Seaweeds as Plant Fertilizer, Agricultural Biostimulants and Animal Fodder

Editors

Leonel Pereira
Department of Life Sciences, Faculty of Sciences and Technology
MARE—Marine and Environmental Sciences Centre
University of Coimbra, Portugal

Kiril Bahcevandziev
CERNAS—Research Centre for Natural Resources,
Environment and Society/IIA—Institute of Applied Research
Polytechnic Institute of Coimbra
Coimbra Agriculture College, Bencanta
Coimbra, Portugal

Nilesh H. Joshi
Fisheries Research Station
Junagadh Agricultural University
Okha Port, Okha, Dist. Dev Bhoomi Dwarka
Gujarat, India

CRC Press
Taylor & Francis Group
Boca Raton   London   New York

CRC Press is an imprint of the
Taylor & Francis Group, an **informa** business

A SCIENCE PUBLISHERS BOOK

Cover credit: Editor – Original from Leonel Pereira (photo) and in public domain (Otto Wilhelm Thomé illustration)

CRC Press
Taylor & Francis Group
6000 Broken Sound Parkway NW, Suite 300
Boca Raton, FL 33487-2742

First issued in paperback 2021

Library of Congress Cataloging-in-Publication Data

Names: Pereira, Leonel, editor. | Bahcevandziev, Kiril, 1959- editor. | Joshi, Nilesh H., 1975- editor.
Title: Seaweeds as plant fertilizer, agricultural biostimulants and animal fodder / editors: Leonel Pereira, Kiril Bahcevandziev, Nilesh H. Joshi.
Description: Boca Raton, FL : CRC Press, 2019. | Includes bibliographical references and index.
Identifiers: LCCN 2019027252 | ISBN 9781138597068 (hardcover)
Subjects: LCSH: Marine algae as fertilizer. | Biofertilizers. | Marine algae as feed.
Classification: LCC S661.2.M3 S33 2019 | DDC 641.6/98--dc23
LC record available at https://lccn.loc.gov/2019027252

Visit the Taylor & Francis Web site at
http://www.taylorandfrancis.com

and the CRC Press Web site at
http://www.crcpress.com

Preface

To keep the soil productive, it is necessary to add nutrients. The types and amounts of nutrients that plants need have been determined and can be supplied by applying to the soil sources that contain these nutrients.

Fertilizers are chemical compounds applied to promote plant growth. Typically, they are applied either to the soil substrate or by foliar feeding. Fertilizers may be organic or inorganic (composed of simple chemicals and minerals). Organic fertilizers are 'naturally' occurring compounds manufactured through natural processes. Inorganic fertilizers are produced through chemical processes using naturally occurring deposits and chemically altering them.

Organic fertilizers contain essential nutrients to improve the health and productivity of soil and encourage plant growth. Organic nutrients increase the soil organisms by providing organic matter and micronutrients. In the long term, inorganic fertilizers may have an adverse impact on the organisms living in soil and on the soil productivity.

The increased use of chemical fertilizers in agriculture during the Green Revolution made many developing countries self-reliant in food production, but on the contrary, it degraded the environment and had harmful impacts on living beings. The excess use of chemical fertilizers in agriculture is costly and among its effects on soils it depletes water holding capacity and soil fertility and causes a disparity in soil nutrients. Hence, the need to develop low-cost, effective and eco-friendly fertilizers that can work without upsetting the natural balance.

Various bacteria, fungi and algae have been used in the past (Chapter 1) as biofertilizers to improve farm lands with regard to nitrogen fixation, solubilization and mobilization of phosphorus, increase in organic carbon content, balance of carbon/nitrogen ratio, plant growth promotion by increasing nutrient absorption, antagonistic activity against plant pathogens, and production of hormones that are beneficial for agriculture. Certain algae species are used that have the unique ability to provide natural products that could serve as a good substitute for chemical fertilizers. Crop performance can be improved with the use of seaweed extracts, which are environmentally friendly alternatives to the fertilizers and biostimulants. Seaweed extracts contain several natural compounds (such as auxins, cytokinins, and gibberellin) that can improve plant development (Chapters 2, 3, 4 and 5).

There are many species of seaweeds and their varied extracts that have been tested to control different plant diseases and insects that cause damage to plants. In general, there are different modes of action by which the seaweeds control plant diseases and insects. Plant pathology studies using seaweed extracts or their isolated compounds show positive results related to induced resistance in plant defense systems against pathogens (Chapter 6).

Seaweeds have also a long history of use as livestock feed. They have a highly variable composition, depending on the species, time of collection and habitat, and on external conditions such as water temperature, light intensity and water nutrient concentration (Chapters 7, 8 and 9).

The results available on the effect of seaweed supplementation on rumen fermentation are controversial, with reports of increase, decrease or absence of effects. If seaweeds are to be extensively used in ruminant feeding, it must be determined after further research whether they impair rumen fermentation efficiency or increase it through a decrease in gas and methane production (Chapter 10).

Leonel Pereira
Kiril Bahcevandziev
Nilesh H. Joshi

Acknowledgements and Credits of Images

This work had the support of Fundação para a Ciência e Tecnologia—FCT, through the strategic project UID/MAR/04292/2019 granted to MARE. It was also co-financed by the European Regional Development Fund through the Interreg Atlantic Area Programme, under the project NASPA.

Credits of Images

Images on Public Domain:

Chapter 1

Figures 2d, l (William H. Harvey, Phycologia Britannica, 1846–1951)

Figures 3a, c (Carl Axel Magnus Lindman, BilderurNordens Flora, 1856–1928)

Figure 3d (Otto Wilhelm Thomé, Flora von Deutschland, Österreich und der Schweiz, 1885)

Figure 4l (Carts laden with vraic collected off the coast at Le Hocq)

Chapter 4

Figures 1-5 (Originals from authors)

Chapter 6

Figure 5 (https://www.wordclouds.com/)

Originals:

Chapter 1

Figures 1a-p

Figures 2a-c, 2e-k

Figure 3b

Figures 4a-k

Leonel Pereira

Chapter 2

Figure 1 (Original from authors)

Chapter 3

Figures 1-4 (after Mori et al. 2017, with permission)

Chapter 5:

Figures 1-4

Carvalho & Castro

Chapter 6:

Figures 1-4 (Machado et al. 2010, Machado 2014, Machado et al. 2014a,with permission)

Chapter 10:

Figure 1 Adapted, with permission, from (Seshadri et al. 2018) under CC BY 4.0 licence.

Figure 2 Reproduced, with permission, from Henderson et al. (2015) under CC BY 4.0 licence.

Figure 3 Reproduced, with permission, from Gerber et al. (2013).

Figure 4 Reproduced, with permission, from Seshadri et al. (2018) under CC BY 4.0 licence.

Figure 5 Reproduced, with permission, from Leahy et al. (2010), and Hill et al. (2016) under CC BY 4.0 licence.

Figure 6 Reproduced, with permission, from Kinley et al. (2016a)

Figure 7 Reproduced, with permission, from Kinley et al. (2016b) under CC BY 4.0 licence.

Figure 8 Reproduced, with permission, from Maia et al. (2016) under CC BY 4.0 licence.

Contents

Historical Use of Seaweed as an Agricultural Fertilizer in the European Atlantic Area

Leonel Pereira and João Cotas[*]

Marine and Environmental Sciences Centre (MARE), Department of Life Sciences,
Faculty of Sciences and Technology, University of Coimbra, 3000-456 Coimbra, Portugal

1 Introduction

The first documented use of seaweed as agricultural fertilizer in the European Atlantic Area occurred among the ancient Romans. L.J.M. Collumella, the most notable Roman writer on agricultural practices, wrote that roots were to be wrapped in seaweed in order to retain the freshness of the seedlings (Battacharyya et al. 2015). In AD 79, Pliny noted the gathering of "margo" (thought to be maerl, a red seaweed) by "peoples of Britain and Gaul" in order to fertilize their soils (Monagail et al. 2017). Seaweed was regularly used by ancient coastal people along the Atlantic to fertilize soil, but only the Romans have left written records of this practice.

The use of seaweed as fertilizer is known in several European maritime countries—in the North Atlantic, Iceland, Scandinavia and Baltic countries, Norway, Denmark, Orkney and Hebrides, Scotland, Ireland, the Aran Islands, the Channel Islands, Brittany, Spain and Mediterranean countries, in the Eastern Adriatic, Kirk Island (Bacelar 1953, Gavazzi 1974, Rasmussen 1974), and also in the Azores and Cape Verde (Medeiros 1967, Ferreira 1968). In Portugal, north of the Douro River it was practiced along the whole coast and was a normal and extremely frequent activity—and in some cases even a fundamental economic category. South of that river seaweed was used on a very small scale, near Peniche and in certain corners of the cliffs of Ericeira and Cape Espichel (Oliveira et al. 1990); on the Alentejo and Algarve coast, where the sea casts up enormous quantities of algae—the limos or the golfo—on to the rocks, they have never been practically used.

The cutting or collection of seaweed growing in the Iberian Peninsula or washed on to the beach—sargaço (see the main species of algae that constitute the sargaço and other traditional marine plants used in agriculture, in section 4), argaço (Basto 1910), or limos (Oliveira et al. 1990)—for fertilization of farm fields was certainly the most important of these agricultural and maritime activities, where tasks contributing to the crop developed in a natural setting that was often foreign to them, and where agriculture and fishing coexisted.

The production of algae depends on certain conditions, which are especially evident precisely in this area of our coast and in southern places—the coast must be fringed with rocks on which algae grow but from which the movement of surface waters during storms and tides can also pull them off;

---

*Corresponding author: jcotas@gmail.com

in fact, it is generally after storms and tides that more sargaço appears on the beaches. In order for this activity to be practicable, the coast must be accessible, and there must be wide sandy areas or available space for the collection and drying of algae—conditions that together are characteristic of the sea shore north of the Douro River, and that do not occur frequently south of this river (Oliveira et al. 1990).

2 Iberian Peninsula

2.1 The sargaço

The algae harvest in the Entre-os-Rios and the Douro regions (North of Portugal) was, from the Middle Ages to the mid-20th century, an economic and socially important activity in mainland Portugal, as is clearly demonstrated by the Dionysian and Manueline provincial laws of Póvoa de Varzim and Maia. In the provincial law published on March 9, 1308, granted to the old regiment of Varzim de Jusão and its terms, King D. Dinis determined that the privilege of collecting sargaço, as an economic activity of great importance at that time for fertilizing the land, belonged to its residents. These provisions were later confirmed in the Manueline Order of 1514 but, since it was a natural resource capable of generating revenue, the Order also established a tax on the sargaço trade (Pereira and Correia 2015).

In light of the close connection between algae and the way of life of a people, it is useful to distinguish the two main mixtures of marine algae traditionally used as fertilizer, that is, moliço and sargaço. Moliço is a mixture of green algae (Chlorophyta) and red algae (Rhodophyta), and also some marine plants. It is composed of specimens of the genus *Enteromorpha* (algae now belonging to the genus *Ulva*), *Chaetomorpha/Rhizoclonium* (Chlorophyta), *Gracilaria* and *Ceramium* (Rhodophyta), and seagrasses (angiosperms) belonging to the genera *Zostera, Ruppia* and *Potamogeton* (Figs. 1, 2 and 3), all harvested in Ria de Aveiro (Aveiro, Portugal) (see also sections 2.2 and 4).

Sargaço (also called "argaço" and "limos" since the time of the first kings of Portugal) is a mixture composed exclusively of several marine macroalgae (*Saccorhiza, Laminaria, Fucus*— Phaeophyceae, *Codium*—Chlorophyta, *Palmaria, Gelidium* and *Chondrus*—Rhodophyta) (Figs. 1 and 2), which grow on the rocks of the coastline (see also section 5). Traditionally, the algae washed on to the rocks on the seashore were collected and spread out on the sands to dry. The equipment used included the carrelo (used to transport the raft), the raft (used to reach the rocks furthest from the beach where the algae cling) (Figs. 4a, b), tools for cutting and collecting (foicinhão, croque, engaceira) (Figs. 4c, d), baskets made of wood splints (to pack and transport them to land), round mesh bags to collect loose sargaço from the water, and ox carts to transport them to the drying ground. Once they were dried, the sargaço was transported in the same baskets to the medas or haystacks (Figs. 4e, f). These haystacks were topped with straw to protect them from rain and sun (Pereira 2010a, Pereira and Correia 2015).

The harvesting of sargaço can be related historically, to a certain extent, with small holdings and intensive crops; it is in fact in these cases that the use of algae as fertilizer was especially effective and viable. And it was precisely these conditions that prevailed in the northern areas of the coast, where this activity had the greatest importance and where, in addition, the search for sargaço and cultivation of farm land were mainly carried out by families using their own labour (Oliveira et al. 1990).

To the north of the river Douro, this activity had characteristic and unique aspects. It was based on distinct sociological and economic conditions, which sometimes related to certain local patterns concerning gender-based division of labour, or the hybrid nature of the activity itself, in which a specific socio-professional evolution or mutation occurred. The gradual transformation of a form of family labour, a subsidiary of one's own agriculture, into an autonomous commercial activity by people of other social categories who were recruited for it, and in the vast dune areas of Aguçadoura

Figure 1. Main species used as agricultural fertilizer: Chlorophyta – a) *Rhizoclonium linum*, b) *Codium tomentosum*, c) *Ulva* sp.; Phaeophyceae – d) *Ascophyllum nodosum*, e) *Bifurcaria bifurcata*, f) *Fucus serratus*, g) *Fucus vesiculosus*, h) *Himanthalia elongata*, i) *Laminaria digitata*, j) *Laminaria hyperborea*, k) *Pelvetia canaliculata*, l) *Saccharina latissima*, m) *Saccorhiza polyschides*; Rhodophyta – n) *Calliblepharis jubata*, o) *Chondrus crispus*, p) *Cryptopleura ramosa*.

Color version at the end of the book

a well-defined movement of exploitation and valorization of the sands, began at the end of the 18th century. Intensive horticulture—the fields of maceira (Fig. 4g) typical of the region—is closely

Figure 2. Main species used as agricultural fertilizer: Rhodophyta – a) *Ceramium* sp., b) *Gelidium corneum*, c) *Gigartina pistillata*. d) *Halarachnion ligulatum*, e) *Lithothamnion corallioides*, f) *Lithothamnion glaciale*, g) *Osmundea pinnatifida*, h) *Palmaria palmata*, i) *Phymatolithon calcareum*, j) *Polyneura bonnemaisonii*, k) *Polysiphonia elongata*, l) *Vertebrata thuyoides*.

Color version at the end of the book

linked to this phenomenon, allowing it on the one hand and being increased by it on the other. In some cases, the activity of collecting and using sargaço was an essential factor of coastal settlement.

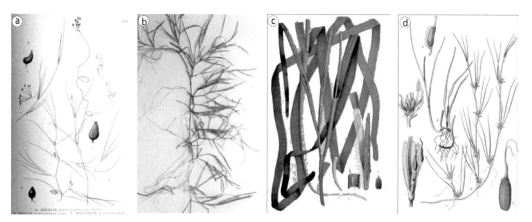

Figure 3. Main species used as agricultural fertilizer: Marine angiosperms – a) *Ruppia* spp., b) *Stuckenia pectinata*, c) *Zostera marina*, d) *Zostera noltei*.

Color version at the end of the book

This task entailed its own tools and techniques and even, in some cases, specific customs and dress, which at times underlined the nature of the different social strata concerned.

Sargaço harvesting in this geographical area was originally carried out primarily by farmers in the coastal zone who, as part of their farming activities, went to the sea to collect sargaço for their own fields and sometimes also for sale. These peasants lived farther inland. The inland landscape revealed villages with large farmhouses with wide façades, open porches and low roofs, scattered along winding roads, with churches and adjacent cemeteries, and pine trees separating neighbouring parishes. On the shores stood simple shacks for shelter and storage of algae, implements, and boats.

Sargaço harvest began practically in May, when the species known as the "folha-de-maio" (May-leaf) (*Laminaria hypeborea*, Phaeophyceae) (Fig. 1j) grows. This alga becomes very hard after drying and decomposes only if the year is rainy (Oliveira et al. 1990, Pereira 2016). The months of greater abundance of sargaço are May and September, because of the tides of the equinoxes.

The most convenient period for this activity is undoubtedly summer—June, July, August, and often also September. In June the folha-de-junho or folha-mimosa (June-leaf or mimosa-leaf) begins to grow, consisting of fine algae that are harvested in July. After September, the catch is generally irregular and hardly profitable; from November onwards, it is practically nil. However, there are often bountiful spells after a few days of rough seas, which pull the algae from the submerged peninsula, near or far, and drag them to the coast. Certain days of exceptional abundance live a long time in the memory of the "sargaceiros" (Guimarães 1916).

2.1.1 Composition and characterization of sargaço

Among the various algae that are collectively referred to as "sargaço" (see also section 4), the most frequently found, on the Minho coastline, belong to the group of brown macroalgae (phylum Ochrophyta, class Phaeophyceae), which in Portuguese is named bodelha (*Fucus vesiculosus*) (Fig. 1g), botelho-bravo (*Pelvetia canaliculata*) (Fig. 1k), cintas (*Saccorhiza polyschides*) (Fig. 1m), cordas or corriolas (*Himanthaliae longata*) (Fig. 1h), folha-de-maio (*Laminaria hypeborea*) (Fig. 1j), and taborro (algae of the genus *Laminaria* and *Saccorhiza*) (Fig. 1m). Among the red algae (phylum Rhodophyta) are found the macroalgae guia or francelha (*Gelidium corneum*) (Fig. 2b), botelho-crespo (*Chondruscrispus*) (Fig. 1o), and argancinho-das-lapas (*Osmundea pinnatifida*) (Fig. 2g) (Pereira 2010a, Pereira and Correia 2015, Pereira 2016). They were all used in agriculture, both as fertilizers and as phytosanitary products (Matos 2018).

The taborro, however, because it has a high water content, is usually separated from the other species and used green immediately after harvesting, without being subjected to the drying and composting process. It is therefore used essentially as meadow fertilizer, or mixed in plowed land, before any seeds are sown.

Red seaweed species, after drying, is used in the pharmaceutical and cosmetic industries and also in the manufacture of agar (E406), carrageenan (E407) and plastics.

Very rich in lime, potassium, phosphoric acid and nitrogen, sargaço is an excellent natural fertilizer. As it dries, the content of those substances rises considerably, and only water decreases by about 60%. Table 1 presents the mineral composition of the fresh and dry sargaço.

Table 1. Composition of fresh and dry sargaço (%) (after Matos 2018)

Composition	Fresh sargaço	Dry sargaço
Water (H$_2$O)	78.0	15.0
Nitrogen (N)	0.35	0.94
Phosphoric acid (H$_3$PO$_4$)	0.35	0.89
Potassium (K)	0.94	2.54
Calcium (Ca)	1.14	3.08

2.1.2 The sargaceiros

The sargaceiros are men and women who, during the tides of the periods between the equinoxes, head towards the sea to collect sargaço. The origin of this ritual is lost in time, but the branquetas or robes customarily worn for this task (Figs. 4a-d) recall a possible connection to the Romans of Puglia (Apulia in Italian), in southern Italy. Armed, in the style of a Roman legion, they fight against the sea on the beaches of Minho (North of Portugal).

Figure 4. Historical use of algae and marine plants in the European North Atlantic: a) the carrelo (used to transport the raft), and the raft (used to reach the rocks furthest from the beach where the algae cling); b) the jangada (raft); c-d) the instruments for cutting and collecting (foicinhão, croque, engaceira); e-f) the medas or haystacks of sargaço; g) Maceiras or fields of maceira; h) landing dock of moliço (ria of Aveiro); i) mural painting representing the harvest of moliço; j) landing of moliço; k) typical moliceiro boat; l) "vraic" or "wrack", the Jersey terms, were used by many writers to cover all types of seaweed but especially those used for agricultural purposes.

Color version at the end of the book

Two types of sargaceiros—peasants and journeymen—are clearly distinguished in the documents that establish different tax regimes for collection of sargaço. In the Provision of the King D. João V of 1742 (which put an end to the old quarrel between the governors of the castle of Póvoa do Varzim and the peasants or villagers, in favour of the latter), farmers are mentioned as picking up sargaço for single fertilizing of their farms, without which they do not produce bread, and journeymen who sell the sargaço and buy bread. (The distinction is also made in the diplomas related to the "Terras da Maia".) The "News of the Village of Póvoa de Varzim", from 1758, refers to the sargaço that peasants collect to use in their fields and that journeymen collect to sell (Silva 1958).

2.1.3 The church and the sargaço

The collection of algae on the Minho coastline (North of Portugal) is undoubtedly a very old tradition, and has frequently been documented since at least the beginning of the 14th century in diplomas of various kinds, often referred to as being practiced from time immemorial (Oliveira et al. 1990) in letters, complaints, judgments and orders. Its importance can be judged by the form, rigor and minutia with which it was regulated, relating to disputes between farmers in the region and local authorities or lords regarding the rights to the sargaço harvested on the beaches under their jurisdiction in certain parish archives, memoirs, souvenirs or customs of parishes, especially those referring to parish rights, tributes or tithes to be paid to the Church for the sargaço collected.

In those older times, sargaço could only be collected by the residents of each parish within their respective areas, under penalty of fine, clearly to avoid conflict. For example, the deputies of Póvoa de Varzim of 1305 and 1308 declared the sargaço of a certain coastal zone could only be collected by the sargaceiros of that area; the same idea, more explicitly, is expressed by King D. João V of 1742, to which we have already referred, summarizing the representation that had been presented to him, in which it was declared that only the inhabitants of that village were to collect algae from his beaches, and they were permitted to use boats and rafts along the coast of the parishes of Amorim and Beiriz, in the permitted season, and that they must not invade the coast of other parishes, under pain of a fine of 5,000 réis (Landolt 1915).

2.1.4 The fields of maceira

Maceiras or fields of maceira constitute a form of agriculture unique in the world (Fig. 4g), and can even be said to be indigenous (as a practice) to Minho, north of Portugal. This form of agriculture, widely used on the coasts of these two countries, was invented, it is said, in the 18th century by the Benedictine monks of the abbey of Tibães. It is now in danger of extinction, due to the aging of the labour force and consequent gradual abandonment of ancestral agricultural practices in favour of intensive and more profitable practices in the short term (Pereira 2010a, Pereira and Correia 2015).

2.2 Moliço

Moliço was used as fertilizer to transform dunes into cultivated land in the context of a scarcely mechanized, subsistence agriculture organized in small family farms.

Moliço was harvested by hand with the help of traditional tools (wood rakes, long-handled rakes and gadanhões, iron-tooth rakes). The grasses and seaweeds were cut and transported to the moliceiro boat (Figs. 4h-j).

This harvesting of the vegetation of the Ria de Aveiro happened initially to fertilize land covered with sandy sediments that was unproductive and low-yielding. Such land was given free to the farmers to cultivate and the moliço mixture was used as fertilizer to increase yields.

The harvested moliço was repressed, accumulated, and discharged through "padiolas", a small wooden staircase with a grid shape. Then it was used fresh or dried. To dry the seaweed, the moliceiro spread it on a slightly sloping area to drain the water. The moliço was repeatedly washed by rainwater to remove the sea salt. Then it was piled in round mounds called "malhadas" that could reach two meters' height. Plants and seaweeds that were washed on to the beach by the tide and wind were called "arrolado".

The beginnings of moliço collection are very poorly documented, but the study of the geographical evolution of the Ria suggests that it began in the 18th century. Harvesting of moliço is mentioned in written sources for the first time in 1758. Parish memoirs indicate that many boats roamed the river in search of agraços or mosses, called "moliços" (Amorim 2008). According to research carried out by Inês Amorim (2008), no previous source mentions the harvest of moliço. This historical source can be compared with a recent environmental study devoted to the topographic evolution of the Ria de Aveiro (Silva 2001), which describes the intrinsic relationship between the moliço and the Ria: the algae composing the mixture develops in brackish water, at salinity between 5 and 30 psu. Thus, the algae present in the moliço could not develop in the water of the Ria till it was brackish, that is to say when the two dunes stretched out to make the Ria an area almost distinct from the sea, in the early 18th century. In a decree dated July 2, 1802, a tax began to be lifted on moliço, proving the existence of the activity before that time.

The harvest and the transport of moliço was the principal activity of the Ria de Aveiro by the late 19th and early 20th century. At that period, moliço harvesting, which was previously done by the farmer himself, had begun to become specialized, and to be a full-fledged profession, because of the increase in demand. Based on the number of moliceiros boats, the average yield of moliço between 1883 and 1967 could reach 274,600 t (Azevedo et al. 2013).

From the late 20th century, environmental causes (increased tidal range and salinity following development of the mouth and pollution) and a possible decrease and disappearance of moliço cannot be invoked as the root causes of the decline of activity, several sources reporting an overabundance of moliço in the 1960s and even in the 1990s. The most plausible cause of the decline was the emigration of the moliceiros from the rural areas.

2.2.1 *The moliceiros and their boats*

The term "moliceiro" is traditionally associated with those who work on board the boat or on the harvest of moliço (Figs. 4h-j). They are essentially a population of poor peasants, forced to supplement their income with occasional fishing in the Ria, small-scale farming and the raising of some livestock along with other seasonal or year-round occupations.

The moliceiro boat is made of pine wood with a flat bottom and small draft. It has broad and low edges, sitting near the water. It has an unmistakable bow, very curved, in a half-moon, ending in a slightly arched back. The traditional means of propulsion is the sail (trapezoidal, made of canvas), the rod, or the tow (Sarmento 2010). Moliceiro boats generally measure 15 m in length and 2,75 m in width (Pilon 1980).

Although the earliest documented references to the moliceiro boat date from the first half of the 18th century, this does not mean that it did not exist before; it merely means that this popular artifact and the practices associated with it were not written about before that.

The principal function of the boat was similar to that of an ox cart, plow or other agricultural equipment, but the boat also served to transport goods, foods, people and livestock within the Ria de Aveiro.

The owner of the boat might be either a farmer collecting seagrass and seaweeds for his own land or a professional vendor of moliço. Most owners had only one moliceiro boat, but the boat itself usually changed hands during its average life of two decades. The life of a moliceiro was financially precarious. It is not surprising, therefore, that the massive emigration that occurred in the region during the 1960s and 1970s took out the majority of moliço workers and brought about the end of the traditional moliço industry.

The function of the moliceiro boat has changed profoundly in the last decades. From an indispensable tool for the economy of the whole region, it has become a simple tourist attraction, a symbol depending on the goodwill and financial well-being of its owner. Pollution, economic evolution and emigration have alienated people from this peculiar lifestyle. Chemical fertilizers replaced the moliço previously used for the fertilization of sandy soils, the salt industry lost much of its importance, and the roads took the place of the boat as the main means of transportation for coastal populations. In 1884, more than half the working boats in Ria de Aveiro were "moliceiro" boats. In 1935, approximately one thousand moliceiro boats were registered in the Captaincy of the Port of Aveiro in 1935. Today, fewer than four dozen survive, mainly as a tourist attraction.

The most unique feature of the moliceiro is the set of four distinct panels that adorn the bow and stern, with characteristic paintings in bright colors (blue, yellow, green, red, black, white), and hand-painted captions. The panels of the prow follow the curvature of the "beak", while those of the stern are more or less rectangular. They have a bright border of several colored bands, made up of flowers and geometric figures. The style of painting varies from crude to elaborate, and a great variety of themes are expressed: humorous panels, panels devoted to moliceiros, Ria and everyday life, religious panels, panels dealing with love, fidelity and marriage, and signs representing personalities. The painter commissioned for the purpose either works on his own inspiration or follows the owner's suggestion. Such painted panels were very rare on workboats, testifying to the particular value of moliceiros.

2.2.2 Cultural heritage

The harvest of moliço, the moliceiros and the moliceiro boat are part of the history of the Ria de Aveiro, its identity, and therefore its heritage. The boats are a surviving symbol of that heritage. In addition to their heritage value, the moliceiro boats today have evolved for tourist use: they are used to ferry tourists along the canals of the city of Aveiro.

The moliceiro boats used for tourism are dissociated from their historical values and memorials. They are presented to tourists as traditional boats, but their story is not transmitted. The boat is therefore only a setting for the experience, a pretty product whose history is hidden.

2.2.3 Agricultural properties of the moliço species

Moliço is a mixture of various species of seagrasses and seaweeds that are autochthonous in the Ria de Aveiro. Some information on species and agricultural properties has been published.

The most notable species in the moliço are *Zostera marina* (Fig. 3c) and *Zostera noltei* (Fig. 3d), since the compounds isolated from these marine angiosperms have antifungal activity, such as zosteric and rosmarinic acid. Zosteric acid demonstrated efficacy in protecting crops against fungal diseases by inhibiting spore adhesion to the leaf surface of plants. Thus, fungi fail to develop and do not cause diseases. Furthermore, it is proven to be non-toxic to the plant itself (Stanley et al. 2002). Rosmarinic acid is a plant-constitutive antimicrobial compound that may be released into the surrounding rhizosphere upon microbe challenge (Bais et al. 2002). This prevents fungal diseases on crops. Moreover, *Zostera* sp. can improve salt tolerance of plants, increasing the productivity of agriculture in dunes (Sadasivam et al. 2017).

For the other seagrasses, *Stuckenia pectinata* (formerly *Potamageton pectinatus*) (Fig. 3b) and *Ruppia* (Fig. 3a), there is a lack of information: only antifungal and antibacterial hypotheses were tested with success (Fareed et al. 2008, Đurđević et al. 2014).

The seaweeds present in moliço are *Gracilaria*, *Ceramium* (Fig. 2a), *Ulva* (Fig. 1c) and *Chaetomorpha/Rhizoclonium* (Fig. 1a), which have a growth-enhancing potential on plants. That potential can be attributed to the compounds that seaweeds naturally have, such as: high content of macro- and microelements, which are necessary to plant development and growth; amino acids; bioactive compounds (such as β-carotene and phenolic compounds) and vitamins. Also, many of the common plant hormones (abscisic acid, auxins, cytokinins, gibberellic acid) can occur in seaweeds. These seaweeds synthesize polysaccharides such as agars (*Gelidium*—Fig.1b, *Ceramium*—Fig. 1a) and ulvans (*Ulva* and *Chaetomorpha/Rhizoclonium*) that improve soil condition, mitigating the effects of drought and protecting seeds and plants from abiotic and biotic factors (Chatzissavvidis and Therios 2014). That last factor, along with the macro- and microelements, was important to the moliceiro and farmers, because the main function of moliço was to turn infertile soil into fertile soil with normal yield for the region.

2.3   The harvest of seaweed in the north of Spain

In Oya, Galicia—the only large village between La Guardia and Cabo Sillero, on the southern edge of the bay of Vigo—the collection of sargaço seems to have had considerable importance. There were in these places a person who, by traditional law, was responsible for announcing the appearance of the sargaço with a repeated cry of "Argazo o mar!" from the heights of La Raiña. At this sign (which could be heard day or night), a representative of each family ran to the sea to collect algae. Then the algae were scattered over the sands, and the sargaceiros who had collected them acquired an inalienable right over them. These algae could lie there for as long as they needed to dry, and no one else would remove them (Bacelar 1953).

We see that the two customs—from Oya, Galicia (Spain) and the Minho coast (Portugal)—beyond their differences of content and purpose, had a very similar external form; both in fact intended to mark the actual beginning of a work—the collection of sargaço—which, although done individually, had collective rules and practices.

2.3.1   Collecting ocle in Asturias (Spain)

In Gozón seaweed was traditionally and generically called "ocle", except in Bañugues, where it was known as "cherba". (Bañugues is one of 13 parishes or administrative divisions in the Gozón council, within the province and autonomous community of Asturias, in northern Spain, near Peñas Cape.) The marine algae of different species, plucked from the rocky bed by storms, were dragged by tides and currents and deposited in the beaches. From there they were collected by the locals to be used, given their richness in salts, nitrates and potassium, as fertilizer for their lands. For centuries, the Cantabrian Sea has shed tons of marine plants that redden a large part of the coast (Fuente 2018, Rodríguez 2018).

In the 20th century, ocle was collected until the appearance of new fertilizers, more effective and less laborious, which replaced the algae blankets that had traditionally covered the farmland and prairies of the council. Even the term "ocle" had a new meaning, more limited, from the early 1950s onward, referring to a single genus of algae cast on the beach, the *Gelidium* (mainly *G. corneum*) (Fig. 2b), which, among all the stranded algae, was collected not for its traditional use as farm fertilizer, but for sale to the chemical industry for obtaining agar (a practice recorded in the council of Gozón from 1951). It could be collected by hand, but it was common to use the tooth, shovel, scribble, rake or other tools, and to pick up the ocle that had dried during low tide (Fuente 2018, Rodríguez 2018).

Collection from the pebbled shore presented an added problem because of the steep slopes and the lack of adequate roads, which made transportation difficult. Loads of algae were drawn up the cliff face by cable, using diesel engines. Once the *Gelidium* was separated from the rest of the algae, it was dried in the sun and wind, normally by women who spread it with their hands or with a brush on the meadows or even the streets, turning it frequently until it dried.

In 1952, a kilogram of dry ocle was sold in Luanco (the capital of the Gozón council) for 3 pesetas (pts) (about 0.02 EUR/0.02281 USD), its value rising to 4 pts between 1957 and 1965 and increasing gradually until reaching its maximum value, 260 pts, in the 1970s. Despite its seasonal nature, the income obtained from its sale during the 1950s to 1970s allowed the families who collected it to prosper in ways that otherwise would not be within their reach (Rodríguez 2018). Synonymous with a way of life, sustainable and with a reduced environmental impact, the collection of ocle ("red gold") has gone from being an occasional activity to one carried out in a regulated and professional manner and having a fundamental importance in the local economy. It is one of the main livelihoods of dozens of families in a council very affected by unemployment and susceptible to the vagaries of the tourism industry (Fuente 2018).

3 British Islands, Channel Islands, Normandy and Brittany

There is a long history of coastal people using seaweeds, especially the large brown seaweeds, to fertilize nearby land. Seaweeds help in stimulating soil bacteria, increase fertility of the soil by humus formation (which feeds on the bacteria), and promote aeration and moisture retention. Wet seaweed is heavy, so it was not usually carried very far inland, although on the west coast of Ireland (see 3.2) it was valued enough to be transported several kilometers from the shore. Generally, drift seaweed or beach-washed seaweed is collected, although in Scotland (see 3.1) farmers sometimes cut *Ascophyllum nodosum* (Fig. 1d) exposed at low tide. In Cornwall (England), the practice was to mix the seaweed with sand, let it rot and then dig it in. For over a few hundred kilometers of the coastline around Brittany (France) (see 3.4), the beach-cast brown seaweed is regularly collected by farmers and used on fields up to a kilometer inland (Sumedha et al. 2016).

3.1 Scotland

The earliest useful accounts of plant use in Scotland were largely written by the "gentleman travelers" who toured the country from the late 17th to the early 19th century. The richest and earliest of these detailed, lucid and often idiosyncratic accounts of life on the west coast and the Hebrides was left by Martin Martin, a naturalist and social historian from Skye Island. Martin's status as a native and his command of Gaelic gave him an advantage over most of the other writers to follow, and his observations on rural Scottish life hold a wealth of valuable ethnobotanical information, including specific descriptions of the uses of seaweeds (Kenicer et al. 2000).

As seaweeds are abundant, nutrient-rich and alkaline, they are particularly suited for use as fertilizer on Scotland's generally acidic soils. It is primarily the large, brown species that are used. In the past these plants were taken from winter storm-cast or cut from the rocks using serrated sickles (Briand 1991), with *Fucus* species sometimes "cultivated" in areas with sandy beaches where they would not otherwise have grown. This was achieved by providing an artificial substrate of large boulders around the half tide mark (approximately the level at which these species grow naturally). The weed would set itself to these boulders and grow to maturity in two seasons or so. The boulders could be turned over after harvesting, allowing a second crop to establish itself for harvesting two seasons later (Martin and Monro 1703, Headrick 1807, Kenicer et al. 2000).

Historically, seaweed fertilizers were an integral part of the farming system in the crofts of the northwest coast (Lightfoot 1777) and were well suited to the cultivation of relatively poor, hilly land. The crofter would dig a series of wide, linear trenches, often running up a hillside. The seaweed was then dug into the earth from the trenches and after around three weeks of composting, the mixture was turned into the trenches again and oats, wheat, barley, potatoes, onions, turnips and brassicas were planted (Martin and Monro 1703). This method of cultivation, known as the "lazy bed" system, remains in use today in the Hebrides, and its legacy can be seen widely across Scotland.

Using the lazy bed system, a crofter with an adequate supply of seaweed would be able to grow two crops of oats in successive years in the same plot without the need for rotation. In addition, the content of glutinous carbohydrates in seaweeds made them particularly suited to land reclamation, as a composted seaweed-soil mix would be relatively resistant to soil runoff. This mixture could be placed directly on rocky ground on hillsides and planted with potatoes, thus allowing land that was otherwise unfit for cultivation to be used (Martin and Monro 1703). If estates only had a small frontage on the sea, the seaweed was generally shared amongst some of the more inland crofts. Alternatively, rights to collection from various plots were rotated amongst the crofters over a biannual cycle (Landsborough 2012).

In the farms of Ayrshire and the east coast, and north to Aberdeen, seaweed was in great demand in the 18th and 19th centuries. There were regional differences in the choice of seaweed species and whether or not they were collected as storm cast, harvested or applied in the form of ash, and these may be attributed to differing soil types and availability of alternative fertilizers (Hendrick 1898, Moffatt 1915).

During the latter half of the 19th and into the 20th century, increasing availability of chemical fertilizers and social changes led to a general decline in use of seaweed as manure. The last significant harvests of raw seaweed for laying on the land were in Ayrshire and around north Berwick in the 1960s. Even then, the abundance of seaweed was such a boon to the farmers in these regions that a proposed alginate processing plant in East Lothian was shelved on the grounds that it would have used up the farmers' supplies. As the 1960s ended, however, most Ayrshire farmers all but stopped taking the weed as it was said to be too sparse and heavily polluted with rubbish (Noble 1975); very few farmers still collect it (Kenicer et al. 2000).

Despite the decline in large-scale farm use of seaweed, many crofters and gardeners (including several National Trust for Scotland properties) still use seaweed fresh from the shore. There are two minor problems associated with using seaweed on the garden. The first of these is the smell, which led to the cessation of its use at Culzean Castle, where it was alleged that staff were becoming physically sick. The second problem is with weeds. Coastal ruderal land plants (*Chenopodium album*, in particular) may be transported as seeds in the seaweed, and then proliferate in the gardens where it is used. These are relatively minor problems, however, and the generally held opinion amongst the amateur gardening community is that seaweed is an excellent fertilizer and soil conditioner (Kenicer et al. 2000).

Certain calcareous red seaweeds can be used on the land as a substitute for lime and are commercially marketed under the name of maerl for this purpose. Deep-water genera such as *Lithothamnion* and *Phymatolithon* (Figs. 2e, f and i), which grow hard calcium-rich "skeletons" and are cast up on beaches after storms (Blunden et al. 1997), are the most important sources. Particularly impressive maerl beaches can be seen on Skye Island. Although the use of maerl in Scotland has never been of great importance, there is a considerable historical basis for this practice in Ireland, Jersey and the Channel Islands. Maerl fertilizer is commercially available in garden centers and is thought to give a better performance than standard lime, although it is marginally more expensive. Maerl is also exploited for use in filtration of drinking water, and as an additive to livestock fodders (Johansen 1981).

3.2 Ireland

Seaweed harvesting has a long tradition in Ireland. A poem, probably dating from the 12th century, describes monks harvesting dillisk (*Palmaria palmata*) (Fig. 2h) from the rocks and distributing it to the poor as one of their daily duties. Dillisk, also known as dulse, a red alga that is eaten on both sides of the North Atlantic, has traditionally been used as both food and medicine. In the 18th century it was used as chewing tobacco, ingested to eliminate worms, and recommended as a remedy for "women's longing".

Throughout the centuries, different types of seaweed were used as fertilizer, enriching the soil with minerals and growth hormones in the impoverished post-glacial soils of the west of Ireland stripped of their coastal bogs (Guiry and Morrison 2013). Seaweed manure was particularly important in areas with poor soil, and families fought over seaweed rights and access. Harvesting rights predate the formation of the Irish State and date back to the British Crown.

Ireland's tradition of kelp harvesting dates back to the 17th century. Coastal communities gathered kelp (*Ascophyllum nodosum*, for example) from the shoreline, especially after a storm, and burnt it in stone circles known as kelp kilns, the ruins of which are still visible along the west coast. The ash that remained contained soda and potash and was used for glazing pottery and for making glass and soap. In the mid-18th century the ash was found to also contain iodine, and this discovery kept the tradition alive until the World War II. Today, small amounts of kelp are harvested for the sea-vegetable industry and as feed for farmed shellfish, specifically urchins and abalone (Blunden 1991, Steele et al. 2010).

3.3 Channel Islands

One of the most interesting and important features of the agriculture of the Channel Islands, and particularly of the agriculture of Jersey before 1900, was the use of seaweed as fertilizer. A habitual feature of the agricultural scene, seaweed was the cause of much litigation and legislation in Jersey, largely because of its supreme value when applied to the sandy soils that cover most of the island. One 19th century writer commented as follows: "Besides his own estate or domain, in the shape of 'terra firma', every islander has a common right of great value, lying on the shore of the barren sea, and belonging to the sea itself. It is true that neither ox nor horse can browse on it, and yet it supplies provender for ox and horse as truly as if it were a field of clover or oats" (Ansted and Latham 1862). Though its efficacy was accepted, and the benefits derived from its use widely enjoyed, it was not till the end of the 19th century that any serious scientific study of the types of seaweed and their chemistry was made (Blench 1966). Seaweed has been used for agricultural purposes in Jersey at least since the 12th century but there are no detailed records that give any details of its application (Gruchy 1957). The main use of vraic is for spreading over potato fields during the winter. It is then plowed into the soil before the potatoes are planted in late winter and spring. Vraic was traditionally gathered by horse and cart in Grouville Bay and St Ouen's Bay, and in smaller quantities at Le Hocq and Havre des Pas (Fig. 4l).

"Vraic" or "wrack", the Jersey terms, were used by many writers to cover all types of seaweed but especially those used for agricultural purposes. The derivation of the terms is obscure, probably being either a corruption of the French "varech", or of the old English "wræc" (Blench 1966). The composition of vraic, after Blench (1966), was as follows: *Ascophyllum nodosum* (Fig. 1d), *Fucus serratus* (Fig. 1f), *F. vesiculosus* (Fig. 1g), *Peveltia canaliculata* (Fig. 1k), *Himanthaliae longata* (Fig.1h), *Laminaria digitata* (Fig. 1i), *Saccharina latissima* (Fig. 1l) (Phaeophyceae), and *Palmaria palmata* (Fig. 2h) (Rhodophyta) (see section 4).

The mineral constituents of algae had a high value in agriculture, especially in the fertilization of poor soils in these compounds. The higher sodium compounds are found especially in weeds, while potassium was extremely useful in clover, grass, tomato or potato cultivation. The highest concentrations of potassium are present in tangle, toothed wrack, and knotted wrack, in that order (see section 4).

Quayle (1815), in his account of the agriculture of the Channel Islands, devotes almost a whole chapter to the value and use of seaweed. He shows that it was still the most important fertilizing agent in use: "the supply of 'vraic' ashes are not equal to the demand; and on these, in the opinion of many, the agriculture of the Island depends for support." After summarizing most of the regulations printed in 1771, he adds one or two further points of interest. Inland farmers were allowed a portion of "mielles" (sand dune areas) for drying of vraic. Another feature first mentioned by him is that many people living near the bays kept a horse and cart so that they could collect seaweed, not for their own use, but for sale: a cartload (wet), the result of two hours' work with an iron-pronged vraic rake, was sold for two livres (Fig. 4l). Four wet loads were equal to one dry load.

3.4 Normandy and Brittany (France)

In France, seaweed gathering dates back to Neolithic times. In these times, the use of seaweed was common along the Atlantic coast. Residues of algae were found in fireplaces during archaeological excavations. Seaweed was used for heating, in mattresses, for cattle and for human food in times of famine. The primary uses have evolved, but some traditional uses are still observed in some coastal areas, for example, the use of seaweed as food for cattle and for soil improvement. This is the case in Brittany, where seaweed gathering has held a prominent place in the history of coastal people (Arzel 1987). Wreck seaweed gathered after storms was spread in fields. Men collected algae from the sea even in winter with large rakes, and women collected algae with litter at the shore. Then, the algae were spread on dunes to be dried for year-round preservation (Arzel 1987). This activity experienced a sharp decline with the advent of chemical fertilizers and the increase of farm size. Soil improvement using fresh seaweed is currently less frequently practiced, except in small private fields, such as on Batz Island in northern Brittany (Mesnildrey et al. 2012).

The primary applications of algae in agriculture are as fertilizers and fodder. Several processes facilitate the production of powders or liquid extracts. Fertilizers are spread on grounds in powder form, as micro-balls, or pulverized in liquid form. The main algae species used are *A. nodosum*, *Fucus* spp. (Phaeophyceae) and maerl, which favor the growth of plants and resistance against diseases. Indeed, seaweed produces defensive substances in response to aggression by gastropods (Pérez 1997). The fucales can also be used as nutritional supplements in animal food for their digestive qualities; they are processed in flours mixed with food (Arzel 1987). The maerl are also used by water-treatment plants for correction of the pH (Mesnildrey et al. 2012).

For Normandy and Brittany, harvesting seaweed is an activity that seems to have been practiced since Roman times (Leclerc and Floc'h 2010). Algae were mainly intended for the amendment of agricultural land. This practice, which lasted until the 1960s, appears to have been vital for local populations, as described by Mauriès, librarian and member of the academic society of Brest (Mauriès 1875).

Biologist Camille Sauvageau (Sauvageau 1920) writes that all marine plants, particularly those cast on shore or cut by coastal inhabitants, have a name that varies with the regions: in the 17th century they were called "varech" or "vraicq" in Normandy, "sar" or "sart" in Aunis, Saintonge and Poitou, and "gouesmon" in Brittany.

The term "varech", according to the dictionaries Ménage of 1650 and of the French Academy of 1798 (5th ed.) is not exclusive to algae but refers to all the debris washing up on the Normandy coasts. It is borrowed from the Norrois word "vagrek" (Norrois was the medieval Scandinavian language spoken in Normandy) and Gaelic "wrack" or "wreck" meaning failed. "Warec" is mentioned in an investigation commissioned by the King of England in 1181 on the archiepiscopal area of "Dol" (between Saint-Malo and Mont St Michel). Though kelp is still associated with Normandy, its meaning has evolved over the centuries. Before 1681, it included everything that ran aground "by turmoil and fortune of the sea, or that arrives so close to the coast that a man on horseback can touch it with a spear". All such matter was considered "gayves" (designating what is cast up by the sea) and called "varech" (Desouches 1972). It is associated with the right of "varech", which allowed the local lords of Normandy to seize all that is cast by the sea on its coasts (Videment 1909, Desouches 1972).

From the 19th century, the term "varech" was limited to marine plants cast on the shore and by extension to other plant debris (Videment 1909). In the beginning of the 20th century, according to C. Sauvageau (Sauvageau 1920), "varech" was associated by botanists with *Fucus* or marine plants such as *Zostera marina*, which was sought after for the horsehair trade for upholstering mattresses and chairs or for the manufacture of packaging. The first official mention of "goémon" is in the Roles of Oléron (nautical code written by Aliénor d'Aquitaine in 1152) (Videment 1909). E. Videment proposes an origin from "gaywon", which itself comes from "gaive" or "gayves", these words designating what was cast up by the sea (sea grass, fish, precious stones, etc.) and, belonging to nobody, became the property of the first person to find it.

4   Main Species Used as Agricultural Fertilizer

The main species of algae and marine angiosperms, constituents of the various mixtures traditionally used as agricultural fertilizer and, in some cases, for the extraction of phycocolloid (agar and carrageenan) and the cosmetics and pharmaceutical industries, in the European Atlantic region, are listed below and described by taxonomic group (Rodrigues 1963, Oliveira et al. 1990, Pereira 2009, 2010b, 2016, Pereira and Correia 2015).

4.1   Green algae (phylum Chlorophyta)

Chaetomorpha/Rhizoclonium spp. Kützing (Fig. 1a)
Common names: Spaghetti algae (English); Limos (Portuguese)
Description: Macroscopic, consisting of unbranched uniseriate filaments, with intercalary growth. It is free-living or attached basally by rhizoids or an often elongate, thick-walled cell. Cells are usually cylindrical to barrel-shaped and have a lamellated wall, with each having a parietal, net-like chloroplast and numerous pyrenoids and nuclei.

Codium tomentosum Stackhouse (Fig. 1b)
Common names: Spongeweed, Velvethorn (English); Chorão, Chorão-do-mar, Candeias, Pingarelhos (Portuguese)
Description: Short stem highly branched dichotomously, slightly flattened, tomentose, dark green, with obtuse ends of segments; thick insertion base.

Ulva spp. Linnaeus (Fig. 1c)
Synonym: *Enteromorpha* spp. Link
Common names: Sea lettuce, Green laver, Gut weed (English); Limo, Limos, Alface-do-mar (Portuguese)
Description: Large, green, translucent, intestine-like or undulating blades.

4.2   Brown algae (phylum Ochrophyta, class Phaeophyceae)

Ascophyllum nodosum (Linnaeus) LeJolis (Fig. 1d)
Common names: Knotted wrack, Asco, Sea whistle, Bladder wrack, Egg wrack (English)
Description: Perennial brown intertidal seaweed species most abundant on sheltered rocky shores in the mid-intertidal zone of the North Atlantic. Olive green in color, *A. nodosum* generally grows upward in the water column, anchoring to hard substrates using a disk-shaped holdfast. Its long, thick, leathery, branching strap-like fronds are typically between 0.5 and 2 m in length, and have large egg-shaped air bladders at regular intervals along their length which keep the plant floating upright when submerged at high tide, and hang downward, draping over intertidal rocks in a thick, tangled, and glistening mat at low tide.

Bifurcaria bifurcata R. Ross (Fig. 1e)
Synonym: *Bifurcaria tuberculata* Stackhouse
Common names: Brown tuning fork weed, Brown forking weed (English); Frosque, Pauzinhos (Portuguese)

Description: Cylindrical, dichotomously branched thallus, olive green in color, with generally unequal ramifications, with yellowish, elongated terminal and simple receptacles, 1.5-5 cm in length. Hermaphrodite concavities with bottle shape, slightly protruding and becoming well defined on dry thalli. Oogonium in the bottom of conceptacle and antheridium in the vicinity of the osteole.

Fucus serratus Linnaeus (Fig. 1f)
Common names: Black wrack, Blackweed, Toothed wrack (English)
Description: Characterized by thalli with serrate edges, of greenish-brown color. The frond surface of this species has numerous pin-pricks with clusters of tiny white hairs. It grows profusely in a wide variety of situations from exposed rocky shores to saline lagoons. It is most commonly found on sheltered, hard, rocky substrata on the lower part of somewhat sheltered coastlines, subject to some degree of disturbance such as from tidal scour. Its growth rate varies considerably depending on environmental conditions but can range from 4 to 12 cm per year. *F. serratus* thalli may become detached and lost to winter storms.

Fucus vesiculosus Linnaeus (Fig. 1g)
Common names: Bladder wrack (English); Botelho, Bodelha, Botelha, Bagão, Borracha, Limo-bexiga, Limo-de-estalo, Trambolho, Esgalhota, Estalos (these Portuguese common names are generically used in the Minho region, north of Portugal and apply to all algae that are collected by hand).
Description: Fronds stipulate, dichotomous, with relatively wide partitions, with a medium vein, whole margin and large spherical or slightly elongated vesicles. Frequent across the Minho coastal shoreline.

Himanthalia elongata (Linnaeus) S.F. Gray (Fig. 1h)
Synonym: *Himanthalia lorea* (Linnaeus) Lyngbye
Common names: Sea thong (English); Cintas, Cordas, Corriolas (Portuguese)
Description: Young frond cylindrical, then obconic or claviform. Mature frond with the upper portion flattened and concave, forming a cup with 2-3 cm height and 3-5 cm width. Receptacles originate from the apex of the frond, long or very long 0.25-3 m, brownish, compressed, with dichotomous branches 5-10 mm wide, regularly spaced, attenuated to the distal end. Dioecious species, with long protruding thongs while young, and with an oosphere in each oogonium.

Laminaria digitata (Hudson) J.V. Lamouroux (Fig. 1i)
Common name: Tangle (English)
Description: Large, tough, glossy kelp, which can grow from 1 to 3 m, and up to 4 m in optimum conditions. Its color ranges from dark brown to golden brown to olive brown to olive green. The broad frond or blade is large, lacks a midrib, and is shaped like the palm of a hand with a number of more or less regular finger-like segments (hence the Latin name for this seaweed). The digits extend almost to the base of the frond and the number of digits varies with amount of exposure. In shelter these are few and short, but with increasing exposure, they are more numerous (up to 10 or 12). The length of the frond varies with season, age of plant and location, and can reach 1-1.5 m in suitable conditions.

Laminaria hyperborea (Gunnerus) Foslie (Fig. 1j)
Synonym: *Laminaria cloustonii* Edmondston
Common names: Oarweed (English); Rabo-negro, Folha, Folha-de-maio, Fitas, Chicote, Taborro-de-pé (Portuguese); Correa (Spanish)
Description: Laminar, cartilaginous, dark olive green, with more or less elongated, cylindrical, rugose, consistent stalk; rounded-backed blade at the base and divided into wide, convergent, sometimes subdivided partitions. Short and strong haptera, joined together by hooks, slightly branched, arranged in vertical series and forming an obconic set radiating from the base of the stipe. Unilocular sporangia forming irregular sori on both sides of the lamina, sometimes developing almost to the base.

Pelvetia canaliculata (Linnaeus) Decaisne & Thuret (Fig. 1k)
Common names: Channeled wrack (English); Botelho-bravo (Portuguese)
Description: Fronds narrow in tuff, 5-15 cm long, deeply fluted on one side and convex on the other, rigid and leathery, green. Regular branching giving the fronds the appearance of corymbiform, simple or bifurcated, more or less numb. Monoecious species with hermaphrodite conceptacles.

Saccharina latissima (Linnaeus) C.E. Lane, C. Mayes, Druehl & G.W. Saunders (Fig. 1l)
Synonym: *Laminaria saccharina* (Linnaeus) J.V. Lamouroux
Common names: Sugar kelp, Sea belt, Sugar wrack (English); Taborrão, Taborra, Rabeiro (Portuguese)
Description: Thallus leathery, dark olive green, with more or less elongated, cylindrical stipe; long basal disc; cuneiform lamina at the base, torn in lacinia, sometimes narrower or broader; surface with

numerous saccharine crystallizations. It can reach 3 m in length. Unilocular sporangia developed on both sides of the border and along its axis, assembled in sori.

Saccorhiza polyschides (Lightfoot) Batters (Fig. 1m)
Common names: Furbelows (English); Caixeira, Carocha, Cintas, Golfe, Golfo, Limo-correia, Limo-corriola (Portuguese)
Description: Species with a distinctive large, bulbous, warty holdfast and a flattened stipe with a frilly margin. The stipe is twisted at the base and widens to form a large flat lamina, which is divided into ribbon-like sections, generally to 3 m long.

4.3 Red algae (phylum Rhodophyta)

Calliblepharis jubata (Goodenough & Woodward) Kützing (Fig. 1n)
Synonyms: *Calliblepharis lanceolata* Batter; *Rhodymenia jubata* (Goodenough & Woodward) Greville
Common names: False eyelash weed (English); Botelho-gordo (Portuguese)
Description: Laminar thalli with reddish fronds, sometimes becoming greenish at the extremity, cartilaginous, flaccid, lanceolate-oblong, in a rhizomatous-like thallus, provided with marginal branches and sometimes clad in thorn-like growths.

Ceramium spp. Roth (Fig. 2a)
Common names: Hornweed, Red hornweed (English)
Description: Cartilaginous, usually bushy, brownish-red to yellowish-green fronds, to 300 mm long. Irregularly dichotomous, clothed with simple to repeatedly dichotomous ramuli. Usually completely corticate, branchlets markedly narrower than main axes, apices normally markedly hooked inwards (but straight in plants bearing tetrasporangia). Highly variable in size, form and color.

Chondrus crispus Stackhouse (Fig. 1o)
Common names: Irish moss, Carrageen moss, Jelly moss (English); Botelho-crespo, Musgo-gordo, Botelho, Botelho-rino, Botelha, Cuspelho, Musgo, Limo-folha, Folha-de-alface, Musgo da Irlanda, Crespo, Musgo-Irlandês; Folhinha (Portuguese); Condrus (Spanish)
Description: Laminar thalli, reddish-pink or brown, iridescent when immersed. Dark, dichotomically branched sub-coriaceous, semicircular outline with a long-crested base, spiny and broad, rounded at the end, slightly overlapping, with outlines of new, slightly bulbous branches. Other times it presents the last branches already formed and more delicate.

Cryptopleura ramosa (Hudson) L. Newton (Fig. 1p)
Synonym: *Nitophyllum laceratum* (S.G. Gmelin) Greville
Common names: Fine-veined crinkle weed (English); Botelho, Botelho-da-pedra (Portuguese)
Description: Laminar, purplish or brownish-red, papyraceous, very thin, but somewhat rigid, sessile or short staple, dichotomous and irregularly branched, semicircular contour, with relatively broad partitions and sometimes terminated in two or more short laciniate parts, appendicular.

Gelidium corneum (Hudson) J.V. Lamouroux (Fig. 2b)
Synonym: *Gelidium sesquipedale* (Clemente) Thuret
Common names: Kanteen (English); Ágar, Cabelo, Cabelo-de-cão, Febra, Francelha, Garagar, Guia, Limo-encarnado, Limo-fino, Limo-preto, Pelinho, Pelo, Ratanho, Sedas (Portuguese); Ocle, Caloca (Spanish).
Description: Cartilaginous, slightly compressed or flattened cartilaginous, linear, pinnate cartilage, with the pinnae attenuated at the base and the linear, short and obtuse ramuli. Red-purple coloration, sometimes lighter and more greenish, especially at the extremities. Appears in winter and also in spring.

Gigartina pistillata (S.G. Gmelin) Stackhouse (Fig. 2c)
Common names: Pestleweed (English); Borracha, Botelho-borriço, Botelho-risso, Corno-de-veado, Pinheirinho, Musgos (Portuguese)
Description: Purple or reddish-brown fronds, densely tufted, undivided at the base, dichotomously branched, constituting a flabelliform whole, with ramifications repeatedly forked, bare or with short horizontal and acute springs; cystocarps solitary or geminate, sessile in the ramuli.

Halarachnion ligulatum (Woodward) Kützing (Fig. 2d)
Synonym: *Halymenia ligulata* (Woodward) C. Agardh
Common names: Sea spider weed (English); Botelho-macio, Botelho-gordo, Argaço-risso (Portuguese)
Description: Laminar thalli, subcartilaginous, purplish-dark, with irregular fronds, sometimes with the broad middle part feathered, spiked and gathered in a basilar portion with a rhizomatous aspect. Sometimes the middle fronds also widen at the edge, bifurcating into equal or unequal partitions. Clings by means of a calloused foot to shells and other hard objects.

Lithothamnion corallioides (P.L. Crouan & H.M. Crouan) P.L. Crouan & H.M. Crouan (Fig. 2e)
Common name: Maerl (English)
Description: An unattached, fragile alga with a calcareous skeleton. Very similar to and often confused with *Phymatolithon calcareum*. Its form is very variable, but it commonly occurs as highly branched nodules forming a three-dimensional lattice. Individual plants may reach 4-5 cm across and are bright pink when alive but white when dead.

Lithothamnion glaciale Kjellman (Fig. 2f)
Common name: Maerl (English)
Description: Bright pink to purplish, minutely white-speckled calcareous crust, becoming very thick, usually with abundant regular or irregular branches, free or attached to substratum.

Osmundea pinnatifida (Hudson) Stackhouse (Fig. 2g)
Synonym: *Laurencia pinnatifida* (Hudson) J.V. Lamouroux
Common names: Pepper dulse (English); Argancinho-das-lapas, Pele-de-lapa, Botelho-preto, Erva-malagueta (Portuguese)
Description: Thallus pale yellow, olive green, or purple, cartilaginous, with alternating, linear integuments, and with the ends of the last rounded-obtuse partitions; markedly compressed and densely branched fronds that grow 2-6 cm tall and arise from a stoloniferous base.

Palmaria palmata (Linnaeus) Weber & Moh (Fig. 2h)
Synonym: *Rhodymenia palmata* (Linnaeus) Greville
Common names: Dulse, Grannogh, Dillisk, Dillesk, Crannogh, Water leaf, Sheep dulse, Dried dulse, Shell dulse, Handed focus (English); Botelho-comprido, Folha (Portuguese)
Description: Laminar brown or reddish thallus, coriaceous or membranous, with a length of 10-40 cm, attenuated at the base, dichotomously round, sometimes subdivided at the base.

Phymatolithon calcareum (Pallas) W.H. Adey & D.L. McKibbin (Fig. 2i)
Common name: Maerl (English)
Description: Fragile, reddish-violet, branched, calcareous fronds; branches 2-3 mm diam.; highly variable in form and difficult to identify with certainty; commonest form resembles stag's horns of irregular diameter.

Polyneura bonnemaisonii (C. Agardh) Maggs & Hommersand (Fig. 2j)
Synonyms: *Nitophyllum hilliae* (Greville) Greville; *Polyneura hilliae* (Greville) Kylin
Common names: Crimson veined weed (English); Asa-de-pito (Portuguese)
Description: Dark red laminar thalli, becoming rosy and then orange, thick, semitransparent, shortly stippled, expanding into a rounded flabelliform fringe, irregularly divided, sometimes almost simple

with few marginal, sometimes deeply cracked, lobes to the base, in a few broad segments; blades sometimes divided in the margins, curly and prolific. Microscopic veins present in younger parts.

Polysiphonia elongata (Hudson) Sprengel (Fig. 2k)
Common names: Lobster horns, Elongate siphon weed (English); Argaço-das-fisgas (Portuguese)
Description: Branched talon, with filamentous ramifications, resembling horsehair, reddish-brown in older parts; naked in winter and densely clothed in spring with numerous attenuated spruces at the base and apex.

Vertebrata thuyoides (Harvey) Kuntze (Fig. 2l)
Synonym: *Polysiphonia thuyoides* Harvey
Common names: Tufted conifer weed (English); Cabelo-de-rabo-negro, Cabelo (Portuguese)
Description: Thallus forming tufts of erect, filamentous ramifications, irregularly branched, with regularly pinnulate, articulate ramblings; oocyte cystocarps, numerous, inserted in the clusters. Sub-cartilaginous consistency; yellowish-green or brownish in color. Epiphyte on other algae (i.e., *Laminaria hyperborea*).

4.4　Marine angiosperms

Ruppia spp. Linnaeus (Fig. 3a)
Common names: Beaked tasselweed, Widgeon grass, Ditch-grass, Tassel pondweed (English); Erva, Erva-do-arganel, Sirgo (Portuguese)
Description: Perennial or annual cycle herbs, rhizomatous, rooted in the substrate. Thin rhizome, very branched. Stems with leaves mainly arranged in the upper half, generally sub-opposed; an inflorescence develops in the crook of one and a vegetative bud in the crook of the other that often develops a new branch. Leaves differentiated in sheath and limb; open sheath, articulate or truncated at the apex, widened at the base, membranous or herbaceous, opaque or translucent; linear or capillary limb, acute or obtuse, weakly or strongly serrated at the apex, flat, with 1 or 3 parallel nerves. Spike-shaped inflorescence.

Stuckenia pectinata (L.) Boerner (Fig. 3b)
Synonym: *Potamogeton pectinatus* Linnaeus
Common names: Fennel pondweed, Sago pondweed (English); Limo-mestre, Rabo (Portuguese)
Description: Submersed plant that grows from a creeping rhizome, with stems slender and flexible. The stems have many branches and may be green, brown, or reddish. The flowers are tiny and greenish and grow in several whorls along the flower spike. Present in fresh, brackish, and saline waters.

Zostera marina Linnaeus (Fig. 3c)
Common names: Common eelgrass, Seawrack (English); Seba, Limo-da-fita, Limo-seval, Fita-do-mar, Moliço (Portuguese)
Description: Angiosperm with true leaves, stems, and rootstocks, not an alga. *Z. marina* leaf blades are characteristically flat and wide (2-12 mm) and can reach up to 3 m in length, although morphology is variable and depends on environmental factors such as substrate type, depth, temperature, and light and nutrient availability. Bright green leaves arise from creeping rhizomes with many hair-like roots.

Zostera noltei Hornemann (Fig. 3d)
Synonyms: *Zostera nana* Roth; *Zostera noltii* Hornemann
Common names: Dwarf eelgrass (English); Cirgo, Musgo, Seba, Cargo (Portuguese)
Description: Grass-like flowering plant with grass-green, long, narrow, ribbon-shaped leaves 6-22 cm long and 0.5-1.5 mm wide with three irregularly spaced veins. The tips of the leaves are blunt, notched, often asymmetric, and become indented in older leaves. Leaves shoot from a creeping rhizome 0.5-2 cm thick, with 1-4 roots per node, which binds the sediment. Leaves shoot in groups

of 2-5, encased in a short, open sheath 0.54 cm long. Several flowers (4-5 male and 4-5 female) occur on a spear-shaped reproductive shoot 2-25 cm long (frequently 10 cm). Seeds are smooth, white, and 1.5-2 mm long (excluding the style). Leaves and rhizomes contain air spaces, lacunae, that aid buoyancy and keep the leaves upright when immersed.

5 Conclusion

Farmers and other agricultural technicians are beginning to recognize the need to reduce the use of chemicals in the soil. They also recognize that the use of seaweeds and other marine plants increases the resistance of crops to diseases and parasites, improves the taste of fruits and vegetables, and favors the precocity and abundance of crops. As they decompose on land, seaweeds considerably increase the temperature of aerobic fermentation, accelerate the biodegradation of manure and dead leaves, improve acidic soils, retain soil humidity, and make sandy and unproductive areas fertile.

A consortium was created within the scope of the NASPA (Interreg) project that integrates several companies and academic institutions from countries along the European Atlantic coast, in search of natural biofungicides and biostimulants, in order to reduce the amount of synthetic chemicals used and thus reduce carbon and nitrogen footprint (www.bionaspa.com).

This project provides knowledge and access to pilot-scale production facilities with the ability to test and validate biostimulants and biofungicides for agricultural use in the Atlantic area. The project also supports and influences a regional strategy to develop eco-products for agriculture using local resources, improving food production and nutritional enrichment and stimulating a greener agriculture in the Atlantic Area.

Acknowledgments

The authors acknowledge financial support from the Portuguese Foundation for Science and Technology (FCT) through the strategic project UID/MAR/04292/2019 granted to MARE. This work is also co-financed by the European Regional Development Fund through the Interreg Atlantic Area Programme, under the NASPA project. The authors are also grateful for the historical bibliographical research done by Dr. Radek de Bragança from the University of Bangor, Wales.

References Cited

Amorim, I. 2008. Porto de Aveiro: Entre a Terra e o Mar. Administração do Porto de Aveiro, S.A., Aveiro, 67 pp.

Ansted, D.T. and R.G. Latham. 1862. The Channel Islands. Wm. H. Allen & Co. London, UK, 396 pp..

Arzel, P. 1987. Les Goémoniers. Chasse marée, Edition de l'Estran, Douarnenez, 305 pp.

Azevedo, A., A.I. Sousa, J.D.L. Silva, J.M. Dias and A.I. Lillebø. 2013. Application of the generic DPSIR framework to seagrass communities of Ria de Aveiro: a better understanding of this coastal lagoon. J. Coast. Res. 65(1): 19–24.

Bacelar, R.L.G. 1953. Costumbres Gallegos – Argazo ó Mar! O el tributo al matrimónio. Faro de Vigo: 3.

Bais, H.P., T.S. Walker, H.P. Schweizer and J.M. Vivanco. 2002. Root specific elicitation and antimicrobial activity of rosmarinic acid in hairy root cultures of *Ocimumbasilicum*. Plant Physiol. Biochem. 40: 983–995.

Basto, C. 1910. Falas e tradições do distrito de Viana do Castelo. Revista Lusitana XIII: 84–88.

Battacharyya, D., M.Z. Babgohari, P. Rathor and B. Prithiviraj. 2015. Seaweed extracts as biostimulants in horticulture. Sci. Hortic. 196: 39–48.

Blench, B.J.R. 1966. Seaweed and its use in Jersey agriculture – the agricultural history review. Agric. Hist. Rev. 14(2): 122–128.

Blunden, G. 1991. Agricultural uses of seaweeds and seaweed extracts. pp. 65–81. *In*: Guiry, M.D. and Blunden, G. (eds.). Seaweed Resources in Europe: Uses and Potential. John Wiley & Sons Ltd., Chichester.

Blunden, G., S.A. Campbell, J.R. Smith, M.D. Guiry, C.C. Hession and R.L. Griffin. 1997. Chemical and physical characterisation of calcified red algal deposits known as maerl. J. Appl. Phycol. 9: 11–17.

Briand, X. 1991. Seaweed harvesting in Europe. pp. 259–308. *In*: Guiry, M.D. and G. Blunden (eds.). Seaweed Resources in Europe: Uses and Potential. John Wiley & Sons Ltd., Chichester.

Chatzissavvidis, C. and I. Therios. 2014. Role of algae in agriculture. pp. 1–37. *In*: Pomin, V.H. (ed.). Seaweeds: Agricultural Uses, Biological and Antioxidant Agents. Nova Science Publishers, New York. ISBN: 978-1-63117-571-8

Desouches, M.-J. 1972. La récolte de goémon et l'ordonnance de la Marine. Annales de Bretagne 79(2): 349–371.

Đurđević, J., S. Vasić, I. Radojević, G. Đelić and L. Čomić. 2014. Antibacterial, antibiofilm and antioxidant activity of Potamogeton nodosus Poir. extracts. Kragujevac Journal of Science 36: 137–144.

Fareed, M.F., A.M. Haroon and S.A. Rabeh. 2008. Antimicrobial activity of some macrophytes from lake Manzalah (Egypt). Pak. J. Biol. Sci. 11(21): 2454–2463.

Ferreira, A.B. 1968. A Ilha Graciosa. *In*: Chorographia – Centro de Estudos Geográficos da Universidade de Lisboa (IAC), Lisboa.

Fuente, A. 2018. El Milagro del Ocle Asturiano: de Desecho del Mar a 'Oro Rojo' de la Alta Cocina. Available online at: https://www.elmundo.es/papel/historias/2018/11/19/5bf154bce2704ea3608b45dc.html (accessed on December 7, 2018)

Gavazzi, M. 1974. Die Nutzniessung des Tangs in den Volksüberlieferungen der Europäischen Meeresküsten. pp. 123–138. *In*: Memoriam António Jorge Dias I, Lisboa.

Gruchy, G.F.B. 1957. Medieval and Tenures in Jersey. Bigwoods, Jersey, 226 pp.

Guimarães, A. 1916. Os Sargaceiros (Litoral Minhoto). Terra Portuguesa: 17–22.

Guiry, M.D. and L. Morrison. 2013. The sustainable harvesting of *Ascophyllum nodosum* (Fucaceae, Phaeophyceae) in Ireland, with notes on the collection and use of some other brown algae. J. Appl. Phycol. 25(6): 1823–1830.

Headrick, J. 1807. View of the Isle of Arran. Archibald Constable & Co., Edinburgh.

Hendrick, J. 1898. The use and value of seaweed as manure. Transactions of the Highland and Agricultural Society of Scotland V(X): 118–134.

Johansen, H.W. 1981. Coralline Algae: A First Synthesis. CRC Press, Boca Raton, FL, 249 pp.

Kenicer, G., S. Bridgewater and W. Milliken. 2000. The ebb and flow of Scottish seaweed use. Bot. J. Scotl. 52(2): 119–148.

Landolt, C. 1915. Folk-Lore Varzino. Póvoa do Varzim.

Landsborough, D. 2012. Arran: Its Topography, Natural History and Antiquities, by the Landsborough's, Father and Son. RareBooksClub.com, 118 pp.

Leclerc, V. and J.-Y. Floc'h. 2010. Les Secrets des Algues. Editions QUAE GIE, Versailles, France, 172 pp.

Lightfoot, J. 1777. Flora Scotica: or, a systematic arrangement, in the Linnaean method, of the native plants of Scotland and the Hebrides. B. White, London, 544 pp.

Matos, M.C. 2018. Caraterização do Sargaço. Grupo dos Sargaceiros da Casa do Povo de Apúlia. Available online at: http://www.sargaceiros.com.pt/caracterizacao.html (accessed on December 6, 2018).

Martin, M. and D. Monro. 1703. A Description of the Western Isles of Scotland Circa 1695. Reprint edited by Macleod, D.J. 1994. Birlinn Ltd., Edinburgh, 392 pp.

Mauriès, M. 1875. Recherches historiques et littéraires sur l'usage de certaines algues. Bulletin de la Société Académique de Brest 2: 1–43.

Medeiros, C.A. 1967. A Ilha do Corvo, pp. 174–180. *In*: Chorographia – Centro de Estudos Geográficos da Universidade de Lisboa (IAC), Lisboa.

Mesnildrey, L., C. Jacob, K. Frangoudes, M. Reunavot and M. Lesuer. 2012. Seaweed Industry in France. Report. Interreg program NETALGAE. Les Publications du Pôle Halieutique Agrocampus Ouest n° 9, 34 pp.

Moffatt, A.A. 1915. Seaweed as a source of potash for agriculture. Transactions of the Highland and Agricultural Society of Scotland V(XXVII): 281–286.

Monagail, M.M., L. Cornish, L. Morrison, R. Araújo and A.T. Critchley. 2017. Sustainable harvesting of wild seaweed resources. Eur. J. Phycol. 52(4): 371–390.

Noble, R.R. 1975. An end to 'wrecking': the decline of the use of seaweed as a manure on Ayrshire coastal farms. Folk Life 8: 13–19.

Oliveira, E.V., F. Galhano and B. Pereira. 1990. Actividades Agro-Marítimas em Portugal. Publicações D. Quixote, Lisboa, 238 pp. ISBN: 972-20-0792-0

Pereira, L. 2009. Guia Ilustrado das Macroalgas – Conhecer e Reconhecer Algumas Espécies da Flora Portuguesa. Imprensa da Universidade de Coimbra, Coimbra, Portugal, 90 pp. ISBN 978-989-26-0002-4

Pereira, L. 2010a. Littoral of Viana do Castelo: Uses in Agriculture, Gastronomy and Food Industry (Bilingual). Câmara Municipal de Viana do Castelo, Portugal, 68 pp. ISBN: 978-972-588-218-4

Pereira, L. 2010b. Littoral of Viana do Castelo: ALGAE (Bilingual). Câmara Municipal de Viana do Castelo, Portugal. 68 pp. ISBN: 978-972-588-217-7

Pereira, L. and F. Correia. 2015. Algas Marinhas da Costa Portuguesa—Ecologia, Biodiversidade e Utilizações. Nota de Rodapé Edições, Paris, 340 pp. ISBN: 978-989-20-5754-5

Pereira, L. 2016. Edible Seaweeds of the World. Science Publishers' (SP), An Imprint of CRC Press/Taylor & Francis Group, Boca Raton, FL, 448 pp. ISBN-13: 978-1498730471

Pérez, R. 1997. Ces Algues Qui Nous Entourent. Conception Actuelle, Rôle dans la Biosphère, Utilisations, Culture. ÉditionIfremer, Plouzané, 272 pp.

Pilon, M. 1980. Les moliceiros, bateaux porteurs de la mémoire et de l'identitéfluviomaritime de la Ria d'Aveiro (Portugal). pp. 1–23. *In*: Heritage, Uses and Representations of the Sea, Porto. Available on line at:
  http://www.citcem.org/encontro/pdf/new_03/TEXTO%202%20-%20Mathilde%20Pilon.pdf (accessed on December 19, 2018).

Quayle, T. 1815. General View of the Agriculture and Present State of the Islands on the Coast of Normandy, Subject to the Crown of Great Britain, 148 pp. Nabu Press, Charleston, South Carolina.

Rasmussen, H. 1974. The Use of Seaweed in the Danish Farming Culture. pp. 385–398. *In*: Memoriam António Jorge Dias I, Lisboa.

Rodrigues, J.E.M. 1963. Contribuição para o conhecimento das Phaeophyceae da costa Portuguesa. Separata das Memórias da Sociedade Broteriana XVI: 1–124.

Rodríguez, L.F. 2018. Historia de Gozón (a través de sus mujeres). Available online at: http://leyendesasturianes.blogspot.com/2012/05/recogida-del-ocle.html (accessed on December 7, 2018)

Sadasivam, V.S., G. Packiaraj, S. Subiramani, S. Govindarajan, K.P. Ganesan, Manju, V. Kalamani, L. Vemuri and J. Narayanasamy. 2017. Evaluation of Seagrass Liquid Extract on Salt Stress Alleviation in Tomato Plants. Asian J. Plant. Sci. Res. 16(4): 172–183.

Sarmento, C. 2010. Cultura Popular Portuguesa e o Discurso do Poder: Práticas e Representações do Moliceiro. e-Cadernos CES 10: 37–69.

Sauvageau, C. 1920. Utilisationdes Algues Marines. Gaston Doi, Paris, 390 pp.

Silva, M. 1958. A evolução de um município. Póvoa do Varzim. Boletim Cultural I: 11–32.

Silva, J.F. and R.W. Duck. 2001. Historical changes of bottom topography and tidal amplitude in the Ria de Aveiro, Portugal – trends for future evolution. Climate Research 18(1): 17–24.

Stanley, M.S., M.E. Callow, R. Perry, R.S. Alberte, R. Smith and J.A. Callow. 2002. Inhibition of fungal spore adhesion by zosteric acid as the basis for a novel, nontoxic crop protection technology. Phytopathology 92: 378–383.

Steele, S., H.J. Lyons, M. Guiry and S. Kraan. 2010. The Seaweed of Ireland's Coastline. The Heritage Council. Available online at:
http://www.seaweed.ie/irish_seaweed_contacts/doc/Seaweed_Poster_2010.pdf (accessed on January 21, 2019)

Sumedha, C., S. Pati, B.P. Dash and A. Chatterji. 2016. Seaweeds – promising organic fertilizers. Science Reporter June: 34–35.

Videment, E. 1909. Les Herbes Marines, Utilisation et Réglementation. Thèse. Université de Rennes. Rennes.

Macroalgae Polysaccharides in Plant Defense Responses

Roberta Paulert[1]* and Marciel João Stadnik[2]

[1] Department of Agronomic Sciences, Federal University of Parana (UFPR), Rua Pioneiro,
 n. 2.153, 85.950-000 Palotina, Paraná, Brazil
[2] Laboratory of Plant Pathology, Agricultural Science Center, Federal University of Santa Catarina
 (UFSC), Rodovia Admar Gonzaga, n. 1.346, 88.034-001 Florianópolis, Santa Catarina, Brazil

1 Introduction

Protecting plants from diseases generally requires a massive use of chemical compounds, such as fungicides. However, some of them cause unintended secondary effects such as occurrence of resistant strains and have an impact on environment and human health. This realization should shift the focus of research from the laboratory to the field and lead to the development of alternative strategies (Ramírez-Carrasco et al. 2017), such as the use of biocontrol products that can activate plant defenses (Gozzo and Faoro 2013) and/or exhibit direct toxicity against pathogens (Hamed et al. 2018). Moreover, modern methods and crop improvement efforts should include special compounds to prime induced resistance in the field (Mauch-Mani et al. 2017) and, above all, to select for induced heritable states in progeny that is primed for defense (Ramírez-Carrasco et al. 2017).

Polysaccharides represent a wide diversity of biomolecules with potential applications in agriculture (Vera et al. 2011, Trouvelot et al. 2014) and are defined as molecules with a degree of polymerization generally higher than 20-25 (Courtois 2009). It seems important to consider the sugar composition, sugar sequences, glycosidic linkage, degree of polymerization, structural variations such as substitution degrees with acetates or sulfates, and the conformation for the possible interactions with the cells (Delattre et al. 2005, Vera et al. 2011).

In this scenario, macroalgae extracts containing polysaccharides are widely studied since different aspects of potential applications in agriculture are well described (Trouvelot et al. 2014, Hamed et al. 2018). Algal extracts have long been used as biofertilizers, biostimulators, and soil conditioners (Selvam and Sivakumar 2013, Nayar and Bott 2014, Sharma et al. 2014, Hamed et al. 2018) and have the ability to protect plants by inducing resistance (Stadnik and de Freitas 2014, Trouvelot et al. 2014). Improved seed germination, enhanced growth and higher yields were observed upon plant treatment with algal extracts (Santos 2016, Vijayanand et al. 2014). Some polymers are composed of rare sugars such as rhamnose and fucose (Klarzynski et al. 2003, Paulert et al. 2010) and may act as elicitors, able to activate defense mechanisms and consequently increase disease resistance in host plants. Thus, seaweed extracts could be used in biocontrol-friendly crop management strategies (Vijayanand et al. 2014).

---
*Corresponding author: roberta@ufpr.br

Seaweed aquaculture is growing rapidly over the last decade owing to a large and diverse array of applications and uses of macroalgal products (Nayar and Bott 2014). Cultivated seaweeds mainly include different communities of red, brown and green macroalgae and improve soil physical and chemical properties (Hamed et al. 2018). The industry is largely based on seaweed extracts such as hydrocolloids for use in fertilizers and bioactive products (Selvam and Sivakumar 2013, Nayar and Bott 2014). The polymers found in the brown algae are presented in several fertilizers and are currently the principal molecules in phytosanitary products (Nayar and Bott 2014, Vijayanand et al. 2014).

In nature, plants are incessantly in contact with pathogenic microorganisms and, unlike mammals, lack mobile defender cells and an adaptive immune system. Instead, they rely on the innate immunity based on systemic signals from infection sites and developed a complex array of recognition and defense reactions to protect themselves against microbial pathogens (Jones and Dangl 2006, Jung et al. 2009).

The innate ability of plants to detect pathogens is essential for their survival and thus they respond to attack with the activation of transcriptional and biochemical pathways that serve to prevent further damage by pathogens (Nürnberger and Brunner 2002). It is well known that the defense events include the expression of a large number of genes that encode a diverse array of proteins and signaling molecules (Lemaître-Guillier et al. 2017). Altogether, these reactions reduce pathogen growth. However, plants are able to defend themselves after the recognition at the cell surface via the perception of inducers or signal molecules, known as elicitors (Trouvelot et al. 2014). First, the plant immune system uses transmembrane pattern recognition receptors (PRRs) that respond to molecular patterns associated with microorganisms (Jones and Dangl 2006). These general signal molecules belong to various biochemical classes including (glyco)proteins, peptides, carbohydrates, and lipids (Nürnberger and Brunner 2002). Second, the plant immune system acts inside the cell, using the nucleotide binding (NB) and leucine-rich repeat (LRR) proteins (Jones and Dangl 2006). Similarities have been revealed between the molecular basis of innate immunity in plants and that known for insects and animals (involving TLR, LRR proteins, MAPK-mediated activation of immune response genes) (Berri et al. 2017), supporting the idea of a common, early evolutionary origin of eukaryotic non-self recognition systems instead of the recognition of general elicitors by plants (Cosse et al. 2009, Nürnberger and Brunner 2002). The pathogen defense and antioxidant processes are ancient mechanisms, which have been conserved during the evolution of eukaryotes, since transcriptional responses in elicited brown algae resemble PAMP-induced patterns in land plants, with similarities in terms of timing, signal transduction pathways and gene induction (Cosse et al. 2009).

In the past years, the general term "elicitor" has been replaced by new ones, such as pathogen- or microbe-associated molecular patterns (PAMPs/MAMPs) (Aziz et al. 2007, Trouvelot et al. 2008). Since elicitors/PAMPs are molecules found in both pathogenic and non-pathogenic microorganisms, the use of the alternative term "microbe-associated molecular pattern" (MAMP) has been suggested and these molecules belong to different families including proteins, carbohydrates and lipids (Varnier et al. 2009).

When the carbohydrates are sprayed on the leaf surface, they have to penetrate through the hydrophobic cuticle to reach epidermal or guard cells (Trouvelot et al. 2014) to be perceived by transmembrane PRRs and trigger signaling events and defense reactions (immune responses) (Jones and Dangl 2006). They can also enter the leaf along the surfaces of the stomatal pores. Microorganisms living in the phyllosphere (bacteria, oomycete, fungi) secrete enzymes to hydrolyze into elicitor-active oligosaccharides. Depending on their structure, the released fragments may induce defense signaling and responses (Trouvelot et al. 2014).

The resistance mechanisms exhibited by plants may be grouped into constitutive and inducible, and efficient morphological, structural barriers accomplish general resistance and chemical defense mechanisms in the plant (Ferreira et al. 2007, Sels et al. 2008). The preformed structural barriers are the cuticle layer and the rigid lignin deposits in the cell wall (Trouvelot et al. 2014). Chemical barriers include the presence of antimicrobial components (Sels et al. 2008). Together, these barriers form a first line of defense and can prevent successful invasion by most of the microorganisms. In addition, plants possess inducible defense mechanisms that are activated upon pathogen attack and include the generation of reactive oxygen species (ROS), accumulation of secondary metabolites like phytoalexins and phenolic compounds, hypersensitive response and the production of a large variety of pathogenesis-related (PR) proteins (Sels et al. 2008, Vera et al. 2011, Trouvelot et al. 2014).

The first line of induced defense in plants is the rapid and transient production of huge amounts of ROS such as superoxide ($\cdot O_2^-$) and hydrogen peroxide (H_2O_2) during the so-called oxidative burst (Wojtaszek 1997). Reactive oxygen species are sensitive indicators (Melcher and Moerschbacher 2016) and H_2O_2 is described as key role in resistance responses against pathogens (Aziz et al. 2003, Trouvelot et al. 2008); it is involved in phytoalexin production, lipid peroxidation and defense-related gene expression (Gauthier et al. 2014). The measurements of H_2O_2 allow detection of new bioactive substances for elicitor and priming agents and can be also used to understand structure-function relationships of derivatives (Melcher and Moerschbacher 2016).

Within hours these events are followed by a broad spectrum of metabolic modifications that include: (a) stimulation of the phenylpropanoid and fatty acid pathways, (b) production of defense-specific chemical messengers such as salicylic acid or jasmonates, and (c) accumulation of components with antimicrobial activities such as phytoalexins and PR proteins (Sels et al. 2008, Gozzo and Faoro 2013). Often, the increase in the production of reactive oxygen intermediates and the resistance reactions are accompanied by programmed host cell death at the sites of microbial attack named hypersensitive response (Wojtaszek 1997, Ortmann et al. 2006). The reaction is thought to benefit the plant by helping to stop microorganisms at the attack site, but evidence shows that hypersensitive responses are not required to trigger plant defense responses by oligo- and polysaccharide from marine macroalgae (Paulert et al. 2010). In contrast with the situation resulting from a treatment with elicitin or chitosan, the hypersensitive response is not observed, by visual and microscopic examination, after seaweed oligo- and polymer infiltration into leaf tissue (Klarzynski et al. 2000, Paulert et al. 2010). Thus, no damage or cell death was induced, suggesting that these responses are not required for the induction of systemic acquired resistance in some plant models. Laminarin-induced resistance in tobacco (*Nicotiana tabacum*) without inducing tissue damage when infiltrated in leaves (Klarzynski et al. 2000). In addition, sulfated fucan oligosaccharides did not cause cell death in tobacco suspension cultures and displayed no symptoms of cell death in leaves (Klarzynski et al. 2003). The expression of a PR gene (*PR10*) and protection of *Medicago truncatula* against *Colletotrichum trifolii* (Fungi, Ascomycota) were observed without necrotizing activity and did not show any phytotoxic effect even when infiltrated into leaves (Cluzet et al. 2004). By contrast, chemically sulfated laminarin induced resistance in grapevine (*Vitis vinifera* cv. Marselan) against *Plasmopara viticola* (Chromista, Oomycota) through priming of defense responses, including hypersensitive response (Trouvelot et al. 2008). High concentration of λ-carrageenan (1 mg/ml) elicited macroscopic changes upon infiltration (necrotic spots) and can elicit an array of plant defense responses, possibly through an effect of its high sulfate content (Mercier et al. 2001).

Pathogenesis-related proteins consist of a large inducible variety of families in many plant species upon infection with oomycetes, fungi, bacteria, viruses, or insect attack (van Loon et al. 2006, Sels et al. 2008). The term "PR proteins" refers to proteins that are not detectable at basal concentrations in healthy tissues, but for which accumulation at the protein level has been demonstrated under pathological conditions (Sudisha et al. 2012). Most PR proteins are induced through the action of the signaling compounds salicylic acid, jasmonic acid, or ethylene; they possess antimicrobial activities and are involved in defense signaling (van Loon et al. 2006). A considerable overlap exists between PR proteins and antifungal proteins. However, many of the now considered 17 families of PR proteins (van Loon et al. 2006, Sels et al. 2008) do not present any known role in antipathogen activity, whereas among the 13 classes of antifungal proteins, most are not PR proteins (Ferreira et al. 2007). Pathogenesis-related proteins are not used to designate all microbe-induced proteins, including enzymes considered the key in resistance mechanisms such as phenylalanine ammonia-lyase (PAL), which are constitutively present but also increase during most infections (Sels et al. 2008). However, it is important to mention that algae polymers and/or oligomers very often increase both PAL activity and the expression of PR proteins. For instance, the heteropolysaccharide from green macroalgae (ulvan) is able to induce PR1a and PR5 in *Arabidopsis thaliana* (Jaulneau et al. 2010) and PR10 in *Medicago truncatula* (Cluzet et al. 2004) and increase PAL activity in olive trees and tomato (El Modafar et al. 2012, Salah et al. 2018).

Most macroalgal and derived polysaccharides activate plant defense responses and protection against a range of pathogens by activating salicylic acid, jasmonic acid and/or ethylene signaling pathways at a systemic level (Jaulneau et al. 2010, Vera et al. 2011, Gozzo and Faoro, 2013). Furthermore, plants have developed diverse mechanisms of defense that potentiate their innate immune system for stronger and/or faster activation of the defense responses against different types of stress; interestingly, some macroalgae polysaccharides/oligosaccharides induce resistance via priming. Nowadays there are numerous studies on the molecular bases of priming (Ramírez-Carrasco et al. 2017).

2 Priming in Plants

The capacity of plant cells to respond to pre-treatment is known as potentiation or priming (Conrath et al. 2015). The response has been known in plants (Conrath et al. 2006) and also in animals (Castro et al. 2006) for many years. Plant resistance priming is equivalent to human vaccines (Martinez-Medina et al. 2016). A prime state is defined as a physiological condition in which plants are able to more strongly and/or more rapidly induce defense responses in many crops to biotic stresses (fungi, bacteria, virus) or abiotic stresses (Varnier et al. 2009, Conrath et al. 2006, Beckers and Conrath 2007). Typically, priming is induced by natural or synthetic compounds (plant activators or priming agents) without elicitor properties (Ton et al. 2005) or used at concentrations that do not stimulate defense responses.

Natural compounds that induce resistance by a priming mechanism include oligogalacturonides, volatile organic compounds, azelaic acid, hexanoic acid and algae carbohydrates (Paulert et al. 2010, Walters et al. 2013, Aranega-Bou et al. 2014). In addition, priming response is usually detected following a pre-treatment of the priming-inducing molecule for at least several hours before the second challenge (Conrath et al. 2006, Ortmann et al. 2006, Beckers and Conrath 2007). A naturally occuring dicarboxylic acid (azelaic acid) has been identified to accumulate and trigger the immune system of *Arabidopsis thaliana*, thereby activating a stronger plant defense response from a secondary infection in distal parts (Jung et al. 2009). When locally infected by pathogens, *Arabidopsis* increases accumulation of the mobile metabolite azelaic acid and the induced AZI1 protein. Since azelaic acid is considered a systemic acquired resistance component, it primes cells to accumulate salicylic acid and confer disease resistance (Jung et al. 2009).

Beckers et al. (2009) reported that the primed state of plants is associated with accumulation of inactive signaling proteins. Thus, benzothiadiazole-primed *Arabidopsis* plants accumulated mRNA and the mitogen-activated protein kinase (MPK3). When plants were then exposed to infection by *Pseudomonas syringae* (Bacteria, Proteobacteria) or to abiotic stress, the MPK3 and MPK6 were more strongly activated in the primed than in the naive cells (Beckers et al. 2009). A proteomic approach showed that primed potato leaves exhibit unique changes in the primary metabolism, associated with selective protein modification via nitric oxide (Arasimowicz-Jelonek et al. 2013).

As described by Trouvelot et al. (2008), a foliar pre-treatment of grapevine (*Vitis vinifera* L.) with sulfated laminarin (with degree of sulfation of 2.4 and degree of polimerization of 25) potentiated the defense responses. Under glasshouse conditions, the polymer efficiently protected a susceptible grapevine cultivar (*Vitis vinifera* cv. Marselan) against downy mildew (*Plasmopara viticola*), induced callose deposition and the jasmonic acid pathway both contributing to the resistance.

Two consecutive treatments of ulvan protected *Medicago truncatula* plants against *Colletotrichum trifolii* more efficiently than one single treatment, in agreement with the *PR10-1* gene expression that persisted for seven days after one treatment and at least 10 days after two consecutive treatments (Cluzet et al. 2004). Paulert et al. (2010) showed that, consistent with this priming activity of ulvan in *M. truncatula*, the pre-treatment of wheat suspension cells increased the chitin-elicited oxidative burst about five- to sixfold. In rice cells, the production of hydrogen peroxide elicited by chitin or chitosan was strongly primed, increasing the burst triggered by the known elicitors alone.

A detailed understanding of the molecular events that take place during a plant–pathogen interaction is an essential goal for disease control in the future. Advances in molecular mechanisms of priming include elevated levels of PRRs, plant heat shock factor as a novel molecular component of primed defense gene activation and systemic acquired resistance activity, elevated levels of dormant signaling enzymes (such as MPKs) and chromatin modifications (Conrath et al. 2015). Studies have shown that epigenetic modifications are mechanisms that enable plant cells to acquire memory and can cause long-term alterations to gene responsiveness (Mauch-Mani et al. 2017).

Priming is a promising alternative approach because it provides a long-lasting, broad-spectrum resistance to biotic and abiotic stress and could provide an effective mechanism for crop protection in the field (Beckers and Conrath 2007, Martinez-Medina et al. 2016). Although many studies are in progress to understand the molecular level of priming (Mauch-Mani et al. 2017), the relevance of the contribution of priming to the extract-induced resistance remains an open question (Martinez-Medina et al. 2016).

3 Oligosaccharides from Algae Elicit Defense Responses and Systemic Resistance

For some decades, active oligosaccharides in plants derived from microbe, plant, or marine macroalgae cell wall structures have been studied for their potential biological activities. Indeed, marine macroalgae constitute an inexpensive possible source of oligomers because they contain a diversity of unique polysaccharides (Courtois 2009).

Many of the elicitors of defense reactions in plants are oligosaccharides (Klarzynski et al. 2003, Delattre et al. 2005) and their biological activity is highly dependent on degree of polymerization, specific sugar sequences and substitution patterns (Trouvelot et al. 2014). The relevance of some of these oligosaccharides as biological signals acting *in vivo* in the defense system is supported by their possible natural occurrence during plant-microbe interactions (Delattre et al. 2005). A pathogen's attack can be mimicked by oligosaccharides derived from marine macroalgae (Potin et al. 1999, Quintana-Rodriguez et al. 2018) and the relevance of some of these oligosaccharides in defense systems is supported by their possible natural occurrence during plant-microbe interactions (Fritig et al. 1998). Evidence has also been provided that macroalgae polysaccharides are sources of oligomers and elicitors of plant defense responses, and enhance plant resistance against pathogens (Klarzynski et al. 2003, Vera et al. 2011). Thus, it is thought that mimicking pathogen attack with such non-specific elicitors might prove useful in the development of alternative strategies for crop protection. Natural extracts and their derivatives bear an as-yet underestimated potential to serve as environmentally friendly vaccines of cultivated plants (Quintana-Rodriguez et al. 2018).

Oligosaccharides are synthesized using glycosyl transferases or by chemical processes, or can be produced from polysaccharides after chemical (acidic) hydrolysis, physical hydrolysis (e.g., microwave irradiation, thermal depolymerization), or enzymatic hydrolysis (from marine bacterium, for example) (Delattre et al. 2005, Courtois 2009). Content between 2 and 10 residues, but degree of polymerization up to 20-25 can also be considered as oligosaccharides or named low molecular weight polysaccharides (Delattre et al. 2005).

The oligosaccharides (oligo-alginates, oligo-carrageenans, oligo-ulvans, oligo-laminarins) prepared by algae polymers are elicitors (Delattre et al. 2005) and induce both early and late defense responses (Aziz et al. 2007). Early responses include extracellular medium alkalinization and release of ROS (among them hydrogen peroxide) acting as secondary messengers. Moreover, a phosphorylation cascade of MAPKs is activated. All this complex cascade of signaling events induces the expression of defense genes. This leads to the induction of PAL, accumulation of salicylic acid and PR proteins, including hydrolytic enzymes (β-1,3-glucanases and chitinases) (Shetty et al. 2009). These defense responses are regulated by phytohormones, such as salicylic acid, jasmonic acid and ethylene. Moreover, plants treated with oligomers can develop systemic acquired resistance (Klarzynski et al. 2003).

In tobacco plants, kappa, lambda and iota oligo-carrageenans induced protection against viral (tobacco mosaic virus, TMV), fungal (*Botrytis cinerea*) and bacterial (*Pectobacterium carotovorum*) infections through activation of PAL activity and the accumulation of phenylpropanoid compounds (Vera et al. 2012). Furthermore, Patier et al. (1995) showed that κ-carrageenan oligomers (hexasaccharide) elicit a rapid glucan hydrolase activity in *Rubus fruticosus* cells.

Laminarin oligomers with an average degree of polymerization 25 and sulfation 2.4 were able to induce PR proteins. On the other hand, smaller oligomers such as laminarin pentaose failed to induce the PR proteins (Ménard et al. 2004). Thus, the structure-activity analysis revealed that a minimum chain length is essential to induce the PR proteins (PR1, PR2, PR3, and PR5) of unsulfated as well as sulfated β-1,3-glucans (Ménard et al. 2005).

In the same way, laminarin oligosaccharides (with a mean degree of polymerization of 10) were perceived by tobacco suspension-cultured cells and induced within a few minutes of alkalinization of the extracellular medium and a transient release of H_2O_2. After a few hours, a strong stimulation of PAL and lipoxygenase activities occurred, as well as accumulation of salicylic acid (Klarzynski et al. 2000). Indeed, structure-activity studies with laminarin and its derivates showed elicitor effects on tobacco with laminaripentaose being the smallest elicitor-active structure. On the other hand, laminarin with degree of polymerization of four showed a weak elicitor activity of PAL and laminarin trimer did not induce the enzyme (Klarzynski et al. 2000).

Oligo-β-1,3-glucans with degree of polymerization (DP) of 6 and 10, obtained from the brown algae *Laminaria digitata* (Ochrophyta, Phaeophyceae), induced the production of H_2O_2 while the β-1,3 glucan of DP of 3 had no effect in grapevine cells (Aziz et al. 2007). The *Botrytis cinerea* (Fungi, Ascomycota) infection in grapevine leaves (*Vitis vinifera* L.) was significantly reduced (by 30-38%) by larger oligomers of β-1,3-glucans (DP >6), whereas oligomers with lower DP had no significant effect (Aziz et al. 2007).

Interestingly, Fu et al. (2011) observed that β-1,3-glucan with low DP played a vital role in the rapid responses (faster extracellular alkalinization and more intensive in stomatal movement), whereas β-1,3-glucan with high DP was responsible for the longer-term effects (better protection against TMV in tobacco).

The treatment of tomato seedlings with the sulfated oligoulvans from *Ulva lactuca* (Chlorophyta) polysaccharide (average DP of 2, prepared from the ulvan using an ulvan lyase) was accompanied by a strong phenolic compound and salicylic acid accumulation. Furthermore, a great stimulation of PAL activity, reaching a maximum activity of 2-2.5 times that of the ulvan polymer, was detected (El Modafar et al. 2012). Consistent with these results, the treatment of tomato seedlings with oligosaccharides significantly reduced (44%) wilt development by *Fusarium oxysporum* f. sp. *lycopersici* (Fungi, Ascomycota) (El Modafar et al. 2012).

Ulvan small oligosaccharides were formed, thereby obtaining alternative products with a modulated bioactivity (Coste et al. 2015), with a disaccharide and a tetrasaccharide as the principal end-products. Xylose-containing oligosaccharides (trisaccharide and pentasaccharide) were also formed, although they were not very abundant (Konasani et al. 2018). Although these oligomers were produced, to our knowledge they were not yet tested in plant biological systems. In the domain structure studies of the first ulvan lyase described, Melcher and co-workers (2017) found, for the first time, that its C-terminal domain is a substrate-binding domain with specificity and high affinity for ulvan. The recombinant ulvan-binding protein is potentially quite powerful for characterization and manipulation.

The elicitor ability of the oligosaccharides to induce natural defense response is influenced by various factors, particularly the monosaccharide composition, sulfate content (Mercier et al. 2001, Klarzynski et al. 2003, Li et al. 2008) and length of the oligosaccharidic chain (Aziz et al. 2007). It seems that at least a DP of 2 is needed to induce biological responses (El Modafar et al. 2012) and DP between 5 and 12 confers high elicitor activities (Klarzynski et al. 2000, Aziz et al. 2007). However, some oligosaccharides with DP higher than 20 are elicitors of plant defense mechanisms (Ménard et al. 2004).

Some macroalgae carbohydrates carry sulfate groups that are important for their biological function. It is well established that the sulfate groups of some oligosaccharides mediate recognition in biochemical and physiological processes occurring in plants (Ménard et al. 2004).

3.1 Sulfate groups and their role in defense activities in plants

The structure and composition of algal compounds were shown to be crucial for their activities on signaling pathways regulating defense mechanism (Mercier et al. 2001, Klarzynski et al. 2003). Oligo- and polysaccharides naturally containing sulfate substituents often play a critical role in major physiological functions in plants and the chemical sulfation can modulate and dramatically change their biological activities in plants (Ménard et al. 2004). The position of sulfate groups is important to the biological activities of algae polysaccharides (Li et al. 2008) and the pattern of sulfate substitution in carrageenans is crucial for recognition (Ménard et al. 2004).

The polysaccharide from red seaweed λ-carrageenan induced plant immune system via jasmonic acid in *Arabidopsis* accompanied by smaller lesions of *Sclerotinia sclerotiorum* (Fungi, Ascomycota) and the resistance was related, at least partly, to the degree of sulfation (Sangha et al. 2010).

For example, sulfated oligosaccharides obtained from fucan and carrageenans were shown to induce the salicylic acid signaling pathway but not the non-sulfated molecules (Mercier et al. 2001, Klarzynski et al. 2003). In addition, the differential expression of the tobacco defense genes by the red algal sulfated carrageenan was related to the sulfate groups and λ-carrageenan was revealed to be a very potent elicitor of salicylic acid (induced four-fold), which suggests a relationship with the higher level of sulfation (Mercier et al. 2001).

In tobacco, the stimulation of PAL activity was accompanied by a systemic resistance to TMV and was associated to the sulfate substituent of the brown alga *Pelvetia canaliculata* (Ochrophyta, Phaeophyceae) oligofucans (Klarzynski et al. 2003), and red and brown algae galactans (Laporte et al. 2007). Furthermore, the treatment of tobacco seedlings with sulfated laminarin was accompanied by total resistance to TMV, whereas treatment with unsulfated laminarin induced a low resistance (Trouvelot et al. 2008).

The biological properties of sulfated laminarin, infiltrated in tobacco leaves, clearly demonstrated that the activity increases with increasing degree of sulfation (DS) and a higher DS than 0.4 is required to trigger acidic and basic PR protein expression. A DS of 1.5 seems to be sufficient to achieve maximal activity (Ménard et al. 2004).

Based on studies of structure-activity relationships, the chemical sulfation of laminarin led to a molecule with improved biological properties (essential for the plant resistance), which were not induced by the native unsulfated molecule and cannot be replaced by other anionic groups (Ménard et al. 2004). The new semisynthetic oligomer increased defense and resistance-eliciting activities in tobacco and *Arabidopsis thaliana* both in tissue culture and in whole plants (Ménard et al. 2004). In tobacco, it induced oxidative burst and caused accumulation of the phytoalexin and PR1, PR2, PR3, and PR5 proteins as well as induction of total immunity to TMV infection. This opens new routes for the development of new compounds suitable for crop protection, since chemical sulfation of naturally occurring polysaccharides can dramatically change their biological activities in plants to increase their defense and resistance-eliciting activities (Ménard et al. 2004).

The desulfation of the ulvan caused a suppression of its elicitor capacity of PAL activity in tomato seedlings (El Modafar et al. 2012). In addition to the sulfate group in ulvans, their high content in rhamnose might be involved in the elicitor capacity as shown with various bacterial lipopolysaccharides (Varnier et al. 2009). On the other hand, the native sulfated ulvan (18.9% of sulfate) and its chemically sulfated derivatives (from 20.9% to 36.6%) similarly reduced the severity of pathogenic fungi infections. Plants of *Arabidopsis thaliana* were sprayed once (i.e., three days before inoculation) with water (control), native ulvan or its chemically sulfated derivatives (1 mg/ml), and all polymers showed the same efficiency in reducing the lesion diameter by 60% and 35% in *Alternaria brassicicola* and *Colletotrichum higginsianum* (Fungi, Ascomycota), respectively. Thus, the authors suggest that ulvan-induced resistance in *A. thaliana* against *A. brassicicola* (necrotroph) and *C. higginsianum* (hemibiotroph fungus) does not depend directly on sulfation degree, at least against these two pathogens (de Freitas et al. 2015).

4 Marine Polysaccharides Involved in Plant Immune System

Naturally occurring carbohydrates formed by chains of sugar residues interconnected by glycosidic linkages have been described with activity in plant immunity, including chitin hexamer, chitosan oligomers, bacterial exopolysaccharides, oligosaccharins, and oligogalacturonides (Ortmann et al. 2006, Courtois 2009, Trouvelot et al. 2014). In this context, marine algae could be interesting sources of active molecules since numerous classes of complex carbohydrates working as priming agents or elicitors have been identified (Vera et al. 2011, Trouvelot et al. 2014). The resulting resistance is rarely complete, with most inducing agents reducing disease by between 20% and 85%. However, if these elicitor applications are successful in reducing pathogen inoculum, less fungicide might be required later in the season (Walters et al. 2013).

The most notable polysaccharides involved in plant immune system and of interest for applications in plant protection are ulvans, fucoidans, laminarin, alginates and carrageenans (Fig. 1) (Mercier et al. 2001, Aziz et al. 2003, Klarzynski et al. 2003, Cluzet et al. 2004, Paulert et al. 2010, Jaulneau et al. 2010).

4.1 Polysaccharides in green seaweeds (Chlorophyta)

Among the three main taxonomic groups of marine macroalgae (Chlorophyta, Phaeophyceae and Rhodophyta), green algae are valuable sources of structurally diverse bioactive compounds and remain largely unexploited to protect plants from pathogens. Scientific attention has been increasingly focused on the research of green biopolymers due to their unique ability to increase resistance toward phytopathogenic organisms (Jaulneau et al. 2010, de Freitas and Stadnik 2015).

The cell wall polysaccharides synthesized by green macroalgae represent 38-54% of the dry algal matter. The major water-soluble polysaccharide is ulvan and represents 8-29% of the algae dry weight. Besides ulvan, there are three other polysaccharide families in the *Ulva* cell wall: the major one is insoluble cellulose and the two minor ones are alkali-soluble linear xyloglucan and an anionic polymer glucuronan (β-(1,4)-polyglucuronic acid) (Redouan et al. 2009, Lahaye and Robic 2007).

Ulvan was widely reported to induce resistance against a broad spectrum of pathogens in many plant species (Cluzet et al. 2004, Jaulneau et al. 2010, Paulert et al. 2010, de Freitas and Stadnik 2012). However, few papers have reported induced resistance of plants or the agricultural uses of glucuronan oligo- and polymer (Redouan et al. 2009, Elboutachfaiti et al. 2009, Abouraïcha et al. 2017). For instance, the severity of blue mold (*Penicillium expansum*) and gray mold (*Botrytis cinerea*) (Fungi, Ascomycota) was reduced by oligoglucuronans from *Ulva lactuca* (average DP of

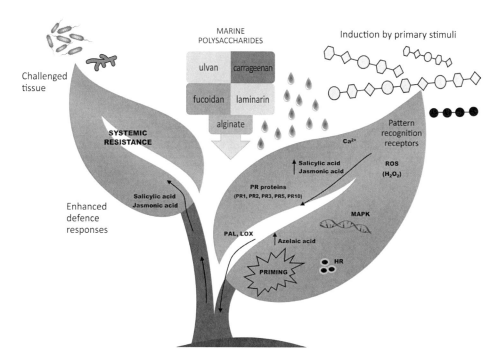

Figure 1. Outcome of sprayed marine polysaccharides at the leaf surface. Carbohydrates have to be perceived by pattern recognition receptors (PRRs) and induce defense signaling events and reactions. The polymers or oligomers induce free cytosolic calcium variation, an oxidative burst, mitogen-activated protein (MAPK) activation and sometimes hypersensitive response. The production of pathogenesis-related (PR) proteins and stimulation of phenylalanine ammonia-lyase (PAL) and lipoxygenase (LOX) are observed. Central regulatory molecules are salicylic acid, jasmonic acid and azelaic acid. The treatment may induce resistance through priming and increases plant protection against future stresses by systemic resistance.

3) on apple fruit. The disease reduction was accompanied by production of hydrogen peroxide and phenolic compounds and by increased activity of PAL and peroxidase (Abouraïcha et al. 2017).

The marine green macroalgae belonging to Ulvaceae family stimulated interest as sources of rare sugars and sulfated polymers with innovative structure and biotechnological applications (Paradossi et al. 2002, Massironi et al. 2019). One of the most popular types of algae offering protection against stress imposed by pathogens is the genus *Ulva* (Order Ulvales, Family Ulvaceae).

The cell wall anionic polysaccharide containing sulfate groups named ulvan is obtained from hot water extraction of *Ulva* and *Enteromorpha* species (Quemener et al. 1997). However, genetic studies revealed that the green seaweeds *Enteromorpha* and *Ulva* are not of distinct genera (Hayden et al. 2003). Thus, ulvan is the major biopolymeric fraction water-soluble, charged heteropolysaccharide extracted from green algae belonging mainly to *Ulva* species (Ulvales, Chorophyta) (Paradossi et al. 2002, Lahaye and Robic 2007, Robic et al. 2009). The common seaweeds used to obtain ulvan are distributed worldwide and include *U. rigida, U. armoricana* (synonym *U. rigida), U. australis* (formerly *U. pertusa), U. lactuca, U. pseudorotundata* (formerly *U. rotundata*), and *U. linza* (formerly *U. fasciata*) (Quemener et al. 1997, Paradossi et al. 1999, Paradossi et al. 2002, Pengzhan et al. 2003, Kirkendale et al. 2013). These green macroalgae sometimes proliferate in an uncontrolled manner because of environmental factors (eutrophication of sea water) and their biomass can be used as flour for polymer extraction (Paradossi et al. 1999). Thus, it is possible to obtain an interesting polysaccharide using a very simple extraction procedure (by extraction in hot water and precipitation by ethanol) with unique chemical versatility and bioactivity that has emerged as a promising candidate for the potential use of elicitor as plant protectant (Vera et al. 2011).

The yield of ulvan varies from 8% to 44% of the algal dry weight, depending on the extraction, purification procedures, and active growth period of the seaweeds (Jaulneau et al. 2011, Lahaye and Robic, 2007, Robic et al. 2009). The different constitutive sugars are rhamnose (16-29%), xylose (30-33%), glucuronic acid (8-16%), iduronic acid (3.7-9%), and smaller amounts of glucose, mannose, galactose, and arabinose arranged in an essentially linear fashion (Quemener et al. 1997, Pengzhan et al. 2003). Seasonal variation of uronic acid content was observed in *Ulva* sp. extracts (Jaulneau et al. 2011). Although different constitutive sugars are described by enzymatic and partial acid hydrolysis, the main structural repeating motif of ulvan is a disaccharide named glucuronorhamnose or aldobiuronic acid (ulvanobiouronic acid A or B), composed of rhamnose 3-sulfate and uronic acid (Alves et al. 2013, Quemener et al. 1997, Lahaye and Robic 2007). The presence of sulfate groups in uronic acids (Jaulneau et al. 2011) and rhamnose 3-sulfate can be linked to xylose and/or to xylose 2-sulfate. Additionally, rhamnose 3-sulfate can be partially branched at C-2 by single side-chains of glucuronic acid (Lahaye et al. 1999). The polymer molecular mass ranges from medium (180-50 kDa) to high (1200-300 kDa) (Robic et al. 2009, Jaulneau et al. 2010).

The properties of ulvan polymer, based on the induced resistance against plant pathogenic fungi, seem to be independent of their sulfation degree (de Freitas et al. 2015). Although the composition varied among batches, all extracts elicited a reporter gene in a transgenic tobacco line and protected cucumber plants against powdery mildew infection (Jaulneau et al. 2011).

For the potential use of elicitors as plant protectants, an ulvan solution obtained from *Ulva* spp. (1 mg/ml) was sprayed twice on plants of *Medicago truncatula*. Nearly complete protection of plants against *Colletotrichum trifolii*, the causal agent of anthracnose disease, was observed. *In vitro* assay showed no effect on *C. trifolii* development (Cluzet et al. 2004). Furthermore, under greenhouse conditions, ulvan (10 mg/ml) was sprayed twice on bean plants (*Phaseolus vulgaris* L.) and reduced the anthracnose (caused by *Colletotrichum lindemuthianum*) severity by 38% to 60%, without direct antifungal activity against the pathogen (de Freitas and Stadnik 2012, Paulert et al. 2007, Paulert et al. 2009). Interestingly, ulvan increased peroxidase and glucanase activity in bean leaves in the first study to report an increase in activity of plant defense-related enzymes by treatment of a polysaccharide before pathogen infection (de Freitas and Stadnik 2012).

Consistent with its strong elicitor activity described by Cluzet et al. (2004), a high polymer ulvan triggered the protection of wheat (*Triticum aestivum* cv. Kanzler) and barley (*Hordeum vulgare* cv. Villa) plants against *Blumeria graminis* f. sp. *tritici* or *hordei* (Fungi, Ascomycota), respectively. The pre-treatment of whole plants with ulvan (0.1 or 1 mg/ml) reduced the symptom severity of powdery mildew by 45% in wheat and by 80% in barley (Paulert et al. 2010). However, ulvan was not able to induce disease resistance in wheat plants against *Puccinia graminis* (Fungi, Basidiomycota) when the disease was evaluated 11 days post-inoculation (R. Paulert, unpublished data).

Similar results were obtained with ulvan from *Ulva lactuca*, when biochemical changes worked as elicitor by inducing resistance in potato (*Solanum tuberosum* L.) against late blight disease caused by *Phytophthora infestans* (Chromista, Oomycota) in field experiments (Ahmed et al. 2016). Ulvan also reduced by 29% the incidence of disease caused by *Verticillium dahliae* (Fungi, Ascomycota) in olive trees (*Olea europaea* L.) with a significant increase in PAL activity (Salah et al. 2018).

In tomato, treatment with *U. lactuca* elicitors (injected at the internodes) significantly reduced wilt development caused by *Fusarium oxysporum* f. sp. *lycopersici* (Fungi, Ascomycota). After 45 days, the percentage of wilting was markedly reduced when plants were treated with oligoulvans, at 44% compared to the control plants. An elicitor activity of ulvan polymer was obtained by the stimulation of PAL, an increase of phenolic compounds content and an induction of salicylic acid in the leaves of tomato seedlings located above and below the elicitation site (El Modafar et al. 2012).

The green algal polysaccharide ulvan, extracted from *U. rigida* (formerly *U. armoricana*), induced defense responses (via the jasmonic acid pathway) in three plant species cultivated in the greenhouse: common bean, grapevine and cucumber, and later inoculated with three powdery mildew pathogens *Erysiphe polygoni*, *E. necator* and *Sphaerotheca fuliginea* (Fungi, Ascomycota) respectively (Jaulneau et al. 2010). In bean plants, the protection level was up to 90% and two successive applications did not increase the activity (Jaulneau et al. 2011). On grapevine, quantification of powdery mildew symptoms showed a reduction of 50-77% and the protection level on cucumber was 75-90% (Jaulneau et al. 2010).

Under greenhouse conditions, ulvan was sprayed twice in bean plants (*Phaseolus vulgaris* L.) and reduced the severity of anthracnose (caused by *Colletotrichum lindemuthianum*) by 38% when plants were inoculated two days after the second application (Paulert et al. 2009). Indeed, a small reduction (20%) of the disease was observed in rice plants sprayed with ulvan solution and infected with *Pyricularia oryzae* (Fungi, Ascomycota) (Paulert et al. 2010).

The mycelial growth of *Alternaria brassicicola* (Fungi, Ascomycota), *in vitro*, was not inhibited by ulvan. On the other hand, the polysaccharide from *U. fasciata* reduced the severity of *A. brassicicola* infection by 90% in *Arabidopsis* plants (foliar spraying) and the induced resistance required functional NADPH oxidase for ROS production (de Freitas and Stadnik, 2015).

Oxidative burst is widely used as a plant defense marker to monitor elicitor or priming responses (Ortmann et al. 2006). Besides acting in plants, ulvan nanoparticles induced the production of ROS and worked as an immunostimulant in fish macrophages (Fernández-Días et al. 2017). Ulvan, as elicitor, induced a biphasic oxidative burst of *Medicago truncatula* cell-suspension cultures and may therefore reflect the detection of multiple ROS or the high-molecular-weight polymer could be cleaved and processed by enzymes secreted by the plant cells during the oxidative burst (Melcher and Moerschbacher 2016). On the other hand, priming activity with dose dependency in the monocot *Oryza sativa* was observed in the oxidative burst measurement in cell-suspension cultures pre-treated with different concentrations of ulvan (0.012-0.2 mg/ml) and then elicited with chitosan. Interestingly, the heteropolysaccahride ulvan acts as an elicitor in dicots (Cluzet et al. 2004, Jaulneau et al. 2010), but as a priming agent in monocots (Paulert et al. 2010). Furthermore, exopolysaccharides from Gram-negative bacteria act as elicitors in dicots, but as priming agents in monocots (Ortmann et al. 2006). To compare dicot and monocot systems directly is highly relevant in studying the signaling molecules underlying immune surveillance, and the detailed understanding of innate immunity in monocotyledons would be worthwhile in view of the large use of cereal crops (Melcher and Moerschbacher 2016). Besides resistance induction, the ulvan polymer was not able to induce hypersensitive response (did not induce cell death) in wheat or bean (Paulert et al. 2010, de Freitas and Stadnik 2012).

When compared with other polysaccharides originating from red and brown algae, ulvan may be considered a late bloomer (Alves et al. 2013). Ulvan could represent a promising candidate (as nanoparticles, for example) in a wide array of applications (Rydahl et al. 2017, Fernández-Días et al. 2017, Massironi et al. 2019) including the management of plant disease (Stadnik and de Freitas, 2014, Trouvelot et al. 2014). Moreover, ulvan is not phytotoxic and is a renewable source of priming agent and MAMPs/elicitors (Cluzet et al. 2004, Jaulneau et al. 2010, Paulert et al. 2010). Because of its low toxicity for humans and its biodegradable nature (Rasyid 2017), the polymer has the potential to be part of alternative strategies in order to reduce or replace pesticides and design environmentally friendly disease control, promoting the valorization of waste algal biomasses and converting these materials into high-value product. Furthermore, the polysaccharides from *Ulva* species represent an underexploited material and are available from abundant and renewable resources (Alves et al. 2013). However, molecular mechanisms involved in ulvan perception by plant cells are still unknown (Jaulneau et al. 2010) and the mechanisms by which ulvan interferes with the different biological systems need to be further studied. To date, there are no reports about ulvan and/or ulvan-derived oligosaccharides promoting abiotic stress tolerance in plants.

4.2 Polysaccharides in brown seaweeds (Phaeophyceae)

The principal cell wall polysaccharides in brown seaweeds are alginates, and storage carbohydrates are laminarin and fucoidans (Sharma et al. 2014, Vera et al. 2011). Alginates are particularly attractive for their hydrogel properties, while fucoidans are famous for the broad range of bioactive effects attributed to their sulfate substitution. The number of studies about fucoidan has increased tremendously over the past few years because of its potential biological properties, including immunomodulatory activities (Lim et al. 2019, Fernando et al. 2019).

4.2.1 Alginates

Alginates are quite abundant and naturally occurring hydrophilic colloidal unbranched polysaccharides present in brown macroalgae, constituting up to 40% of the dry matter (Zhang et al. 2013). They are linear, consisting mainly of residues of β-1,4-linked D-mannuronic acid and α-1,4-linked guluronic acid, arranged either in heteropolymeric and/or homopolymeric blocks depending on the seaweed source (Zhang et al. 2015, Fernando et al. 2019). Samples collected from different localities have marked differences in the ratio of sugars (Chandía et al. 2004); properties are influenced by the uronic acid composition and by the distribution of the different blocks along the chain.

Alginates in extracts can induce oxidative burst in plants (Vera et al. 2011, Sharma et al. 2014) and the elicitor activity of the polymannuronic acid fraction from Chile increased the PAL and peroxidase activities in wheat plants (Chandía et al. 2004).

Accordingly, in addition to the alginate activity as described by Chandía et al. (2004), the oligomers demonstrated elicitor activity by accumulating phytoalexin and inducing PAL activity in soybean cotyledon bioassays (An et al. 2009). The alginate-degraded products with DP of an average of 6, 7, 8, and 11 were active in increasing the phytoalexin accumulation, although the oligomers with DP of 7 showed the maximal elicitor properties (An et al. 2009). In the same way, the oligomers (DP of 6 and 7) induced the enzyme activities of PAL, peroxidase, and catalase and decreased by 39% the infection caused by *Magnaporthe grisea* (Fungi, Ascomycota) in rice plants (Zhang et al. 2015).

Increasing attention has been paid to alginate oligomers as a new functional material. Thus, the oligosaccharides are produced from an acid-hydrolyzed or heat-degraded alginate, and lyases are also used (Zhang et al. 2015). Indeed, the treatment with alginate-derived oligosaccharides (composed by a mixture of mainly di- and trisaccharides) promoted drought stress resistance in tomato (*Lycopersicon esculentum* Miller) seedlings by increasing the contents of free proline, total soluble sugars, abscisic acid, and the activities of several enzymes. The biomass and dry weight of tomato seedlings can also be promoted significantly by the application of alginate oligomers by foliar spray (Liu et al. 2009).

Concerning abiotic stress, alginate-derived oligosaccharides, when sprayed on the seedling leaf surface, promoted water stress tolerance in cucumber (*Cucumis sativus* L.) by stimulating abscisic acid signaling pathway. Treatment of plants with marine alginate-derived oligosaccharides significantly improved the fresh weight and photosynthesis and transpiration rates. Additionally, it induced the expression of genes involved in abscisic acid and drought resistance genes (*CsRAB18, CsABI5, CsRD29A* and *CsRD22*), thus reversing the effects of drought stress (Li et al. 2018). Furthermore, wheat plants showed enhanced tolerance to drought stress through abscisic acid-dependent signal pathway, and *P5CS* gene was up-regulated when treated with the alginate oligomers (Liu et al. 2013). Wheat plants increased the generation of nitric oxide, which promoted the formation and elongation of roots in a dose-dependent manner by treatment with oligo-alginate (Zhang et al. 2013).

The spraying of alginate oligosaccharides (500 μg/ml of around 20 units of polyguluronic or polymannuronic acid fraction, prepared by chemical depolymerization) obtained from *Lessonia trabeculata* Villouta et Santelices (Phaeophyceae) and *Lessonia flavicans* (formerly *Lessonia vadosa*) (Phaeophyceae) induced both stimulation of growth and defense responses in tobacco plants. For this, tobacco leaves were sprayed with oligosaccharides once a week, for one month. Seven days after the final spraying, the plants treated gained height by 49% and an increase in defense against TMV corresponding to decreases in the number of necrotic lesions of 22% (Laporte et al. 2007).

4.2.2 Fucoidans

The term "fucoidan" is now used, as well as fucan, sulfated fucan, or fucosan. It is commonly applied to a family of sulfated fucose-rich polysaccharides often extracted from algae (Ale and Meyer, 2013), whereas the term "fucan" is reserved for that extracted from marine animals (Holtkamp et al. 2009). Thus, fucoidans are major structural polysaccharides of brown seaweed cell wall containing high percentages of L-fucose (central stem content 34-44%) with sulfate groups in positions C2 and/ or C4, representing 5-20% of the algal dry weight (Holtkamp et al. 2009). Some of them are simple, mainly being composed of fucose and sulfate, and some are heterogeneous and complex because they contain other monosaccharides (such as mannose, galactose, glucose, xylose, arabinose, and rhamnose) and uronic acids (glucuronic acid) (Duarte et al. 2001, Li et al. 2008, Lim et al. 2019). Some are highly branched at position 2. Furthermore, the content and structures of fucoidans show a wide variation between algae species, related to the place of growth (Zvyagintseva et al. 2003, Fernando et al. 2019); thus, the molecular weight varies from 13 to 950 kDa (Li et al. 2008, Holtkamp et al. 2009). For example, fucoidan purified from the alga *Pelvetia canaliculata* contains about 32% fucose and 33% of sulfate groups (Klarzynski et al. 2003) and that obtained from *Fucus vesiculosus* is composed of 44% fucose and 26% sulfate (Li et al. 2008). The fucoidan content can vary from about 5% to 10% (Zvyagintseva et al. 2003).

Reported fucoidan structure can be obtained from different brown seaweed species of the order Chordariales, Laminariales, Ectocarpales and Fucales (Ale and Meyer 2013), such as the following: *Ascophyllum nodosum, Fucus vesiculosus, F. evanescens, Lessonia flavicans* (formerly *Lessonia vadosa*), *Saccharina japonica* (formerly *Laminaria japonica*, an important economic alga in China), *Macrocystis pyrifera, Ecklonia kurome, Stoechospermum polypodioides* (formerly *Stoechospermum marginatum*), *Pelvetia canaliculata, Sargassum stenophyllum,* and *Cladosiphon okamuranus* (Duarte et al. 2001, Wang et al. 2011, Li et al. 2008, Holtkamp et al. 2009, Vera et al. 2011, Lim et al. 2019).

Under greenhouse, carrot plants treated (sprayed) with *Ascophyllum nodosum* extract increased enzyme activity and transcript levels of PAL and chitinase, resulting in reduction of disease severity caused by *Alternaria radicina* and *Botrytis cinerea* (Fungi, Ascomycota) (Jayaraj et al. 2008).

For the past decade, fucoidans isolated from different species have been extensively studied for their varied biological activities because, compared with other sulfated polysaccharides, they are widely and cheaply available (Wang et al. 2011, Klarzynski et al. 2003, Li et al. 2008).

Oligofucans with an average molecular weight less than 10 kDa have been obtained by marine enzyme digestion (endo-fucanase) from *Pelvetia caniculata* and it is important to note that fucan oligosaccharides with elicitor activity are estimated with an average polymerization degree of 10 fucose residues (Klarzynski et al. 2003).

Enzymes catalyzing partial cleavage of seaweed polysaccharides have been proposed to be useful tools for investigating the function relationship of fucoidan polysaccharides. Although bacterial enzymes catalyzing modifications of fucoidan polysaccharides (including sulfatases, fucoidanases and fucosidases) are described (Ale and Meyer 2013), only a few works related with oligomers in plant defense are found.

A fucoidan polymer from *Lessonia flavicans* (formerly *Lessonia vadosa*) containing fucose and about 34% of sulfate groups increased the defense enzymes PAL and lipoxygenase in tobacco plants (Chandía and Matsuhiro 2008). In the same way, fucan oligosaccharides (0.2 mg/ml) induced a rapid alkalinization of the extracellular medium and the release of hydrogen peroxide in tobacco suspension cell culture followed by increased activity of PAL and lipoxygenase. Tobacco leaves accumulated salicylic acid and after 48 hours accumulated PR proteins (PR-1, PR-2, PR-3 and PR-5). Furthermore, treatment with oligofucans induced a strong reduction of both the number and size of lesions induced by TMV. Thus, oligofucans both locally and systemically induced a broad range of late defense responses in tobacco (including markers of systemic acquired resistance), without cell death (Klarzynski et al. 2003).

Potato virus (PVX) infection was reduced in *Datura stramonium* L. when leaves were treated with fucoidan (from *Fucus evanescens*), and less accumulation of viral particles was observed in comparison to the control. The polymer treatment induced intracellular lytic processes causing virus destruction (Reunov et al. 2009). The spread of infection induced by TMV was reduced by fucoidan in the leaves of two cultivars of tobacco. When the leaves were treated with fucoidan before inoculation with TMV, its antiviral activity was not so strong. However, the number of local necrotic lesions decreased by 62% to 90% when a mixture of the virus and the polysaccharide was used as inoculum and electron microscopic analysis showed agglutinated particles (Lapshina et al. 2006). After four days, in the infected and treated leaves of tobacco, specific intracellular granular inclusions related to early stages of the virus reproduction were observed. Thus, fucoidan affected not only the plant but the virus as well and demonstrated that the algae carbohydrate delayed the development of the TMV-induced infection (Lapshina et al. 2007).

4.2.3 Laminarin

Laminarin is a reserve/storage carbohydrate of *Laminaria digitata* and other Laminariales species such as *Saccharina cichorioides* (formerly *Laminaria cichorioides*) (Zvyagintseva et al. 2003, Nayar and Bott 2014). It is a water-soluble linear β-1,3-glucan with an average DP of 25-33 glucose units and with occasional β-(1 → 6) linked branches (Read et al. 1996). The content and structure vary depending on the season, place of growth and age of the seaweed (Zvyagintseva et al. 2003). For example, the content of laminarin in *Saccharina japonica* (formerly *Laminaria japonica*), the most widely distributed brown algae in China, is less than 1% compared with the 10% of the dry weight of *L. digitata* in Europe (Zvyagintseva et al. 2003).

In animals, β-1,3 glucan has been reported to play an important role in immune-stimulatory activities (Kim et al. 2011). Laminarin has been proved as plant elicitor via recognition as MAMPs or PAMPs and thus induced defense reactions and resistance in different plants, such as tobacco (Klarzynski et al. 2000, Ménard et al. 2004, Fu et al. 2011), alfafa (Kobayashi et al. 1993), grapevine (Aziz et al. 2003, Gauthier et al. 2014, Garde-Cerdán et al. 2017) and *Arabidopsis* (Ménard et al. 2004, Wu et al. 2016). As a major component, laminarin has been packed as a commercial product in France in the past few decades. However, the polymer is still far from being fully understood with respect to induced resistance in plants (Fu et al. 2011).

The elicitor activity of the polysaccharides (derivatives of the laminarin) in an extract of *Ascophyllum nodosum* was demonstrated by the increased activity of β-1,3-glucanase in *Rubus fructus* (blackberry) cell-suspension cultures (Patier et al. 1993).

The laminarin (DP of 25 to 40) induced nitric oxide release and H_2O_2 burst in tobacco cell suspension and inhibition of stomatal opening and demonstrated the best inhibition against TMV at 200 μg/ml (Fu et al. 2011). Furthermore, laminarin (200 μg/ml) protected tobacco plants against TMV and the inhibition rate was about 67% based on ethylene-dependent basic plant PR proteins (PR2 and PR3) (Fu et al. 2011). It also protected tobacco against the soft rot pathogen *Erwinia carotovora* (Bacteria, Proteobacteria) (Klarzynski et al. 2000). In a similar manner, Aziz and associates (2003) found that this glucan was an effective elicitor in reducing downy mildew (*Plasmopara viticola*—Chromista, Oomycota) and gray mold (*Botrytiscinerea*—Fungi, Ascomycota) infection of grapevine.

Klarzynski et al. (2000) showed that laminarin (purified from *Laminaria digitata*) elicits a variety of defense reactions in tobacco, conferring resistance to *Erwinia carotovora* (Bacteria, Proteobacteria). Treatment of tobacco leaves with laminarin (infiltrated) causes no tissue damage or cell death, but the pre-treatment induced a strong restriction of soft rot disease symptoms when compared with control leaves. Consistent with its strong elicitor activity, laminarin also induced the accumulation of four PR proteins (PR1, PR2, PR3 and PR5).

In 2004, Ménard and co-workers observed that laminarin induced only a weak resistance in plants. After chemical sulfation, however, laminarin induced the salicylic acid signaling pathway in tobacco and induced strong immunity against TMV infection. Moreover, it possesses elicitor activity by inducing the expression of different families of PR proteins (acidic PR1) in *Nicotiana tabacum*. The activities of laminarin sulfates infiltrated in tobacco leaves clearly demonstrated that a degree of sulfation of 0.4 to 1.5 is required to trigger PR protein expression and the activity increases with increasing the sulfation (Ménard et al. 2004).

Interestingly, enhanced tolerance to abiotic stress (salt and heat) was observed when *Arabidopsis* plants were treated with laminarin. The plants increased fresh weight transcriptome analysis, indicating that defensin-like gene (*DEFL202*) was over expressed and resulted in increased chloroplast stability under stress (Wu et al. 2016).

Laminarin and its sulfated derivative induce resistance in grapevine against downy mildew (caused by the oomycete *Plasmopara viticola*) (Aziz et al. 2003, Trouvelot et al. 2008). Laminarin elicits classical defenses such as oxidative burst, PR proteins and phytoalexin production (Aziz et al. 2003), while the chemically sulfated polymer triggered grapevine resistance via priming (Gauthier et al. 2014). Thus, sulfated laminarin triggered a significantly higher expression of the well-established ROS-related gene (respiratory burst oxidative homolog D, *RbohD*) that correlated with H_2O_2 production and hypersensitive response-like cell death (Trouvelot et al. 2008, Gauthier et al. 2014). It also primed the biosynthesis of salicylic acid (SA) and the expression of SA-marker genes in grapevine plants challenged with *Plasmopara viticola*. It induced a stronger resistance by reducing the *P. viticola* sporulating area by 84% (Gauthier et al. 2014) without effect on sporangia viability and no significant direct toxic effect against zoospores (Trouvelot et al. 2008).

When primed, plants do not trigger notable defense responses, but they respond to very low levels of a stimulus in a faster and stronger manner (Conrath et al. 2015). Acting as a priming agent like ulvan (Paulert et al. 2010), the sulfated derivative of laminarin was unable to elicit ROS production, cytosolic Ca^{2+} concentration variations or mitogen-activated protein kinase activation (Gauthier et al. 2014).

Under greenhouse conditions, sulfated laminarin was sprayed in grapevine (*Vitis vinifera* L. cv. Marselan) and elicited the emission of volatile organic compounds and induction of methyl salicylate, monoterpenes and sesquiterpenes (such as α-farnesene). It is well documented that volatile terpenes are part of the direct or indirect plant defense against pathogens (Chalal et al. 2015). The resistance induced in grapevine by the sulfated polysaccharide is associated with the activation of the primary metabolism especially on amino acid and carbohydrate pathways. Based on the proteomic approach, few proteins such as the 12-oxophytodienoate reductase 11 (OPR-like) could be considered as useful markers of induced resistance (Lemaître-Guillier et al. 2017).

Sulfated laminarin induced higher resistance in the adult leaves of grapevine against *Plasmopara viticola*, the causal agent of grape downy mildew. Thus, the production of hydrogen peroxide, phytoalexins and deposition of phenolics were more abundant in mature than in young leaves. In addition, there was significantly reduced stomatal colonization by zoospores only in mature leaves. Why a young organ is less responsive to resistance inducers still remains to be uncovered (Steimetz et al. 2012). However, the polymer was able to penetrate the leaf cuticle (diffusion via stomata, anticlinal cell walls and trichomes) of *Arabidopsis thaliana* and grapevine only when formulated with a highly ethoxylated surfactant (Paris et al. 2016).

4.3   Polysaccharides in red macroalgae (Rhodophyta)

Polysaccharides in red macroalgae include agars, carrageenans, xylans, and floridean starch (Fernando et al. 2019). Among them, carrageenans are receiving attention for their interesting functional properties in plant defense responses.

Little research on the application of agar polysaccharides from *Pyropia yezoensis* is described, although it is a commercially important edible red alga in Southeast Asia. For instance, polysaccharides (galactopyranose units linked to galactose or galactose 6-sulfate) with different molecular weights were prepared and the results suggested that the low-molecular-weight polysaccharides (3.2 kDa) could regulate antioxidant enzyme activities and modulate intracellular ion concentration, thereby protecting wheat seedlings from salt stress damage (Zou et al. 2018).

Carrageenans are a collective family of sulfated polysaccharides with a linear backbone of D-galactopyranose (galactans) linked to anhydrogalactose units in some cases. They represent the major cellular constituents present in many red seaweeds (from 30% to 75% of the algal dry weight) (Mercier et al. 2001, Vera et al. 2011, Shukla et al. 2016).

Natural carrageenans differ considerably from particular seaweed species and geographic districts. They are mixtures of different sulfated galactans and traditionally split into six basic types: iota (ι)-, kappa (κ)-, lambda (λ)-, mu (μ)-, nu (υ)-, and theta (θ)-carrageenans (Shukla et al. 2016). Two of them, kappa and iota, are the most widely distributed in algae and used in the food industry (Tuvikene et al. 2006). For instance, lambda carrageenan is formed by a D-galactose sulfated in C2 linked to a D-galactose sulfated in C2 and C6 and obtained from the genera *Gigartina* and *Chondrus* (Vera et al. 2011, Shukla et al. 2016). Kappa carrageenan is constituted by a D-galactose sulfated in C4 linked to anhydrogalactose and is commercially extracted from the red alga *Kappaphycus alvarezii* (Shukla et al. 2016). Iota carrageenan, obtained mainly from *Eucheuma denticulatum*, is a sulfated galactose (C4) linked to an anhydrogalactose sulfated in C2 (Vera et al. 2011). Sulfate content varies from 20% in κ-carrageenan to 33% in ι-carrageenan and to 41% in λ-carrageenan (Shukla et al. 2016).

Patier et al. (1995) showed that κ-carrageenan oligomers (hexasaccharide) elicit a rapid glucan hydrolase activity in *Rubus fruticosus* cells. Some studies have shown the biological activity of these polysaccharides and of their oligomers as elicitors of defense responses, enhancing plant immunity, against various plant pathogens or herbivores and as promoters of plant growth by regulating metabolic processes (Sangha et al. 2011, Vera et al. 2011, Castro et al. 2012, Shukla et al. 2016). Probably the activity depends on the degree of sulfation (Sangha et al. 2010, Sangha et al. 2011). Thus, carrageenans hold promise as low-cost alternatives to chemical products. Furthermore, *Arabidopsis* plants treated with carrageenans modulated the resistance against the herbivore *Trichoplusia ni* (cabbage looper). In particular, ι-carrageenan elicited resistance by inducing defense mechanisms including jasmonic acid and SA-dependent pathways and induced several genes involved in indole glucosinolate biosynthesis (Sangha et al. 2011). Glucosinolates have essential roles because they can be easily converted to volatile compounds important against various herbivores. Furthermore, λ-carrageenan induced jasmonic acid-dependent defenses in *Arabidopsis* and tomatoes against a necrotrophic pathogen (*Sclerotinia sclerotiorum*) and tomato chlorotic dwarf viroid (TCDVd) (Sangha et al. 2010, Sangha et al. 2015).

Mercier et al. (2001) reported that λ-carrageenan, when infiltrated into the mesophyll of tobacco leaves, efficiently induced defense genes (ACC oxidase, lipoxygenase, sesquiterpene cylase, basic chitinase and type II proteinase inhibitor) with a few visible necrotic spots and later induced systemic resistance, highlighting that the signaling pathways were mediated by ethylene, jasmonic acid and salicylic acid (Mercier et al. 2001).

From *Kappaphycus alvarezii* the sulfated κ-carrageenan was obtained that induced the expression, through foliar applications, of the PR proteins without toxicity to the chili plants (*Capsicum annuum*). The expression level of peroxidase, PR1 and defensin gene 1.2 (*PDF1.2*) was significantly up-regulated. Consistently, the biopolymer (0.3%) was sprayed on 40-day-old chili plants and induced resistance against anthracnose disease caused by *Colletotrichum gloeosporioides*. An induction of SA- and jasmonic-dependent PR genes by carrageenan reinforced their efficacy in promoting plant defense. Furthermore, the polysaccharide showed fungistatic effect on mycelial growth of *Colletotrichum gloeosporioides* (Mani and Nagarathnam 2018).

Nagorskaya et al. (2010) reported that the development of potato virus X (PTX) in *Datura stramonium* leaves was reduced because carrageenan induced the protein synthesis and the formation of laminar structures able to bind viral particles blocking intracellular survival. Additionally, Reunov et al. (2004) reported the inhibitory effect of carrageenan on the accumulation of TMV in tobacco leaves. The number of necrotic lesions on the leaves inoculated with the mixture of carrageenan (1 mg/ml) and TMV was reduced by 87%, compared to the leaves inoculated only with the virus.

After the leaves were sprayed once a week for a month, tobacco plants treated with oligomers of sulfated galactan (around 20 units) showed an increase in height of 23% and an increase in defense against TMV corresponding to decreases in the number of necrotic lesions of 74% (Laporte et al. 2007). Indeed, the TMV was significantly reduced and fewer spots were observed when the plants were treated with κ-carregeenan obtained from *Hypnea musciformis*. The polymer showed elicitor activity able to activate defense mechanisms by both SA and jasmonic acid or ethylene pathways (Ghannam et al. 2013).

In this manner, the treatment of carrageenans elicits jasmonic acid signaling pathway inducing *PDF1.2, PR3, Def1.2* genes in most plant models studied, although the SA signaling pathway was also observed in some plants inducing *PR1, PR2* and *PR5* genes (Shukla et al. 2016). It is important to add that the antiviral activity of carrageenan may be explained by its action not only on the plant but also on the virus itself when applied together with TMV (Nagorskaya et al. 2008).

5 Conclusion

Marine algae represent an abundant and inexpensive source of bioactive compounds that can be used in agriculture. There is increasing interest in the mechanisms of action of marine oligo- and polysaccharides in plant immune system. The potential roles of marine macroalgae in plant protection improvement are still being studied. Although marine macroalgae have been used as soil conditioners and biofertilizers for a long time, only over the last few decades has the research focused on the capacity of their carbohydrates to induce plant defense responses. However, their biological activities are so attractive that much research is being done on their structures and bioactivities every year.

Seaweed extracts containing oligo- or polysaccharides can be integrated into biocontrol-friendly crop management strategies since seaweed aquaculture is rapidly growing. Although rare reports are related to algae polymer inducing resistance against bacteria, macroalgal products, mainly including brown and green species, represent promising candidates in the management of plant disease caused by fungi or virus. In this context, ulvan represents an underexploited material that is available from abundant biomass, with potential for converting waste materials into high-value products based on its elicitor activity. The heteropolysaccharide ulvan can be used to efficiently protect crop plants against powdery mildew diseases, although plant protection experiments were mainly evaluated in laboratory conditions.

References Cited

Abouraïcha, E.F., Z.E. Alaoui-Talibi, A. Tadlaoui-Ouafi, R. El Boutachfaiti, E. Petit, A. Douira, B. Courtois and C. El Modafar. 2017. Glucuronan and oligoglucuronans isolated from green algae activate natural defense responses in apple fruit and reduce postharvest blue and gray mold decay. J. Appl. Phycol. 29: 471–480.

Ahmed, S.M., S.R. El-Zemity, R.E. Selim and F.A. Kassem. 2016. A potential elicitor of green alga (*Ulva lactuca*) and commercial algae products against late blight disease of *Solanum tuberosum* L. Asian J. Agric. Food Sci. 4: 86–95.

Ale, M.T. and A.S. Meyer. 2013. Fucoidans from brown seaweeds: an update on structure, extraction techniques and use of enzymes as tools for structural elucidation. RSC Adv. 3: 8131–8141.

Alves, A., R.A. Sousa and R.L. Reis. 2013. A practical perspective on ulvan extracted from green algae. J. Appl. Phycol. 25: 407–424.

An, Q.-D., G.-L. Zang, H.-T. Wu, Z.-C. Zhang, G.-S. Zheng, L. Luan, Y. Murata and X. Li. 2009. Alginate-deriving oligosaccharide production by alginase from newly isolated *Flavobacterium* sp. LXA and its potential application in protection against pathogens. J. Appl. Microbiol.106: 161–170.

Aziz, A., B. Poinssot, X. Daire, M. Adrian, A. Bézier, B. Lambert, J.M. Joubert and A. Pugin. 2003. Laminarin elicits defense responses in grapevine and induces protection against *Botrytis cinerea* and *Plasmopara viticola*. Mol. Plant. Microbe Interact. 16: 1118–1128.

Aziz, A., A. Gauthier, B. Bézier, B. Poinssot, J.M. Joubert, A. Pugin, A. Heyraud and F. Baillieul. 2007. Elicitor and resistance-inducing activities of beta-1,4 cellodextrins in grapevine, comparison with beta-1,3 glucans and alpha-1,4 oligogalacturonides. J. Exp. Bot. 58: 1463–1472.

Aranega-Bou, P., M. de la O. Leyva, I. Finiti, P. García-Agustín and C. González-Bosch. 2014. Priming of plant resistance by natural compounds. Hexanoic acid as a model. Front. Plant. Sci. 5: 1–12.

Arasimowicz-Jelonek, M., A. Kosmala, L. Janus, D. Abramowski and J. Floryszak-Wieczorek. 2013. The proteome response of potato leaves to priming agents and S-nitrosoglutathione. Plant Sci. 198: 83–90.

Berri, M., M. Olivier, S. Holbert, J. Dupont, H. Demais, M.L. Goff and P.N. Collen. 2017. Ulvan from *Ulva armoricana* (Chlorophyta) activates the PI3K/Akt signalling pathway *via* TLR4 to induce intestinal cytokine production. Algal Res. 28: 39–47.

Beckers, G.J. and U. Conrath. 2007. Priming for stress resistance: from the lab to the field. Curr. Opin. Plant Biol. 10: 425–431.

Beckers, G.J., M. Jaskiewicz, Y. Liu, W.R. Underwood, S.Y. He, S. Zhang and U. Conrath. 2009. Mitogen-activated protein kinases 3 and 6 are required for full priming of stress responses in *Arabidopsis thaliana*. Plant Cell 21: 944–953.

Castro, R., M.C. Piazzon, I. Zarra, I. Leiro, M. Noya and J. Lamas. 2006. Stimulation of turbot phagocytes by *Ulva rigida* C. Agardh polysaccharides. Aquaculture 254: 9–20.

Castro, J., J. Vera, A. González and A. Moenne. 2012. Oligo-carrageenans stimulate growth by enhancing photosynthesis, basal metabolism, and cell cycle in tobacco plants (var. Burley). J. Plant Growth Regul. 31: 173–185.

Chalal, M., J.B. Winkler, K. Gourrat, S. Trouvelot, M. Adrian, J.-P. Schnitzler, F. Jamois and X. Daire. 2015. Sesquiterpene volatile organic compounds (VOCs) are markers of elicitation by sulfated laminarine in grapevine. Front. Plant Sci. 6: 1–9.

Chandía, N.P., B. Matsuhiro, E. Mejías and A. Moenne. 2004. Alginic acids in *Lessonia vadosa*: partial hydrolysis and elicitor properties of the polymannuronic acid fraction. J. Appl. Phycol. 16: 127–133.

Chandía, N.P. and B. Matsuhiro. 2008. Characterization of a fucoidan from *Lessonia vadosa* (Phaeophyta) and its anticoagulant and elicitor properties. Int. J. Biol. Macromol. 42: 235–240.

Cluzet, S., C. Torregrosa, C. Jacquet, C. Lafitte, J. Fournier, L. Mercier, S. Salamagne, X. Briand, M.-T. Esquerré-Tugayé and B. Dumas. 2004. Gene expression profiling and protection of *Medicago truncatula* against a fungal infection in response to an elicitor from green algae *Ulva* spp. Plant Cell Environ. 27: 917–928.

Conrath, U., G.J.M. Beckers, V. Flors, P. García-Agustín, G. Jakab, F. Mauch, M.-A. Newman, C.M.J. Pieterse, B. Poinssot, M.J. Pozo, A. Pugin, U. Schaffrath, J. Ton, D. Wendehenne, L. Zimmerli and B. Mauch-Mani. 2006. Priming: getting ready for battle. Mol. Plant Microbe Interact. 19: 1062–1071.

Conrath, U., G.J.M. Beckers, C.J.G. Langenbach and M.R. Jaskiewicz. 2015. Priming of enhanced defense. Annu. Rev. Phytopathol. 53: 97–119.

Courtois, J. 2009. Oligosaccharides from land plants and algae: production and application in therapeutics and biotechnology. Curr. Opin. Microbiol. 12: 261–273.

Cosse, A., P. Potin and C. Leblanc. 2009. Patterns of gene expression induced by oligoguluronates reveal conserved and environment-specific molecular defense responses in the brown alga *Laminaria digitata*. New Phytol. 182: 239–250.

Coste, O., E.-J. Malta, J.C. López and C. Fernández-Díaz. 2015. Production of sulfated oligosaccharides from seaweed *Ulva* sp. Using a new ulvan-degrading enzymatic bacterial crude extract. Algal Res. 10: 224–231.

De Freitas, M.B. and M.J. Stadnik. 2012. Race-specific and ulvan-induced defense responses in bean (*Phaseolus vulgaris*) against *Colletotrichum lindemuthianum*. Physiol. Mol. Plant Pathol. 78: 8–13.

De Freitas, M.B. and M.J. Stadnik. 2015. Ulvan-induced resistance in *Arabidopsis thaliana* against *Alternaria brassicicola* requires reactive oxygen species derived from NADPH oxidase. Physiol. Mol. Plant Pathol. 90: 49–56.

De Freitas, M.B., L.G. Ferreira, C. Hawerroth, M.E.R. Duarte, M.D. Noseda and M.J. Stadnik. 2015. Ulvans induce resistance against plant pathogenic fungi independently of their sulfation degree. Carbohydr. Res. 133: 384–390.

Delattre, C., P. Michaud, B. Courtois and J. Courtois. 2005. Oligosaccharides engineering from plants and algae application in biotechnology and therapeutics. Minerva Biotecnol. 17: 107–117.

Duarte, M.E.R., M.A. Cardoso, M.D. Noseda and A.S. Cerezo. 2001. Structural studies on fucoidans from the brown seaweed *Sargassum stenophyllum*. Carbohydr. Res. 333: 281–293.

Elboutachfaiti, R., C. Delattre, E. Petit, M. El Gadda, B. Courtois, P. Michaud, C. El Modafar and J. Courtois. 2009. Improved isolation of glucuronan from algae and the protection of glucuronic acid oligosaccharides using a glucuronan lyase. Carbohydr. Res. 344: 1670–1675.

El Modafar, C., M. Elgadda, R. El Boutachfaiti, E. Abouraicha, N. Zehhar, E. Petit, B. Courtois and J. Courtois. 2012. Induction of natural defence accompanied by salicylic acid-dependant systemic acquired resistance in tomato seedlings in response to bio-elicitors isolated from green algae. Sci. Hortic. 138: 55–63.

Fernández-Díaz, C., O. Coste and E.-J. Malta. 2017. Polymer chitosan nanoparticles functionalized with *Ulva ohnoi* extracts boost in vitro ulvan immunostimulant effect in Solea senegalensis macrophages. Algal Res. 26: 135–142.

Fernando, I.P.S., D. Kim, J.-W. Nah and Y.-J. Jeon. 2019. Advances in functionalizing fucoidans and alginates (bio)polymers by structural modifications: a review. Chem. Eng. J. 355: 33–48.

Ferreira, R.B., S. Monteiro, R. Freitas, C.N. Santos, Z. Chen, L.M. Batista, J. Duarte, A. Borges and A.R. Teixeira. 2007. The role of plant defence proteins in fungal pathogenesis. Mol. Plant Pathol. 8: 677–700.

Fritig, B., T. Heitz and M. Legrand. 1998. Antimicrobial proteins in induced plant defense. Curr. Opin. Immunol. 10: 16–22.

Fu, Y., H. Yin, W. Wang, M. Wang, H. Zhang, X. Zhao and Y. Du. 2011. β-1,3-glucan with different degree of polymerization induced different defense responses in tobacco. Carbohydr. Polym. 86: 774–782.

Garde-Cerdán, T., V. Mancini, M. Carrasco-Quiroz, A. Servili, G. Gutiérrez-Gamboa, R. Foglia, E.P. Pérez-Alvarez and G. Romanazzi. 2017. Chitosan and laminarin as alternatives to copper for *Plasmopara viticola* control: effect on grape amino acid. J. Agri. Food Chem. 65: 7379–7386.

Gauthier, A., S. Trouvelot, J. Kelloniemi, P. Frettinger, D. Wendehenne, X. Daire, J.-M. Joubert, A. Ferrarini, M. Delledonne, V. Flors and B. Poinssot. 2014. The sulfated laminarin triggers a stress transcriptome before priming the SA- and ROS-dependent defenses during grapevine's induced resistance against *Plasmopara viticola*. PLoS One 9: e88145.

Ghannam, A., A. Abbas, H. Alek, Z. Al-Waari and M. Al-Ktaifani. 2013. Enhancement of local plant immunity against tobacco mosaic virus infection after treatment with sulphated-carrageenan from red alga (*Hypnea musciformis*). Physiol. Mol. Plant Pathol. 84: 19–27.

Gozzo, F. and F. Faoro. 2013. Systemic acquired resistance (50 years after discovery): moving from the lab to the field. J. Agric. Food Chem. 61: 12473–12491.

Hamed, S.M., A.A.A. El-Rhman, N. Abdel-Raouf and I.B.M. Ibraheem. 2018. Role of marine macroalgae in plant protection and improvement for sustainable agriculture technology. J. Basic Appl. Sci. 7: 104–110.

Hayden, H.S., J. Blomster, C.A. Maggs, P.C. Silva, M.J. Stanhope and J.R. Waaland. 2003. Linnaeus was right all along: *Ulva* and *Enteromorpha* are not distinct genera. Eur. J. Phycol. 38: 277–294.

Holtkamp, A.D., S. Kelly, R. Ulber and S. Lang. 2009. Fucoidans and fucoidanases – focus on techniques for molecular structure elucidation and modification of marine polysaccharides. Appl. Microbiol. Biotechnol. 82: 1–11.

Jaulneau, V., C. Lafitte, C. Jacquet, S. Fournier, S. Salamagne, X. Briand, M.-T. Esquerré-Tugayé and B. Dumas. 2010. Ulvan, a sulfated polysaccharide from green algae, activates plant immunity through the jasmonic acid signaling pathway. J. Biomed. Biotechnol. 2: 1–11.

Jaulneau, V., C. Lafitte, M.-F. Corio-Costet, M.J. Stadnik, S. Salamagne, X. Briand, M.-T. Esquerré-Tugayé and B. Dumas. 2011. An *Ulva armoricana* extract protects plants against three powdery mildew pathogens. Eur. J. Plant Pathol. 131: 393–401.

Jayaraj, J., A. Wan, M. Rahman and Z.K. Punja. 2008. Seaweed extract reduces foliar fungal diseases on carrot. Crop Prot. 27: 1360–1366.

Jung, H.W., T.J. Tschaplinski, L. Wang, J. Glazebrook and J.T. Greenberg. 2009. Priming in systemic plant immunity. Science 324: 89–91.

Jones, J.D.G. and J. Dangl. 2006. The plant immune system. Nature 444: 323–329.

Klarzynski, O., B. Plesse, J.-M. Joubert, J.-C. Yvin, M. Kopp, B. Kloareg and B. Fritig. 2000. Linear β-1,3 glucans are elicitor of defense responses in tobacco. Plant Physiol. 124: 1027–1037.

Klarzynski, O., V. Descamps, B. Plesse, J.-C. Yvin, B. Kloareg and B. Fritig. 2003. Sulfated fucan oligosaccharides elicit defense responses in tobacco and local and systemic resistance against tobacco mosaic virus. MPMI 16: 115–122.

Kim, H.S., J.T. Hong, Y. Kim and S.-B. Han. 2011. Stimulatory effect of β-glucans on immune cells. Immune Netw. 11: 191–195.

Kirkendale, L., G.W. Saunders and P. Winberg. 2013. A molecular survey of *Ulva* (Chlorophyta) in temperate Australia reveals enhanced levels of cosmopolitanism. J. Phycol. 49: 69–81.

Kobayashi, A., A. Tai, H. Kanzaki and K. Kawazu. 1993. Elicitor active oligosaccharides from algal laminaran stimulate the production of antifungal compounds in alfalfa. Z. Naturforsch. 48c: 575–579.

Konasani, V.R., C. Jin, N.G. Karlsson and E. Albers. 2018. A novel ulvan lyase family with broad-spectrum activity from the ulvan utilization loci of *Formosa agariphila* KMM3901. Sci. Rep. 8: 1–11.

Lahaye, M., E.A.-C. Cimadevilla, R. Kuhlenkamp, B. Quemener, V. Lognoné and P. Dion. 1999. Chemical composition and ^{13}C NMR spectroscopic characterisation of ulvans from *Ulva* (Ulvales, Chlorophyta). J. Appl. Phycol. 11: 1–7.

Lahaye, M. and A. Robic. 2007. Structure and functional properties of ulvan, a polysaccharide from green seaweeds. Biomacromolecules 8: 1765–1774.

Laporte, D., J. Vera, N.P. Chandía, E. Zuñiga, B. Matsuhiro and A. Moenne. 2007. Structurally unrelated oligosaccharides obtained from marine macroalgae differentially stimulate growth and defense against TMV in tobacco plants. J. Appl. Phycol. 19: 79–88.

Lapshina, L.A., A.V. Reunov, V.P. Nagorskaya, T.N. Zvyagintseva and N.M. Shevchenko. 2006. Inhibitory effect of fucoidan from brown alga *Fucus evanescens* on the spread of infection induced by tobacco mosaic virus in tobacco leaves of two cultivars. Russ. J. Plant Physiol. 53: 246–251.

Lapshina, L.A., A.V. Reunov, V.P. Nagorskaya, T.N. Zvyagintseva and N.M. Schevchenko. 2007. Effect of fucoidan from brown alga *Fucus evanescens* on a formation of TMV-specific inclusions in the cells of tobacco leaves. Russ. J. Plant Physiol. 54: 111–114.

Lemaître-Guillier, C., A. Hovasse, C. Schaeffer-Reiss, G. Recorbet, B. Poinssot, S. Trouvelot, X. Daire, M. Adrian and M.-C. Héloir. 2017. Proteomics towards the understanding of elicitor induced resistance of grapevine against downy mildew. J. Proteomics 156: 113–125.

Li, B., F. Lu, X. Wei and R. Zhao. 2008. Fucoidan: structure and bioactivity. Molecules 13: 1671–1695.

Li, J., X. Wang, X. Lin, G. Yan, L. Liu, H. Zheng, B. Zhao, J. Tang and Y.-D. Guo. 2018. Alginate-derived oligosaccharides promote water stress tolerance in cucumber (*Cucumis sativus* L.). Plant Physiol. Biochem. 130: 80–88.

Lim, S.J., W.M.W. Aida, S. Schiehser, T. Rosenau and S. Böhmdorfer. 2019. Structural elucidation of fucoidan from *Cladosiphon okamuranus* (Okinawa mozuku). Food Chem. 272: 222–226.

Liu, R., X. Jiang, H. Guan, X. Li, Y. Du, P. Wang and H. Mou. 2009. Promotive effects of alginate-derived oligosaccharides on the inducing drought resistance of tomato. J. Ocean Univ. China 8: 303–311.

Liu, H., Y.-H. Zhang, H. Yin, W.-X. Wang, X.-M. Zhao and Y.-G. Du. 2013. Alginate oligosaccharides enhanced *Triticum aestivum* L. tolerance to drought stress. Plant Physiol. Biochem. 62: 33–40.

Mani, S.D. and R. Nagarathnam. 2018. Sulfated polysaccharide from *Kapphaphycus alvarezzi* (Doty) ex P.C. Silva primes defense responses against anthracnose disease of *Capsicum annuum* Linn. Algal Res. 32: 121–130.

Martinez-Medina, A., V. Flors, M. Heil, B. Mauch-Mani, C.M.J. Pieterse, M.J. Pozo, J. Ton, N.M. van Dam and U. Conrath. 2016. Recognizing plant defense priming. Trends Plant Sci. 21: 818–822.

Massironi, A., A. Morelli, L. Grassi, D. Puppi, S. Braccini, G. Maisetta, S. Esin, G. Batoni, C. Della Pina and Federica Chiellini. 2019. Ulvan as novel reducing and stabilizing agent from renewable algal biomass: application to green synthesis of silver nanoparticles. Carbohydr. Polym. 203: 310–321.

Mauch-Mani, B., I. Baccelli, E. Luna and V. Flors. 2017. Defense priming: an adaptive part of induced resistance. Annu. Rev. Plant Biol. 68: 485–512.

Melcher, R.L.J. and B.M. Moerschbacher. 2016. An improved microtiter plate assay to monitor the oxidative burst in monocot and dicot plant cell suspension cultures. Plant Methods 12: 1–11.

Melcher, R.L.J., M. Neumann, J.P.F. Wener, F. Gröhn and B.M. Moerschbacher. 2017. Revised domain structure of ulvan lyase and characterization of the first ulvan binding domain. Sci. Rep. 7: 1–9.

Mercier, L., C. Lafitte, G. Borderies, X. Briand, M.-T. Esquerré-Tugayé and J. Fournier. 2001. The algal polysaccharide carrageenans can act as an elicitor of plant defence. New Phytol. 149: 43–51.

Ménard, R., S. Alban, P. de Ruffray, F. Jamois, G. Franz, B. Fritig, J.-C. Yvin and S. Kauffmann. 2004. β-1,3 glucan, induces the salicylic acid signaling pathway in tobacco and *Arabidopsis*. Plant Cell 16: 3020–3032.

Ménard, R., P. de Ruffray, B. Fritig, J.-C. Yvin and S. Kauffmann. 2005. Defense and resistance-inducing activities in tobacco of the sulfated β-1,3 glucan PS3 and its synergistic activities with the unsulfated molecule. Plant Cell Physiol. 46: 1964–1972.

Nagorskaya, V.P., A.V. Reunov, L.A. Lapshina and I.M. Yermak. 2008. Influence of κ/β-carrageenan from red alga *Tichocarpus crinitus* on development of local infection induced by tobacco mosaic virus in Xanthi-nc tobacco leaves. Biol. Bull. 35: 310–314.

Nagorskaya, V.P., A.V. Reunov, L.A. Lapshina, I.M. Ermak and A.O. Barabanova. 2010. Inhibitory effect of κ/β-carrageenan from red alga *Tichocarpus crinitus* on the development of a potato virus X infection in leaves of *Datura stramonium* L. Biol. Bull. 37: 653–658.

Nayar, S. and K. Bott. 2014. Current status of global cultivated seaweed production and market. World Aquacult. 45: 32–37.

Nürnberger, T. and F. Brunner. 2002. Innate immunity in plants and animals: emerging parallels between the recognition of general elicitor and pathogen-associated molecular patterns. Curr. Opin. Plant Biol. 5: 1–7.

Ortmann, I., U. Conrath and B.M. Moerschbacher. 2006. Exopolysaccharides of *Pantoea agglomerans* have different priming and eliciting activities in suspension-cultured cells of monocots and dicots. FEBS Lett. 580: 4491–4494.

Paradossi, G., F. Cavalieri, L. Pizzoferrato and A.M. Liquori. 1999. A phyco-chemical study on the polysaccharide ulvan from hot water extraction of the macroalga *Ulva*. Int. J. Biol. Macromol. 25: 309–315.

Paradossi, G., F. Cavalieri and E. Chiessi. 2002. A conformational study on the algal polysaccharide ulvan. Macromolecules 35: 6404–6411.

Paris, F., Y. Krzyzaniak, C. Gauvrit, F. Jamois, F. Domergue, J. Joubès, V. Ferrières, M. Adrian, L. Legentil, X. Daire and S. Trouvelot. 2016. An ethoxylated surfactant enhances the penetration of the sulfated laminarin through leaf cuticule and stomata, leading to increased induced resistance against grapevine downy mildew. Physiol. Plant. 156: 338–350.

Patier, P., J.-C. Yvin, B. Kloareg, Y. Liénart and C. Rochas. 1993. Seaweed liquid fertilizer from *Ascophyllum nodosum* contains elicitors of plant D-glycanases. J. Appl. Phycol. 5: 343–349.

Patier, P., P. Potin, C. Rochas, B. Kloareg, J.-C. Yvin and Y. Liénart. 1995. Free or silica-bound oligokappa-carrageenans elicit laminarinase activity in *Rubus* cells and protoplasts. Plant Sci. 110: 27–35.

Paulert, R., A. Smânia Júnior, M.J. Stadnik and M.G. Pizzolatti. 2007. Antimicrobial properties of extracts from the green seaweed *Ulva fasciata* Delile against pathogenic bacteria and fungi. Algol. Stud. 123: 123–130.

Paulert, R., V. Talamini, J.E.F. Cassolato, M.E.R. Duarte, M.D. Noseda, A. Smânia Jr and M.J. Stadnik. 2009. Effects of sulfated polysaccharide and alcoholic extracts from green seaweed *Ulva fasciata* on anthracnose severity and growth of common bean (*Phaseolus vulgaris* L.). J. Plant. Dis. Prot. 116: 263–270.

Paulert, R., D. Ebbinghaus, C. Urlass and B.M. Moerschbaher. 2010. Priming of the oxidative burst in rice and wheat cell cultures by ulvan, a polysaccharide from green macroalgae, and enhanced resistance against powdery mildew in wheat and barley plants. Plant Pathology 59: 634–642.

Pengzhan, Y., Z. Quanbin, L. Ning, X. Zuhong, W. Yanmei and L. Zhi'en. 2003. Polysaccharides from *Ulva pertusa* (Chlorophyta) and preliminary studies of their antihyperlipidemia activity. J. Appl. Phycol. 15: 21–27.

Potin, P., K. Bouarab, F. Küpper and B. Kloareg. 1999. Oligosaccharide recognition signals and defence reactions in marine plant-microbe interactions. Curr. Opin. Microbiol. 2: 276–283.

Quemener, B., M. Lahaye and C. Bobin-Dubigeon. 1997. Sugar determination in ulvans by a chemical-enzymatic method coupled to high performance anion exchange chromatography. J. Appl. Phycol. 9: 179–188.

Quintana-Rodriguez, E., D. Duran-Flores, M. Heil and X. Camacho-Coronel. 2018. Damage-associated molecular patterns (DAMPs) as future plant vaccines that protect crops from pests. Sci. Hortic. 237: 207–220.

Ramírez-Carrasco, G., K. Martínez-Aguilar and R. Alvarez-Venegas. 2017. Transgenerational defense priming for crop protection against plant pathogens: a hypothesis. Front. Plant Sci. 8: 1–8.

Rasyid, A. 2017. Evaluation of nutritional composition of the dried seaweed *Ulva lactuca* from Pameungpeuk waters, Indonesia. Trop. Life Sci. Res. 28: 119–125.

Read, S.M., G. Currie and A. Bacic. 1996. Analysis of the structural heterogeneity of laminarin by electrospray-ionisation-mass spectrometry. Carbohydr. Res. 281: 187–201.

Redouan, E., D. Cedric, P. Emmanuel, E.G. Mohamed, C. Bernard, M. Philippe, E.M. Cherkaoui and C. Josiane. 2009. Improved isolation of glucuronan from algae and the production of glucuronic acid oligosaccharides using a glucuronan lyase. Carbohydr. Res. 344: 1670–1675.

Reunov, A.V., V.P. Nagorskaya, L.A. Lapshina, I.M. Yermak and A.O. Barabanova. 2004. Effect of κ/β-carrageenan from red alga *Tichocarpus crinitus* (Tichocarpaceae) on infection of detached tobacco leaves with tobacco mosaic virus. J. Plant Dis. Protect. 111: 165–172.

Reunov, A., L. Lapshina, V. Nagorskaya, T. Zvyagintseva and N. Shevchenko. 2009. Effect of fucoidan from the brown alga *Fucus evanescens* on the development of infection induced by potato virus X in *Datura stramonium* L. leaves. J. Plant Dis. Protect. 116: 49–54.

Robic, A., J.-F. Sassi, P. Dion, Y. Lerat and M. Lahaye. 2009. Seasonal variability of physicochemical and rheological properties of ulvan in two *Ulva* species (Chlorophyta) from the Brittany coast. J. Phycol. 45: 962–973.

Rydahl, M.G., S.K. Kracun, J.U. Fangel, G. Michel, A. Guillouzo, S. Génicot, J. Mravec, J. Harholt, C. Wilkens, M.S. Motawia, B. Svensson, O. Tranquet, M.-C. Ralet, B. Jorgensen, D.S. Domozych and W.G.T. Willats. 2017. Development of novel monoclonal antibodies against starch and ulvan – implications for antibody production against polysaccharides with limited immunogenicity. Sci. Rep. 7: 1–13.

Sangha, J.S., S. Ravichandran, K. Prithiviraj, A.T. Critchley and B. Prithiviraj. 2010. Sulfated macroalgal polysaccharides λ-carrageenan and ι-carrageenan differentially alter *Arabidopsis thaliana* resistance to *Sclerotinia sclerotiorum*. Physiol. Mol. Plant Pathol. 75: 38–45.

Sangha, J.S., W. Khan, X. Ji, J. Zhang, A.A.S. Mills, A.T. Critchley and B. Prithiviraj. 2011. Carrageenans, sulphated polysaccharides of red seaweeds, differentially affect Arabidopsis thaliana resistance to *Trichoplusia ni* (cabbage looper). PLoS ONE 6: 1–11.

Sangha, J.S., S. Kandasamy, W. Khan, N.S. Bahia, R.P. Singh, A.T. Critchley and B. Prithiviraj. 2015. λ-carrageenan suppresses tomato chlorotic dwarf viroid (TCDVd) replication and symptom expression in tomatoes. Mar. Drugs 13: 2865–2889.

Santos, P.H.Q.P. 2016. Biotechnological evaluation of seaweeds as bio-fertilizer. M.S. Thesis, University of Coimbra, Portugal, 98 pp.

Salah, I.B., S. Aghrouss, A. Douira, S. Aissam, Z. El Alaoui-Talibi, A. Filali-Maltouf and C. El Modafar. 2018. Seaweed polysaccharides as bio-elicitors of natural defenses in olive trees against verticillium wilt of olive. J. Plant Interact. 13: 248–255.

Sels, J., J. Mathys, B.M.A. De Coninck, B.P.A. Cammune and M.F.C. De Bolle. 2008. Plant pathogenesis-related (PR) proteins: a focus on PR peptides. Plant Physiol. Biochem. 46: 941–950.

Selvam, G.G. and K. Sivakumar. 2013. Effect of foliar spray from seaweed liquid fertilizer of *Ulva reticulate* (Forsk.) on *Vigna mungo* L. and their elemental composition using SEM – energy dispersive spectroscopic analysis. Asian Pac. J. Reprod 2: 119–125.

Sharma, H.S.S., C. Fleming, C. Selby, J.R. Rao and R. Martin. 2014. Plant biostimulants: a review on the processing of macroalgae and use of extracts for crop management to reduce abiotic and biotic stresses. J. Appl. Phycol. 26: 465–490.

Shetty, N.P., J.D. Jensen, A. Knudsen, C. Finnie, N. Geshi, A. Blennow, D.B. Collinge and H.J.L. Jorgensen. 2009. Effects of β-1,3-glucan from *Septoria tritici* on structural defence responses in wheat. J. Exp. Bot. 60: 4287–4300.

Shukla, P.S., T. Borza, A.T. Critchley and B. Prithiviraj. 2016. Carrageenans from red seaweeds as promoters of growth and elicitors of defense response in plants. Front. Mar. Sci. 3: 42–9.

Stadnik, M.J. and M.B. de Freitas. 2014. Algal polysaccharides as source of plant resistance inducers. Trop. Plant Pathol. 39: 111–118.

Steimetz, E., S. Trouvelot, K. Gindro, A. Bordier, B. Poinssot, M. Adrian and X. Daire. 2012. Influence of leaf age on induced resistance in grapevine against *Plasmopara viticola*. Physiol. Mol. Plant Pathol. 79: 89–96.

Sudisha, J., R.G. Sharathchandra, K.N. Amruthesh, A. Kumar and H.S. Shetty. 2012. Pathogenesis related proteins in plant defense response. pp. 379–403. *In*: Mérillon J. and Ramawat K. (eds.). Plant Defence: Biological Control. Progress in Biological Control, vol 12. Springer, Dordrecht.

Ton, J., G. Jakab, V. Toquin, V. Flors, A. Iavicoli, M.N. Maeder, J.P. Métraux and B. Mauch-Mani. 2005. Dissecting the beta-aminobutyric acid-induced priming phenomenon in *Arabidopsis*. Plant Cell 17: 987–999.

Trouvelot, S., A.-L. Varnier, M. Allègre, L. Mercier, F. Baillieul, C. Arnould, V. Gianinazzi-Pearson, O. Klarzynski, J.-M. Joubert, A. Pugin and X. Daire. 2008. A β-1,3 glucan sulfate induces resistance in grapevine against *Plasmopara viticola* through priming of defense responses, including HR-like cell death. MPMI 21: 232–243.

Trouvelot, S., M.-C. Héloir, B. Poinssot, A. Gauthier, F. Paris, C. Guillier, M. Combier, L. Trdá, X. Daire and M. Adrian. 2014. Carbohydrates in plant immunity and plant protection: roles and potential application as foliar sprays. Front. Plant Sci. 5: 1–14.

Tuvikene, R., K. Truus, M. Vaher, T. Kailas, G. Martin and P. Kersen. 2006. Extraction and quantification of hybrid carrageenans from the biomass of the red algae *Furcellaria lumbricalis* and *Coccotylus trunctus*. Proc. Estonian Acad. Sci. Chem. 55: 40–53.

van Loon, L.C., M. Rep and C.M. Pieterse. 2006. Significance of inducible defense-related proteins in infected plants. Annu. Rev. Phytopathol. 44: 135–162.

Varnier, A.-L., L. Sanchez, P. Vatsa, L. Boudescocque, A. Garcia-Brugger, F. Rabenoelina, A. Sorokin, J.-H. Renault, S. Kauffmann, A. Pugin, C. Clement, F. Baillieul and S. Dorey. 2009. Bacterial rhamnolipids are novel MAMPs conferring resistance to *Botrytis cinerea* in grapevine. Plant Cell Environ. 32: 178–193.

Vera, J., J. Castro, A. Gonzalez and A. Moenne. 2011. Seaweed polysaccharides and derived oligosaccharides stimulate defense responses and protection against pathogens in plants. Mar. Drugs 9: 2514–2525.

Vera, J., J. Castro, R.A. Contreras, A. González and A. Moenne. 2012. Oligo-carrageenans induce a long-term and broad-range protection against pathogens in tobacco plants (var. Xanthi). Physiol. Mol. Plant Pathol. 79: 31–39.

Vijayanand, N., S.S. Ramya and S. Rathinavel. 2014. Potential of liquid extracts of *Sargassum wightii* on growth, biochemical and yield parameters of cluster bean plant. Asian Pac. J. Reprod. 3: 150–155.

Zhang, Y., H. Liu, H. Yin, W. Wang, X. Zhao and Y. Du. 2013. Nitric oxide mediates alginate oligosaccharides-induced root development in wheat (*Triticum aestivum* L.). Asian Pac. J. Reprod. 71: 49–56.

Zhang, S., W. Tang, L. Jiang, Y. Hou, F. Yang, W. Chen and X. Li. 2015. Elicitor activity of algino-oligosaccharide and its potential application in protection of rice plant (*Oryza sativa* L.) against *Magnaporthe grisea*. Biotechnol. Biotechnol. Equip. 29: 646–652.

Zou, P., X. Lu, C. Jing, Y. Yuan, Y. Lu, C. Zhang, L. Meng, H. Zhao and Y. Li. 2018. Low-molecular-weight polysaccharides from *Pyropia yezoensis* enhance tolerance of wheat seedlings (*Triticum aestivum* L.) to salt stress. Front. Plant Sci. 9: 1–16.

Zvyagintseva, T.N., N.M. Shevchenko, A.O. Chizhov, T.N. Krupnova, E.V. Sundukova and V.V. Isakov. 2003. Water-soluble polysaccharides of some far-eastern brown seaweeds: distribution, structure, and their dependence on the developmental conditions. J. Exp. Mar. Biol. Ecol. 294: 1–13.

Walters, D.R., J. Ratsep and N.D. Havis. 2013. Controlling crop diseases using induced resistance: challenges for the future. J. Exp. Bot. 64: 1263–1280.

Wang, J., Q. Zhang, Z. Zhang, Y. Hou and H. Zhang. 2011. *In-vitro* anticoagulant activity of fucoidan derivatives from brown seaweed *Laminaria japonica*. Chin. J. Oceanol. Limnol. 29: 679–685.

Wojtaszek, P. 1997. Oxidative burst: an early plant response to pathogen infection. Biochem. J. 322: 681–692.

Wu, Y.-R., Y.-C. Lin and H.-W. Chuang. 2016. Laminarin modulates the chloroplast antioxidant system to enhance abiotic stress tolerance partially through the regulation of the defensin-like gene expression. Plant Sci. 247: 83–92.

3

Marine Macroalgae: A Potential Source of Plant Growth Regulators

Emad A. Shalaby

Professor of Biochemistry, Biochemistry Department, Faculty of Agriculture,
Cairo University, Giza, Egypt
E-mail: dremad2009@yahoo.com

1 Introduction

Algae are aquatic plants that lack the leaves, stem, roots, vascular systems, and sexual organs of the higher plants. They range in size from microscopic phytoplankton to giant kelp 200 feet long (Shalaby 2011). They live in temperatures ranging from hot spring to arctic snows, and they come in various colors, mostly green, brown and red. There are about 25,000 species of algae compared to 250,000 species of land plants. Algae make up in quantity what they lack in diversity, for the biomass of algae is immensely greater than that of terrestrial plants (Lowenstein 1986).

1.1 Chemical contents of algae

The current application of chemical compounds isolated from diverse classes of algae is enormous. Since 1975, three areas of research in aquatic natural products emerged: toxins, byproducts and chemical ecology. Over 15,000 novel compounds were chemically determined. Focusing on bio-products, some trends in drug research from natural sources suggested that algae are a promising group to furnish novel biochemically active substances (Singh et al. 2005, Blunt et al. 2005). To survive in a competitive environment, freshwater and marine algae developed defense strategies that resulted in a significant level of structural-chemical diversity from different metabolic pathways (Barros et al. 2005). The exploration of these organisms for pharmaceutical purposes revealed important chemical prototypes for the discovery of new agents and stimulated the use of sophisticated physical techniques and new syntheses of compounds with biomedical application. Moreover, algae were promising organisms for providing both novel biologically active substances and essential compounds for human nutrition (Mayer and Hamann, 2004). Therefore, an increasing supply of algae was needed for algal extracts, fractions or pure compounds for the economic sector (Dos Santos et al. 2005). In this regard, both primary and secondary metabolites were studied as a prelude to future rational economic exploitation as shown in Fig. 1.

Algal products have been used in the food, cosmetic, agriculture and pharmaceutical industries (Fig. 2). An expanding market for these products is a reality and faces a new challenge of growing algae on a large scale without further harming the marine environment. Micro- and macroalgae are essential to the development of aquaculture since they provide the main micronutrients to many aquatic organisms, including vitamins, nitrogen-containing compounds, sterols, and specific fatty

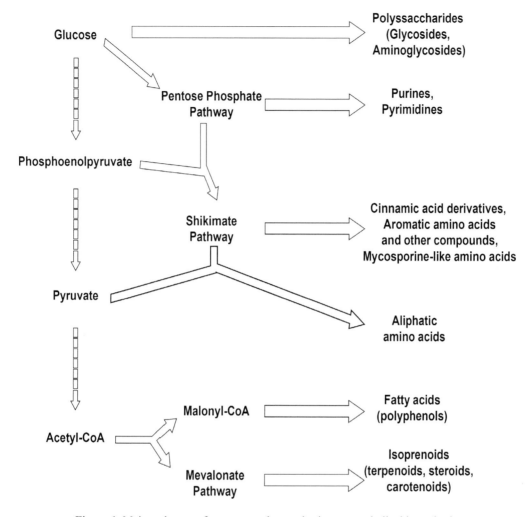

Figure 1. Main pathways of some secondary and primary metabolite biosynthesis.

acids. Total aquaculture production in 2000 was reported to be 45.71 million metric tons (mmt) by weight, valued at US$ 56.47 billion, with production up by 6.3% by weight and 4.8% by value since 1999 (Cardozo et al. 2006).

1.2 Biological activity of macroalgae (seaweeds)

Algae have mainly been used in western countries as raw material to extract alginates (from brown algae) and agar and carrageenans (from red algae). However, algae have also been found to contain a multitude of bioactive compounds that might have antioxidant, antibacterial, antiviral, and anticarcinogenic properties (Plaza et al. 2008).

1.2.1 Algae as potential plant growth stimulators

Crouch and Staden (1992) revealed that seaweed concentrate (SWC) prepared from *Ecklonia maxima* (Osbeck) Papenfuss improved the growth of tomato seedlings when applied as a soil drench but their foliar application in the form of spray had no effect on young plants. However, in a second experiment (as drench), SWC-treated plants exhibited early fruit ripening and total fruit fresh

Figure 2. Some macroalgae products (agriculture, foods, pharmaceuticals).

Color version at the end of the book

weight increase by 17%. The number of harvested fruits was also increased by about 10%. These activities may be due to the algae content from the essential and bioactive forms of the five classical phytohormones—auxin, ABA, CKs, GAs, and ET—as reported in Table 1 (Lu and Xu 2015).

Johansen (1993) reported that the activities of soil algae were thought to enhance soil formation and water retention, stabilize soil, increase the availabilit0y of nutrients of plants growing nearby, and reduce soil erosion. Because of their benefit to agriculture, they were suggested for use as biofertilizers. Bograh et al. (1997) found an increase in pigments (chlorophylls and total carotenoids)

Table 1. Existence of phytohormones in cyanobacteria and algae (Lu and Xu 2015)

Phytohormone	Cyanobacteria	Diatoms	Eustigmatophytes	Brown algae (multicellular)	Red algae (multicellular)	Green algae
Auxin	*Synechocystis* sp., *Chroococcidiopsis* sp., *Anabaena* sp., *Phormidium* sp., *Oscillatoria* sp., *Nostoc* sp.	N/A	N/A	*Ectocarpus siliculosus*	*Prionits lanceolata*, *Porphyra* sp., *Gelidium* sp., *Gracilaria* sp., *Gracilariopsis* sp., *Chondracanthus* sp., *Hypnea* sp.	*Scenedesmus armatus*, *Chlorella pyrenoidosa*. *Chlorella minutissima*
ET	*Synechococcus* sp., *Anabaena* sp., *Nostoc* sp., *Calothrix* sp., *Scytonema* sp., *Cylindrospermum* sp.	N/A	N/A	*Padina arborescens*, *Ecklonia maxima*	*Porphyra tenera*	*Chlorella pyrenoidosa*
ABA	*Synechococcus leopoliensis*, *Nostoc muscorum*, *Trichormus variabilis*, *Anabaena variabilis*	*Coscinodiscus granii*	*Nannochloropsis oceanica*	*Ascophyllum nodosum*	*Porphyra* sp., *Gelidium* sp., *Gracilaria* sp., *Gracilariopsis* sp., *Chondracanthus* sp., *Hypnea* sp.	*Chlamydomonas reinhardtii*. *Dunaliella* sp. *Draparnaldia mutabilis*, *Chlorella minutissima*
CK	*Synechocystis* sp., *Chroococcidiopsis* sp., *Anabaena* sp., *Phormidium* sp., *Oscillatoria* sp., *Calothrix* sp., *Chlorogloeopsis* sp., *Rhodospirillum* sp.	*Ecklonia* sp.	*Nannochloropsis oceanica*	*Ecklonia maxima*, *Laminaria pallida*	*Porphyra* sp., *Gelidium* sp., *Gracilaria* sp., *Gracilariopsis* sp., *Chondracanthus* sp., *Hypnea* sp., *Gigartina dathrata*. *Hypnea* sp.	*Chlorella minutissima*
GA	*Anabaenopsis* sp., *Cylindrospermum* sp., *Phormidium foveolarum*	N/A	*Nannochloropsis oceanica* (Y. Lu et al., unpublished)	*Ecklonia radiata*	*Hypnea musciformis*	*Chlorella* sp., *Chlamydomonas reinhardtii*

Abbreviation: N/A, no reports available.

and carbohydrate production in *Lupinus* leaves pretreated with algal filtrate of *Cylindrospermum* (Cyanobacteria). Adam (1999) found that algal filtrate of the cyanobacterium *Desmonostoc muscorum* (formerly *Nostoc muscorum*) increased germination of wheat seeds as well as their growth parameters and nitrogen compounds, compared to controls. Also, Lozano et al. (1999) stated that the application of an extract from algae to soil or foliage increased ash, protein and carbohydrate contents of potato tubers (*Solanum tuberosum*). In field experiments, Ghallab and Salem (2001) studied the effect of some biofertilizer treatments—"Cerealin" (*Azospirillum* spp.) and Nemales (*Serratia* spp.)—on wheat plant. They found that the two biofertilizers increased growth characters (plant height and weight) and nutrients, sugar, amino acids and growth regulators (IAA, GA$_3$ and cytokinin) (the chemical structures are shown in Fig. 3), and crude protein content in the plants. On the other hand, Abdel-Monem et al. (2001) reported that fertilization with *Azospirillum brasilense* or commercial biofertilizer "Cerealin" improved the growth and yield of maize in rotation with wheat as affected by irrigation regime.

Ascophyllum nodosum (Ochrophyta, Phaeophyceae) extracts (at 0, 1%, 5% and 10%) were reported to improve germination, root growth, flower production, fruit set, and crop quality and increase yield as well as enhancing stress and disease resistance of cabbage and tomato plant dry weight with the 5% treatment, followed by decreases at 10% (Carolyn et al. 2001). Also, Zaccaro et al. (2001) documented that the algal biofertilizers were likely to assume greater importance as complement and/or supplement to chemical fertilizers in improving the nutrient supplies to cereal crops because of high nutrient turnover in the cereal production system, exorbitant cost of fertilizers and greater consciousness of environmental protection.

The current work will focus on macroalgae (seaweed) contents from plant growth regulator compounds and methods of determination and pilot experiment for applications of seaweed extracts on plant and crops.

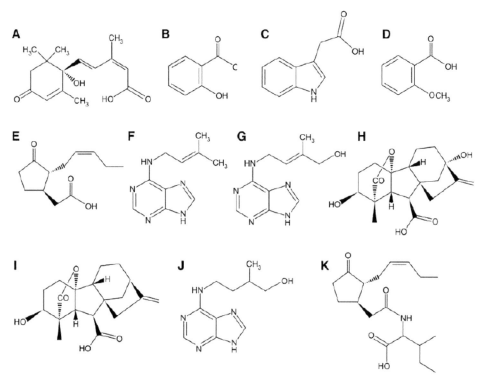

Figure 3. Structure of phytohormones. (A) Abscisic acid, (B) salicylic acid, (C) indole-3-acetic acid, (D) ortho-anisic acid, (E) jasmonic acid, (F) isopentenyladenine, (G) trans-zeatin, (H) gibberellin A1, (I) gibberellin A4, (J) dihydrozeatin, (K) jasmonoyl-leucine (Mori et al. 2017).

2. Application of Algae as Source for Phytohormones

2.1 Extraction and quantitative determination of endogenous phytohormones

2.1.1 Extraction

Ten grams of the dried algal materials (*Asparagopsis taxiformis, Sargassum vulgare, Gelidium corneum, Corallina officinalis, Ulva intestinalis*) collected from Marsa Matrouh (Egypt) for the first species and from Alexandria beach (Egypt) for the remaining species were homogenized, extracted twice with 200 ml methanol (96%), then twice with 200 ml methanol (40%), each for 24 h. The combined methanolic extract was evaporated in a rotary evaporator (100 rpm at 40°C) to an aqueous solution. The aqueous solution was adjusted to pH 2.6-2.8 and extracted four times with ethyl acetate (50 ml each). The ethyl acetate extract was dried on anhydrous sodium sulfate (10 g/100 ml), then filtered and evaporated in a rotary evaporator to dryness; the residue was dissolved in 4 ml absolute methanol. This methanolic solution was used for the determination of gibberellic acid, abscisic acid and indole-3-acetic acid by gas-liquid chromatography (GLC) (Vogel 1975).

Also, solid phase extraction technique can be used for extraction of phytohormones from algal samples as shown in Fig. 4.

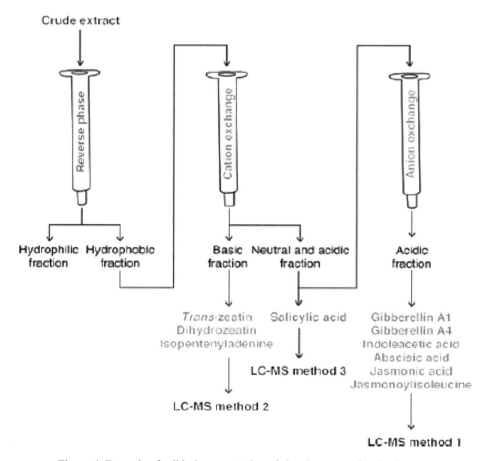

Figure 4. Example of solid-phase extraction of phytohormones for simultaneous analysis (after Mori et al. 2017).

2.1.2 Determination

The GLC analyses were carried out with a Pro-GC gas chromatography, with a dual flame ionization detector. The glass column (1.5 m × 4 mm) was packed with 1% OV-17. Temperatures of injector and detector were 250°C and 300°C, respectively. The column was maintained at 200°C for 3 min and then programmed from 200 to 240°C (at rate 10°C/min). Nitrogen as a carrier gas, hydrogen and air gas flow rates were 30, 33 and 330 ml/min.

The identification of phytohormones was accomplished by comparing the peak retention times with retention of authentic substances (ABA, IAA and GA_3). The quantity of individual algal hormones was determined by comparing the peak area produced by known weight of the algal material with standard curves of the authentic substances that expressed the relation between the different concentrations and their peak area.

2.2 Growth stimulator activity

The study was conducted during the two successive winter seasons at the Experimental Nursery of the Botany Department, Faculty of Agriculture, Cairo University, Giza.

2.2.1 Plant material

Seeds of broad bean (*Vicia faba* L.) cultivar Assiut 86 were obtained from Legume Research Division, Field Crop Research Institute, Agricultural Research Center, Giza.

2.2.2 Experimental procedure

Seeds were planted in pots 25 cm in diameter, filled with sandy soil. Ten seeds per pot were sown, and pots were irrigated. Two dry grams of each type of algae studied (*Asparagopsis taxiformis, Sargassum vulgare, Gelidium corneum, Corallina officinalis, Ulva intestinalis*) were incorporated with the potting medium before planting. Nitrogen was added in the form of ammonium sulfate (20.5% N), while phosphorus (P) was applied as calcium (Ca) super phosphate (15.5% P_2O_5). Seedlings were thinned to two plants per pot after three weeks from sowing date.

2.2.3 Recorded data

Vegetative characters including plant height (cm, measured from the soil surface to the highest point of the uppermost leaf whose tip is pointing down, using a meter), stem length (cm, measured from the centre of the head tube to an imaginary vertical line drawn through the bottom bracket), number of internodes of the main stem, main stem diameter (mm, determined by using a clipper), number of lateral branches/plant, and number of leaves/plant were recorded at 25, 50, 75 and 100 days after sowing (DAS). Also, leaf area (cm^2), shoot fresh and dry weight/plant (g), root fresh weight/plant (g) and root dry weight/plant (g) were recorded using digital balance at 50 DAS. For determination of leaf area, five full mature leaves on the fifth node of the main stem were taken for each replicate. The average area/leaf was determined by using portable leaf area meter Licor (Model Li-3000). The yield characters, including number of pods/plant, number of seeds/plant, seed yield/plant (g), and specific seed weight (g) were recorded at maturity (130 DAS).

2.2.4 Treatments

The plants received the following treatments.

Treatments	Source
Asparagopsis taxiformis (Rhodophyta)	Marsa Matrouh
Sargassum vulgare (Phaeophyceae)	Alexandria

Gelidium corneum (Rhodophyta)	Alexandria
Corallina officinalis (Rhodophyta)	Alexandria
Ulva intestinalis 1 (Chlorophyta)	Alexandria
Ulva intestinalis 2 (Chlorophyta)	Alexandria
Control (−)	Untreated plants
Control (+)	Plants received 5 g/L super phosphate and 2 g/L ammonium sulfate

2.2.5 Statistic analysis

Data were subjected to an analysis of variance, and the means were compared using the least significant difference (LSD) test at the 0.05 and 0.01 levels, as recommended by Snedecor and Cochran (1982).

3 Plant Growth Bioregulators

The concentration of plant growth bioregulators (IAA, GA_3 and ABA) of five algal species is presented in Fig. 5. Red algae *Gelidium* sp., *Corallina* sp. and *Asparagopsis* sp. collected through two seasons (spring and summer) showed variations in the concentration of plant growth bioregulators. The total auxin concentration among these macroalgae species was 0.0035%, 0.062% and 0.019% respectively. The green alga *Ulva* sp. collected during the summer had the greatest concentration of auxin content (0.34%). Gibberellic acid was found in *Corallina* sp. (0.005%). In addition, the brown alga *Sargassum* sp., collected during the spring, was found to have the highest concentration of gibberellic acid (0.6%). However, *Corallina* sp. had the highest concentration of abscisic acid (cis 0.22% and trans 0.18%) when compared with all five species of macroalgae under study. Zaccaro et al. (2001) documented that biofertilizers are likely to assume greater importance as complement and/or supplement to chemical fertilizers in improving the nutrient supplies to cereal crops because of the high nutrient turnover in the cereal production system. Bentley (1960) was the first to suggest the presence of gibberellin-like substances in algae. She found two non-identical components in acidified extracts of phytoplankton.

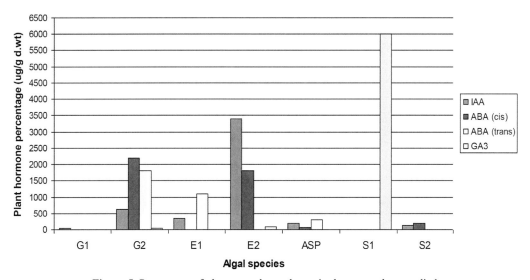

Figure 5. Percentage of plant growth regulators in the macroalgae studied.

Color version at the end of the book

4 Algae as Plant Growth Stimulator (Organic Fertilizer)

4.1 Vegetative characters

4.1.1 Plant height (cm)

Data in Table 2 shows that plant height of *Vicia faba* cv. Assuit 86 was increased at all ages in first season (2006/2007) with significant differences at the ages of 75 and 100 DAS only by using algae as an organic fertilizer compared with both untreated and chemically fertilized plants, except E_2, G and S_1 compared with chemical fertilization treatment. In the second season (2007/2008), plant height was significantly increased in most algae treatments at all ages. Maximum increase was obtained at 75 and 100 DAS by using ASP, measuring 62.4% over untreated control in the second season, and 49.6% over chemical fertilization treatment at 25 DAS in the same season.

Table 2. Effect of chemical and bio-fertilization on plant height (cm) of *Vicia faba* L. during two seasons

	First season (2006/2007)				Second season (2007/2008)			
	DAS				DAS			
Treatments	25	50	75	100	25	50	75	100
Cont. (−)	17.00	29.67	49.60	52.17	19.21	29.87	47.55	55.50
Cont. (+)	19.67	32.50	62.00	67.50	20.13	34.50	67.81	83.14
E_1	21.54	32.00	50.71	59.52	30.05	41.22	62.55	73.16
E_2	22.00	35.17	63.67	70.67	27.43	38.19	60.15	72.54
G	21.50	33.33	55.80	63.00	25.13	36.48	58.44	69.61
C	21.83	32.33	58.30	60.50	24.77	36.05	58.95	65.11
S_1	22.00	33.50	60.90	69.67	27.22	39.18	73.65	86.01
ASP	22.31	35.01	65.24	72.33	30.11	48.25	77.20	90.15
$LSD_{0.05}$	NS	NS	4.75	6.53	4.11	3.50	7.05	5.66

DAS: days after sowing. LSD: least significant difference.
E_1: *Ulva* collected from Alexandria during spring; E_2: *Ulva* collected from Alexandria during summer; S_1: *Sargassum* sp. collected from Alexandria during spring; ASP: *Asparagopsis* sp. collected from Marsa Matrouh; G: *Gelidium* sp. collected from Alexandria; C: *Corallina* sp. collected from Alexandria during spring.

4.1.2 Stem length (cm)

The data on stem length shown in Table 3 reveal that *Vicia faba* stems were generally taller in the second season than in the first one. The differences between treatments were all significant in both seasons except at 25 and 50 DAS in the first season, in addition, cont. (+) and E_2 at 75 DAS as well as S_1 at 100 DAS in the same season, E_1 at 25 DAS and S_1 at 100 DAS in the second season. The maximum increases in stem length were 87.2% and 57.2% over cont. (−) and cont. (+), respectively, compared with ASP biofertilizer, in the second season.

4.1.3 Number of internodes of the main stem

Results in Table 4 show a superiority of chemically fertilized plants in number of internodes of the main stem compared with the other treatments, especially at 75 and 100 DAS in both seasons. ASP treatment gave the highest number of internodes followed by the positive control without

Table 3. Effect of chemical and bio-fertilization on stem length/plant (cm)
of *Vicia faba* L. during two seasons

Treatments	First season DAS				Second season DAS			
	25	50	75	100	25	50	75	100
Cont. (−)	14.21	24.05	39.43	41.16	14.56	24.22	37.43	44.22
Cont. (+)	16.56	27.11	50.11	53.10	17.33	29.16	48.96	72.15
E₁	18.15	28.55	38.25	44.51	26.33	35.44	46.55	66.55
E₂	19.50	29.32	49.96	51.30	23.61	33.62	42.33	63.55
G	17.44	28.17	44.55	48.61	20.11	31.19	42.10	60.05
C	17.36	26.66	46.32	50.33	20.66	31.45	44.03	55.43
S₁	19.17	27.32	49.13	55.63	22.50	34.36	53.40	76.31
ASP	19.41	29.67	52.33	57.13	27.25	40.33	60.13	79.83
LSD.₀.₀₅	NS	NS	3.11	3.51	3.41	3.71	4.15	5.62

DAS: days after sowing. LSD: least significant difference.
E₁: *Ulva* collected from Alexandria during spring; E₂: *Ulva* collected from Alexandria during summer; S₁: *Sargassum* sp. collected from Alexandria during spring; ASP: *Asparagopsis* sp. collected from Marsa Matrouh; G: *Gelidium* sp. collected from Alexandria; C: *Corallina* sp. collected from Alexandria during spring.

Table 4. Effect of chemical and bio-fertilization on number of internodes of the main stem of *Vicia faba* L. in two seasons

Treatments	First season DAS				Second season DAS			
	25	50	75	100	25	50	75	100
Cont. (−)	5.00	7.50	13.00	14.17	5.32	8.11	13.56	15.10
Cont. (+)	5.33	8.00	18.50	22.00	5.67	8.90	19.22	24.16
E₁	5.50	8.60	13.00	17.33	6.17	9.23	14.50	18.22
E₂	5.67	8.50	16.70	20.50	5.84	9.40	15.81	19.05
G	5.83	8.50	15.00	18.67	5.91	9.60	15.50	17.65
C	6.17	8.60	15.33	18.33	6.32	9.21	16.13	18.66
S₁	5.67	8.50	15.47	19.50	6.51	9.75	16.35	20.31
ASP	6.00	8.80	16.80	21.67	5.95	9.45	17.94	22.75
LSD₀.₀₅	NS	NS	2.76	2.11	NS	NS	2.55	2.05

DAS: days after sowing. LSD: least significant difference.
E₁: *Ulva* collected from Alexandria during spring; E₂: *Ulva* collected from Alexandria during summer; S₁: *Sargassum* sp. collected from Alexandria during spring; ASP: *Asparagopsis* sp. collected from Marsa Matrouh; G: *Gelidium* sp. collected from Alexandria; C: *Corallina* sp. collected from Alexandria during spring.

any significant difference. The maximum significant increases were 55.3% and 60.0% for positive control and 52.9% and 50.7% for ASP over the negative control in the two consecutive seasons, respectively.

4.1.4 Main stem diameter (mm)

Data in Table 5 indicate a significant effect of bio-fertilization treatments on diameter of the 5th internode of the main stem in comparison with untreated control as well as most of those treatments with positive control plant at all ages except that of 25 days in the two successive seasons. Meanwhile, these differences were insignificant at 25 DAS in both seasons. The thickest stems (7.76 and 7.22 mm) were obtained from plants that received ASP organic fertilizer at 100 DAS in both seasons, respectively. The increasing ratios were 66.2% and 28.7% over the negative control. These ratios reached 28.5% and 15.9% over untreated plants in comparison with positive control plants.

Table 5. Effect of chemical and bio-fertilization on main stem diameter (mm) of *Vicia faba* L. during 2005/2006 and 2006/2007 seasons

	First season				Second season			
	DAS				DAS			
Treatments	25	50	75	100	25	50	75	100
Cont. (−)	4.02	4.37	4.47	4.67	4.95	5.32	5.56	5.61
Cont. (+)	4.47	4.90	5.80	6.00	5.56	5.91	6.31	6.50
E_1	4.65	5.00	5.60	6.10	5.82	6.32	6.65	6.82
E_2	4.60	5.13	5.38	5.88	5.51	6.10	6.45	6.68
G	4.60	5.23	6.20	6.40	5.16	5.75	6.11	6.41
C	4.70	5.02	5.83	6.73	5.42	5.92	6.25	6.58
S_1	4.60	4.75	6.03	6.20	5.80	6.41	6.75	6.91
ASP	4.92	5.80	6.73	7.76	5.92	6.60	6.96	7.22
$LSD_{0.05}$	NS	0.33	0.35	0.41	NS	0.43	0.32	0.34

DAS: days after sowing. LSD: least significant difference.
E_1: *Ulva* collected from Alexandria during spring; E_2: *Ulva* collected from Alexandria during summer; S_1: *Sargassum* sp. collected from Alexandria during spring; ASP: *Asparagopsis* sp. collected from Marsa Matrouh; G: *Gelidium* sp. collected from Alexandria; C: *Corallina* sp. collected from Alexandria during spring.

4.1.5 Number of lateral branches/plant

Data of number of lateral branches/plant at consecutive ages in two seasons given in Table 6 proved that there were significant differences between bio-fertilization treatments and either negative or positive control. ASP treatment was the highest in number of lateral branches, which ranged between 1-2.5 and 1-3 branches in both seasons, respectively. These numbers ranged between 0.0-1 and 0.0-1.5 branches for untreated plants, and between 0.0 and 2 for positive control, in the two successive seasons, respectively. The maximum increase ratio was 300% for ASP over negative control at 50 DAS in the second season, while such ratio reached 200% for positive control at the same age.

4.1.6 Number of leaves/plant

Results of number of leaves per plant at successive ages in two seasons are given in Table 7. Number of leaves increased gradually from 4 to 22 leaves in the first season, and from 5 to 23 leaves in the second season at 25, 50, 75 and 100 DAS. Such increment was insignificant at the first and second ages in the first season, and at the first age only in the second season. The differences

Table 6. Effect of chemical and bio-fertilization on number of lateral branches/plant of *Vicia faba* L. over two seasons

Treatments	First season				Second season			
	DAS				DAS			
	25	50	75	100	25	50	75	100
Cont. (–)	0.0	0.0	1.0	1.0	0.0	0.5	1.0	1.5
Cont. (+)	0.0	1.0	2.0	2.0	0.0	1.5	2.0	2.0
E_1	0.5	0.5	2.0	2.5	0.5	1.0	2.0	2.5
E_2	1.0	1.0	2.0	2.5	1.0	1.5	2.5	2.5
G	0.5	0.5	2.0	2.5	0.5	1.0	2.0	2.5
C	1.0	1.0	2.0	2.0	1.0	1.5	2.0	2.0
S_1	1.5	1.5	1.50	2.0	1.0	1.5	2.0	2.0
ASP	1.0	1.5	2.50	2.50	1.0	2.0	3.0	3.0
$LSD_{0.05}$	0.20	0.25	0.52	0.20	0.23	0.22	0.20	0.20

DAS: days after sowing. LSD: least significant difference.
E_1: *Ulva* collected from Alexandria during spring; E_2: *Ulva* collected from Alexandria during summer; S_1: *Sargassum* sp. collected from Alexandria during spring; ASP: *Asparagopsis* sp. collected from Marsa Matrouh; G: *Gelidium* sp. collected from Alexandria; C: *Corallina* sp. collected from Alexandria during spring.

Table 7. Effect of chemical and bio-fertilization on number of leaves/plant of *Vicia faba* L. over two seasons

Treatments	First season				Second season			
	DAS				DAS			
	25	50	75	100	25	50	75	100
Cont. (–)	4.00	9.33	13.00	14.17	5.10	10.33	13.91	15.20
Cont. (+)	4.17	11.50	18.50	22.00	5.32	11.96	16.50	21.91
E_1	4.80	10.67	13.00	17.33	5.72	12.10	16.30	18.41
E_2	4.70	11.67	16.67	21.67	5.96	13.43	17.85	21.50
G	4.60	9.80	15.00	18.70	5.82	10.72	15.42	18.95
C	4.83	11.17	15.17	18.30	5.81	13.11	16.05	19.41
S_1	4.77	10.67	15.50	19.50	5.43	12.51	17.33	20.16
ASP	5.27	12.83	16.83	20.67	6.42	14.26	18.23	23.17
$LSD_{0.05}$	NS	NS	2.75	1.61	NS	0.94	1.50	2.11

DAS: days after sowing. LSD: least significant difference.
E_1: *Ulva* collected from Alexandria during spring; E_2: *Ulva* collected from Alexandria during summer; S_1: *Sargassum* sp. collected from Alexandria during spring; ASP: *Asparagopsis* sp. collected from Marsa Matrouh; G: *Gelidium* sp. collected from Alexandria; C: *Corallina* sp. collected from Alexandria during spring.

were significant between untreated plants and all treated plants. Meanwhile, most bio-fertilization treatments significantly differed with chemically fertilized plants at 75 and 100 DAS in the first season as well as at 50 to 100 DAS in the second season. The maximum increasing ratios were

45.9% and 52.4% for ASP and 55.3% and 44.1% for positive control over untreated plants at 75 and 100 DAS in the two successive seasons, respectively (Mostafa, 2004), and inoculation of sandy soil with dry *Spirulina* recorded the highest number of leaves of 30 and 34.13/plant for 2.0 g/pot (dry *Spirulina*) and chemical fertilizer, respectively.

4.1.7　Leaf area (cm²)

The leaf area in *Vicia faba* plants was considerably affected, in both seasons, by the different fertilization treatments: chemical and bio-fertilization (Table 8). Untreated plants had significantly smaller leaves (16.26 and 18.37 cm²) in both seasons, respectively, than all other treatments. NP fertilization significantly increased leaf area, 24.76 and 28.43 cm², in both seasons, respectively, compared to the untreated control. Treating plants with organic fertilizers resulted in larger leaves, especially when ASP was used: 30.66 and 36.17 cm² in both seasons, respectively. The maximum increase in leaf area was obtained by ASP followed by G and then by E_1 with insignificant differences in both seasons. The significant increase ratios of ASP over the negative control were 88.6% and 96.9%; they were 52.3% and 54.8% over the positive control in the first and second seasons, respectively.

Table 8. Effect of chemical and bio-fertilization on means of some morphological characters of *Vicia faba* L. at 50 days after sowing over two seasons

Characters	Leaf area (cm²)		SFW (g)		SDW (g)		RFW (g)		RDW (g)	
Treatments	FS	SS	FS	SS	FS	SS	FS	SS	FS	SS
Cont. (−)	16.26	18.37	8.47	9.22	0.85	0.90	16.03	16.95	2.53	2.59
Cont. (+)	24.76	28.43	11.38	13.28	1.05	1.25	18.35	21.17	2.17	2.48
E_1	30.37	33.29	16.27	19.11	1.72	1.75	15.34	17.55	1.89	1.94
E_2	23.47	28.11	17.68	18.40	1.74	1.80	15.32	18.30	2.49	2.55
G	30.56	35.95	14.27	16.21	1.49	1.54	18.36	20.41	2.41	2.46
C	29.67	30.05	14.73	16.33	1.78	1.83	10.10	15.05	1.76	1.80
S_1	28.36	31.44	15.30	17.45	1.55	1.65	13.62	15.21	2.57	2.65
ASP	30.66	36.17	20.43	23.90	2.47	2.76	20.29	24.05	3.75	3.95
$LSD_{0.05}$	1.75	3.14	2.11	3.33	0.35	0.55	2.13	2.56	0.31	0.32

LSD: least significant difference; FS: first season; SS: second season; SFW: stem fresh weight; SDW: stem dry weight; RFW: root fresh weight; RDW: root dry weight.

E_1: *Ulva* collected from Alexandria during spring; E_2: *Ulva* collected from Alexandria during summer; S_1: *Sargassum* sp. collected from Alexandria during spring; ASP: *Asparagopsis* sp. collected from Marsa Matrouh; G: *Gelidium* sp. collected from Alexandria; C: *Corallina* sp. collected from Alexandria during spring.

4.1.8　Shoot fresh weight (SFW)/plant (g)

The herb fresh weight/plant was significantly increased in all bio-fertilization treatments of both seasons compared to the negative or positive control (Table 8). Moreover, the most effective treatment in this regard was ASP, giving the highest herb fresh weight, 20.43 g and 23.90 g, in the first and second seasons, respectively, followed by E_1 or E_2, in the first and second seasons, respectively. The significant increases for ASP over the negative control were 141.2% and 159.2%, while they were 34.4% and 44.0% over positive control in both seasons, respectively. Ordeg (1999) documented that the suspension of extract of cyanobacteria and microalgae contain a special set of biologically active compounds including plant growth regulators, which can be used for treatment to decrease senescence and transcription as well as to increase leaf chlorophyll, protein content and root and shoot development.

4.1.9 Shoot dry weight (SDW)/plant (g)

The data in Table 8 show that the untreated control plants had the lowest herb dry weight/plant, the same as herb fresh weight compared with chemically or bio-fertilized plants. The highest herb dry weights/plant (2.47 and 2.76 g) were obtained from plants that received ASP bio-fertilization followed by C (1.78 and 1.83 g), then by E_2 (1.74 and 1.80 g), with significant differences in the two successive seasons, respectively. The highest increase in shoot dry weight per plant was 190.6% in the first season and 206.7% in the second season over the negative control. Meanwhile, such ratios were 23.5% and 38.9% over positive control in the first and second seasons, respectively.

4.1.10 Root fresh weight/plant (g)

Regarding fresh weight of root (Table 8), results showed that untreated plants had a slightly higher root fresh weight (RFW) per plant than E_1 and E_2 in the first season as well as C and S_1 in the second season, but such increase was significant with C and S_1 in the first season. Chemical fertilizer caused significant increase in RFW over negative control, E_1, E_2, C and S_1 in both seasons in addition to ASP treatment in the second season. Among all bio-fertilization treatments, ASP realized the highest increasing ratios of RFW, measuring 26.6% and 41.9% over untreated plants as well as 10.6% and 13.6% over chemically fertilized plants in the first and second seasons, respectively. These data were in agreement with the results obtained by Smith and Staden (1983). They reported that greenhouse tests were conducted to determine the effects of a commercially available seaweed (derived from *Ecklonia maxima*) concentrate (Kelpak 66) on the growth of tomato plants (*L. esculentum* Mill.) "Kelpak 66" at a dilution of 1:500 improved the growth of tomato plants significantly, irrespective of whether it was applied as a foliar spray at regular intervals, or whether the soil in which the tomatoes were planted was flushed once with the diluted seaweed concentrate. Root growth was significantly improved whenever the seaweed concentrate was applied.

4.1.11 Root dry weight/plant (g)

Data presented in Table 8 clarify that root dry weight (RDW) obtained from untreated plants was 2.53 g in the first season and 2.59 g in the second. There were insignificant differences between negative control and each of E_2, G and S_1 bio-fertilization treatments in both seasons in addition to positive control in the second, while such differences were significant with E_1, C and ASP in both seasons as well as positive control in the first season. ASP algae gave the heaviest RDW, 3.75 and 3.95 g/plant, with increasing ratios 48.2% and 52.5% over negative control as well as 72.8% and 59.3% over positive control. These data were in agreement with the results obtained by Carolyn et al. (2001). Seaweeds have been used in agriculture for many centuries as fertilizers, animal feeds and soil conditioners. Extracts from *Ascophyllum nodosum* (0.1%, 5% and 10%) have been reported to improve germination, stimulate root growth, flower production, fruit set, and crop quality, and increase yield as well as enhancing stress and disease resistance. Greenhouse results indicated a slight increase in cabbage and tomato plant dry weight in the 5% treatment, followed by decreases at the higher rates (Carolyn et al. 2001).

4.2 Yield characters

4.2.1 Number of pods/plant

It is clear from Table 9 that all adopted treatments of chemical and bio-fertilization significantly increased number of pods per *Vicia faba* plant in both seasons. The maximum increase in number

Table 9. Yield and yield components of *Vicia faba* L. as affected by chemical and
bio-fertilization in two successive seasons

Characters	No. of pods/plant		No. of seeds/plant		Seed yield/ plant (g)		Specific seed weight (g)	
Treatments	FS	SS	FS	SS	FS	SS	FS	SS
Cont. (−)	3.0	4.1	6.6	8.2	2.75	3.56	41.21	43.41
Cont. (+)	9.0	13.0	24.1	32.8	12.85	18.20	53.32	55.49
E_1	6.5	9.5	17.3	26.2	8.93	14.07	51.62	53.70
E_2	7.5	11.0	21.2	24.5	11.07	12.96	52.22	52.90
G	6.0	9.0	17.5	25.4	8.66	13.03	49.49	51.30
C	7.0	10.0	15.3	23.7	7.45	11.66	48.69	49.20
S_1	7.0	10.4	19.1	26.1	9.72	13.47	50.89	51.61
ASP	8.0	11.5	22.7	29.5	11.99	15.96	52.82	54.10
$LSD_{0.05}$	1.5	2.3	5.0	6.0	2.25	2.55	1.93	2.21

LSD: least significant difference. FS: first season. SS: second season.
E_1: *Ulva* collected from Alexandria during spring; E_2: *Ulva* collected from Alexandria during summer; S_1: *Sargassum* sp. collected from Alexandria during spring; ASP: *Asparagopsis* sp. collected from Marsa Matrouh; G: *Gelidium* sp. collected from Alexandria; C: *Corallina* sp. collected from Alexandria during spring.

of pods was obtained by positive control followed by ASP in both seasons, measuring 200.0% and 166.7% over untreated plants in the first season as well as 217.1% and 180.5% over negative control in the second one, respectively. This data was in agreement with the results obtained by Meeting et al. (1988), suggesting that plant growth regulator may be produced by the Cyanobacteria and be responsible for enhancing crop yield.

4.2.2   Number of seeds/plant

Data presented in Table 9 clearly show that all chemical and bio-fertilization treatments significantly increased number of seeds/plant of *Vicia faba* in both seasons over unfertilized plants. Chemically fertilized plants had the maximal number of seeds per plant among all treatments, at 24.1 and 32.8 seeds in both seasons, respectively. ASP had the second rank without any significant difference with positive control, at 22.7 and 29.5 seeds in the first and second seasons, respectively.

4.2.3   Seed yield/plant (g)

Data on seed yield per plant of *Vicia faba* as affected by chemical and bio-fertilization in two seasons are given in Table 9. The obtained results revealed that all treatments followed the same trend as in number of seeds per plant. Chemical fertilization ranked first, at 372.4% and 411.2%, followed by ASP, at 340.8% and 348.3%, with insignificant differences between them, over negative control in both seasons, respectively.

4.2.4   Specific seed weight (g)

Results of Table 9 indicate that weight of 100 seeds of *Vicia faba* cultivar Assiut 86 showed a significant increase due to chemical and bio-fertilization treatments over untreated plants. The percentage of increases in 100-seed weight ranged between 18.2% and 29.4% for C and positive control, respectively, in the first season, and between 13.3% and 27.8% for the same treatments in the second season, over negative control. This data was in agreement with the results obtained by Adam (1999), which found that algal filtrate of *Desmonostoc muscorum* significantly increased germination of wheat seeds as well as their growth parameters and nitrogen compounds, compared to controls. Also, Lozano et al. (1999) stated that the application of an extract from algae to soil or foliage increased ash, protein and carbohydrate contents of potato (tubers) (*Solanum tuberosum*). Ghallab and Salem (2001) studied the effect of some biofertilizer treatments—"Cerealin" (*Azospirillum* spp.) and Nemales (*Serratia* spp.)—on wheat plant in field experiments and found that the two biofertilizers increased growth characters and nutrients, sugar, amino acids and growth regulators (IAA, GA_3 and cytokinin) and crude protein content in the plant. On the other hand, Abdel-Monem et al. (2001) reported that fertilization with *Azospirillum brasilense* or commercial biofertilizer "Cerealin" improved the growth and yield of maize in rotation with wheat as affected by irrigation regime.

4.3 NPK contents in plant cells

The NPK contents in *Vicia faba* (Assiut 86) after treatment with macroalgae (dry matter) as biofertilizer are shown in Table 10. The results indicated that the nitrogen content increased in plants treated with algae and reached to four-fold over negative control and two-fold over positive control in *Sargassum* sp. The same result occurred with the other elements (P and K). Lean and Nalewajko (1976) reported that seaweeds are rich in nitrogen and potash, but poor in phosphates. Seaweed addition is best suited for light sandy soils generally deficient in potash. Physical condition of the soil improves by seaweed application because of their gelatinous nature. Seaweeds are also valuable as a source of trace elements and other organic substances such as amino acids, auxins, gibberellins and vitamins.

Table 10. NPK content in plant cells after cultivation with algae as organic fertilizer

Treatment	N%	P%	K%
Cont.+	0.11	0.17	1.22
ASP	0.10	0.14	1.62
G	0.19	0.26	1.77
S_1	0.20	0.33	1.76
E_1	0.04	0.04	0.51
E_2	0.06	0.07	0.60
C	0.18	0.12	1.52
Cont.-	0.05	0.07	0.64

Cont. + (positive control); Cont. – (negative control).
E_1: *Ulva* collected from Alexandria during spring; E_2: *Ulva* collected from Alexandria during summer; S_1: *Sargassum* sp. collected from Alexandria during spring; ASP: *Asparagopsis* sp. collected from Marsa Matrouh; G: *Gelidium* sp. collected from Alexandria; C: *Corallina* sp. collected from Alexandria during spring.

5 Conclusion

From the reported data, it could be concluded that marine macroalgae can be used as organic fertilizer for improved vegetative character as well as structural characteristics of plants and subsequently yield character in comparison with untreated and chemically fertilized plants. These results suggest major opportunities for a national project to utilize extracts from macroalgae as organic fertilizer for different plants.

References Cited

Abdel-Monem, M.A.S., H.E. Khalifa, M. Beider, I.A. El-Ghandour and Y.G. Galal. 2001. Using biofertilizer for maize production: response and economic return under different irrigation treatments. J. Sustainable Agric. 19(2): 41–48.

Adam, M.S. 1999. The promotive effect of the cyanobacterium *Nostocmuscorum* on the growth of some crop plants. Acta Microbiol. Polonica 48(2): 163–171.

Barros, M.P., E. Pinto, T.C.S. Sigaud-Kutner, K.H.M. Cardozo and P. Colepicolo. 2005. Rhythmicity and oxidative/nitrosative stress in algae. Biol. Rhythm Res. 36(1-2): 67–82.

Bentley, J.A. 1960. Plant hormones in marine phytoplankton, zooplankton and sea water. J. Mar. Biol. Ass. U.K. 39(03): 433–444.

Blunt, J.W., B.R. Copp, M.H.G. Munro, P.T. Northcote and M.R. Prinsep. 2005. Marine natural products. Nat. Prod. Rep. 22(1): 15–61.

Bograh, A., Y. Gingras, R. Tajmir and R. Carpentier. 1997. The effect of spermine and spermidine on the structure of photosystem II proteins in relation to inhibition of electron transport. FEBS Letters 402(1): 41–44.

Cardozo, K.H.M., T. Guaratini, M.P. Barros, V.R. Falcao, A.P. Tonon, N.P. Lopes, S. Campos, M.A. Torres, A.O. Souza, P. Colepicolo and E. Pinto. 2006. Metabolites from algae with economical impact. Comp. Biochem. Physiol. C. Toxicol. Pharmacol. 146(1-2): 60–78.

Carolyn, P., C. Claude and N. Jeffery. 2001. Seaweed extract residue as a soil amendment. Hortscience 36: 436.

Crouch, I.J. and V. Staden. 1992. Effect of seaweed concentrates on the establishment and yield of greenhouse tomato plants. J. Appl. Phycol. 4(4): 291–296.

Dos Santos, M.D., T. Guaratini, J.L.C. Lopes, P. Colepicolo and N.P. Lopes. 2005. Plant cell and microalgae culture. pp. 143–172. *In*: Taft C.A. (ed.). Modern Biotechnology in Medicinal Chemistry and Industry. Research Signpost, Kerala, India.

Ghallab, A.M. and S.A. Salem. 2001. Effect of biofertilizer treatments on growth, chemical composition and productivity of wheat plants grown under different levels of NPK fertilization. Annals of Agril. Sci. Cairo 46: 485–509.

Johansen, J.R. 1993. Cryptogamic crusts of semiarid and arid lands of North-America. J. Phycol. 29(2): 140–147.

Lean, D.R.S. and C. Nalewajko. 1976. Phosphate exchange and organic phosphorus excretion by freshwater algae. J. Fish Res. Bd. Can. 33(6): 1312–1323.

Lowenstein, J. 1986. The secret life of seaweeds. Oceans 19: 72–75.

Lozano, M.S., J.V. Star, R.K. Maiti, C.A. Oranday, R.H. Gaona, H.A. Aranda and G.M. Rojas. 1999. Effect of an algal extract and several plant growth regulators on the nutritive value of potatoes (*Solanum tuberosum* L. var. *gigant*). Arch. Latinoam. Nutr. 49(2): 166–170.

Lu, Y. and J. Xu. 2015. Phytohormones in microalgae: a new opportunity for microalgal biotechnology? Trends Plant Sci. 20(5): 273–282.

Mayer, A.M.S. and M.T. Hamann. 2004. Marine pharmacology in 2000: marine compounds with antibacterial, anticoagulant, antifungal, anti-inflammatory, antimalarial, antiplatelet, antituberculosis, and antiviral activities; affecting the cardiovascular, immune, and nervous system and other miscellaneous mechanisms of action. Mar. Biotechnol. 6(1): 37–52.

Meeting, D., W.R. Rayburn and P.A. Reynaud. 1988. Algae and agriculture. pp. 335–370. *In*: Lembi, C.A. and Waaland, J.R. (eds.). Algae and Human Affairs. Cambridge University Press, Cambridge.

Mori, I.C., Y. Ikeda, T. Matsuura, T. Hirayama and K. Mikami. 2017. Phytohormones in red seaweeds: a technical review of methods for analysis and a consideration of genomic data. Bot. Mar. 60(2): 153–170.

Ordog, V. (1999). Beneficial effects of microalgae and cyanobacteria in plant/soil-system, with special regard to their auxin- and cytokinin-like activity. International workshop and training course on Microalgal Biology and Biotechnology Mosonmagyarovar, Hungary, June: 13–26.

Plaza, M., A. Cifuentes and E. Ibáñez. 2008. In the search of new functional food ingredients from algae. Food Sci. Tech. 19(1): 31–39.

Shalaby, E.A. 2011. Algal biomass and biodiesel production. pp. 111–132. *In*: Stoytcheva, M. (ed.). Biodiesel – Feedstocks and Processing Technologies. InTechOpen, London, U.K.

Singh, S., B.N. Kate and U.C. Banerjee. 2005.Bioactive compounds from cyanobacteria and microalgae: an overview. Crit. Rev. Biotechnol. 25(3): 73–95.

Smith., F.B.C. and V.J. Staden. 1983. The effect of seaweed concentrate on the growth of tomato *Lycopersicon-esculenum* plants in nematode-infested soil. Sci. Hortic. 20(2): 137–146.

Snedecor, G.W. and W.G. Cochran. 1982. Statistical Methods. The Iowa State Univ. Press., Ames., Iowa, USA, 507 pp.

Vogel, A.I. 1975. A text book of practical organic chemistry, 3rd Edition. English Language Book Society and Longman Group Ltd., London, 1165 pp.

Zaccro, M.C., C. Salazer, Z. Caire, S. Cans and A.M. Stella. 2001. Lead toxicity in cyanobacteria porphyrin metaboilsm. Environ. Toxicol. and Water Quality, 16: 61–67.

Role of Secondary Metabolites from Seaweeds in the Context of Plant Development and Crop Production

Hossam S. El-Beltagi[1,2*], **Heba I. Mohamed**[3] and **Maged M. Abou El-Enain**[3]

[1] Cairo University, Faculty of Agriculture, Biochemistry Department, P.O. Box 12613,
  Gamma St, Giza, Cairo, Egypt
[2] King Faisal University, College of Agriculture and Food Sciences,
  Agriculture Biotechnology Department, P.O. Box 420, Alhassa 31982, Saudi Arabia
[3] Ain Shams University, Faculty of Education, Biological and Geological Sciences Department,
  El Makres St, Roxy, Cairo 1575, Egypt

1 Introduction

Seaweeds (or macroalgae) are aquatic photosynthetic organisms belonging to the domain Eukarya and to kingdoms Plantae (green and red algae) and Chromista (brown algae). Although classification systems have varied greatly over time and according to the authors, the following are generally agreed:

(a) The green algae are included in the phylum Chlorophyta and their pigmentation is identical to that of terrestrial plants (chlorophyll a, b and carotenoids).
(b) The red algae belong to the phylum Rhodophyta and their photosynthetic pigments are chlorophyll a, phycobilins (R-phycocyanin and R-phycoerythrin) and carotenoids, mostly β-carotene, lutein and zeaxanthin.
(c) The brown algae are included in the phylum Ochrophyta (or Heterokontophyta), class Phaeophyceae, and their pigments include chlorophylls a, c and carotenoids, dominated by fucoxanthin (Pereira 2009, Pereira 2018).

These organisms are often regarded as an under-utilized bioresource, although many species have been used as sources of food and industrial gums, and in therapeutic and botanical applications for centuries (Khan et al. 2009). Many studies have been carried out on the plant growth-stimulating effects of marine algae (Russo and Berlyn 1990). Seaweeds have been proven as a source of antioxidants, plant growth hormones, osmoprotectants, mineral nutrients and many other organic compounds including novel bioactive molecules (Ramarajan et al. 2013, Ismail and El-Shafay 2015, Pacholczak et al. 2016). The use of seaweeds as biofertilizer was considered to compensate for the deficiency of N, P and K in soils (Tuhy et al. 2015, Singh et al. 2016, Vyomendra and Kumar 2016).

Seaweed extracts are known to contain a wide range of bioactive compounds. Several studies have shown that the stimulating effects are due to a variety of major constituents within the seaweed

---

*Corresponding author: helbeltagi@agr.cu.edu.eg; lbltg@yahoo.com

extracts, such as diverse plant nutrients, phytohormones or betaines (Alam et al. 2014, Divya et al. 2015, Shahbazi et al. 2015, Al-Hameedawi 2016, Mirparsa et al. 2016). On the other hand, organic matter contained in seaweed biofertilizers is known to stimulate plant growth due to its nutrient content (Davari et al. 2012). It was also shown that the addition of different seaweeds as organic fertilizers in adequate quantities improved soil condition and growth parameters in field crops (Badar et al. 2015). Seaweeds improve the level of soil nutrients such as N, P and K and other minerals necessary for plant growth (Sethi 2012, Mirparsa et al. 2016).

Furthermore, seaweeds are rich in proteins, fiber, fat, cellulose, hemicelluloses, lignin, vitamins, bromine, and iodine (Mohammadi et al. 2013, Heltan et al. 2015). Also, they contain a diverse range of organic compounds, at least 17 amino acids (Yoo 2003, Shevchenko et al. 2007). A higher composition of minerals was found in marine seaweeds than in land vegetables (Manivannan et al. 2008, Kumar et al. 2009).

Seaweeds are generally known as a rich source of a high diversity of natural molecules. Seaweeds can produce high amounts of secondary metabolites, including terpenes, lipid-, steroid-, and aromatic-like compounds, acetogenins, amino acid-derived products, phlorotannin and other polymeric substances (Zbakh et al. 2012, Thinakaran and Sivakumar 2013). Marine algae also produce bioactive metabolites in response to microbial activities (Alam et al. 2014, Michalak et al. 2015), insects (Abbassy et al. 2014), and viruses (Mendes et al. 2010). It was revealed that red macroalgae show the best production of halogenated compounds (Pereira and Teixeira 1999). One study revealed that polysaccharides, fatty acids, phlorotannins, pigments, lectins, alkaloids, terpenoids and halogenated compounds isolated from green, brown and red algae have potent antimicrobial activities (Perez et al. 2016).

2 Antioxidant Activity

Marine algae, like other photosynthesizing plants, are exposed to a combination of light and oxygen that leads to the formation of free radicals and other strong oxidizing agents. However, the absence of oxidative damage in the structural components of macroalgae (i.e., polyunsaturated fatty acids) and their stability to oxidation during storage suggest that their cells have protective antioxidative defense systems (Matsukawa et al. 1997). In fact, algae have protective enzymes (superoxide dismutase, peroxidase, glutathione reductase, catalase) and antioxidative molecules (e.g., phlorotannins, ascorbic acid, tocopherols, carotenoids, phospholipids, chlorophyll-related compounds, bromophenols, catechins, mycosporine-like amino acids, polysaccharides) that are similar to those of vascular plants (Rupérez et al. 2002, Yuan et al. 2005).

Fresh seaweeds are known to contain reactive antioxidant molecules, such as ascorbate and glutathione (GSH), as well as secondary metabolites, including carotenoids (α- and β-carotene, fucoxanthin, astaxanthin), mycosporine-like amino acids (mycosporine-glycine) and catechins (e.g., catechin, epigallocatechin), gallate, phlorotannins (e.g., phloroglucinol), eckol and tocopherols (α-, χ-, δ-tocopherols) (Yuan et al. 2005). Terpenoids, polyphenols, phenolic acids, anthocyanins, hydroxycinnamic acid derivatives and flavonoids form other important antioxidant molecules from macroalgae (Bandoniene and Murkovic 2002). Many phenolic compounds from seaweeds have demonstrated high antioxidant activity, for example, stypodiol, isoepitaondiol, taondiol, and terpenoids (Nahas et al. 2007). Polyphenols constitute important antioxidant molecules of plant origin. Polyphenolics such as catechin, epicatechin and gallate showing antioxidant activity have been isolated from *Halimeda* sp. (Chlorophyta) (Devi et al. 2008). Phlorotannins isolated from *Sargassum pallidum* and *Fucus vesiculosus* (Ochrophyta, Phaeophyceae) have also shown significant antioxidant properties (Díaz-Rubio et al. 2011). Sulfated polysaccharides are another group of compounds isolated from seaweeds having antioxidant properties. Sulfated polysaccharides such as fucoidan, alginic acid and laminaran from *Turbinaria* (Ochrophyta, Phaeophyceae) have demonstrated antioxidant activity (Chattopadhyay et al. 2010). Many other sulfated polysaccharides extracted from seaweeds—such as sulfated galactans, galactans, sulfated glycosaminoglycan, and porphyran—have also shown significant radical scavenging properties (Luo et al. 2009, Costa et al. 2010, Barahona et al. 2011).

3 Secondary Metabolites

Secondary metabolites are classified according to the biosynthetic pathway from which they are derived. Biosynthetic and genetic studies have revealed that a limited number of core biosynthetic pathways, generalized in Fig. 1, are responsible for the production of most of the natural products.

Seaweeds (macroalgae) are not only the source of primary metabolites but also an extensive prolific source of secondary metabolites and many of these compounds were extensively evaluated using bioassays and pharmacological studies (del Val et al. 2001). During the log or exponential phase, organisms produce a variety of substances that are essential for their growth, such as nucleotides, nucleic acids, amino acids, proteins, carbohydrates, and lipids, or by-products of energy-yielding metabolism, such as ethanol, acetone, or butanol; this phase is described as tropophase, and the products are usually called primary metabolites. Organisms produce several products other than the primary metabolites with the objective of protecting themselves and to survive in their own environment; these are known as secondary metabolites (Selvin and Lipton 2004). Many secondary metabolites are produced from intermediates and end products of secondary metabolism are grouped into terpenes (a group of lipids), phenolics (derived from carbohydrates) and alkaloids (derived from amino acids, the building blocks of proteins) (Jiang and Gerwick 1991). More than 600 secondary metabolites have been isolated from nearly 3600 seaweeds (Fig. 2) (Renn 1993). It is reported that of the total marine algae so far evaluated, 25% showed one or the other biological activity. Seaweed resources have been used for a variety of major metabolites such as polysaccharides, lipids, proteins, carotenoids, vitamins, sterols, enzymes, antibiotics and many other fine chemicals (Ibtissam et al. 2009).

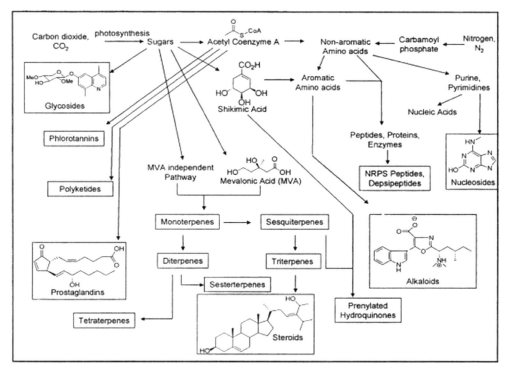

Figure 1. Biosynthetic origin of the major classes of natural products.

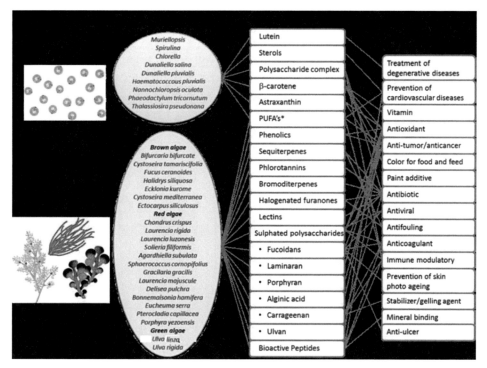

Figure 2. Components of secondary metabolites of marine algae and their possible application.

Color version at the end of the book

3.1 Terpenes

Terpenes are a large and diverse class of compounds produced by a wide variety of organisms, though plants are an especially prolific source. Terpenes are hydrocarbons (composed only of carbon and hydrogen). More than half of the reported secondary metabolites from macroalgae are isoprenoids. Terpenes, steroids, carotenoids, prenylated quinines, and hydroquinones make up the isoprenoid class, which is understood to derive from either the classical mevalonate pathway, or the mevalonate-independent pathway (Stratmann et al. 1992). Mevalonic acid (MVA) is the first committed metabolite of the terpene pathway (Fig. 3). Dimethylallyl pyrophosphate (DMAPP) and its isomer isopentenyl pyrophosphate are intermediates of the MVA pathway and exist in nearly all life forms (Fig. 3) (Humphrey and Beale 2006).

3.1.1 Osmoprotective compounds

Seaweeds living in marine waters are exposed to a continuous salt stress because of the high salinity of the marine environment. Under hyperosmotic conditions, organic compounds such as proline, betaines, as well as polyamines and sorbitol are synthesized and accumulated at high concentrations that are intimately involved in the osmotic stress adjustment in macroalgae (Van Alstyne et al. 2003, El Shoubaky and Salem 2016). It also has been revealed that many marine algae produce 3-dimethylsulfoniopropionate (DMSP), a potent osmoprotective compound; its degradation product dimethylsulfide plays a central role in the biogeochemical S-cycle (Pichereau et al. 1998, Summers et al. 1998). Tertiary sulfonium compounds (DMSP) and quaternary ammonium compounds (such as glycine betaine) were reported as efficient osmoprotectants for agricultural bacteria and plants as well (Asma et al. 2006, Rezaei et al. 2012, Manaf 2016).

In order to make the best use of particularly osmoprotective substances of seaweeds, many experiments have been performed to restore plant growth under saline conditions by using different

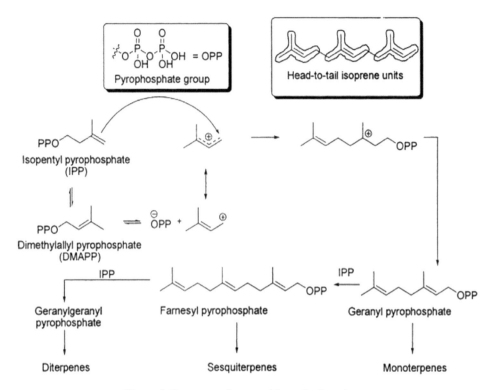

Figure 3. Summary of terpene biosynthetic pathway.

algal extracts. It was shown that, e.g., *Ulva lactuca* (Chlorophyta) was able to restore the leaf area and pigment content in soy bean under saline stress conditions (Ramarajan et al. 2013). Furthermore, maize growth under cold stress was restored after addition of seaweed extract (Bradacova et al. 2016). It was also demonstrated that the addition of extracts of *Sargassum vulgare* (Ochrophyta, Phaeophyceae) improved the germination behavior of two bean cultivars (*Phaseolus vulgaris*) under salt stress (Latique et al. 2014). Other examples of successful use of seaweed extracts are application of seaweed extracts of *Ascophyllum nodosum* (Ochrophyta, Phaeophyceae) to *Amaranthus tricolor* enhancing flowering and its chemical constituents under high salinity conditions (Abdel Aziz et al. 2011), the use of seaweed extracts to stimulate germination and growth of tomato (*Lycopersicon* spp.) seedlings under salt stress (Alalwani et al. 2012), and the application of liquid extracts of *U. lactuca* in high salinity conditions to restore growth of durum wheat (*Triticum durum*) (Nabti et al. 2010). Finally, *Durvillaea potatorum* (Ochrophyta, Phaeophyceae) application substantially improved growth and yield of bean plants under water suppression (Bastos et al. 2016).

3.2 Betaines

Ascophyllum nodosum (Ochrophyta, Phaeophyceae) extracts contain various betaines and betaine-like compounds (Blunden et al. 1986). In plants, betaines serve as a compatible solute that alleviates osmotic stress induced by salinity and drought stress; however, other roles have also been suggested (Blunden and Gordon 1986), such as enhancing leaf chlorophyll content of plants following their treatment with seaweed extracts (Blunden et al. 1997). This increase in chlorophyll content may be due to a decrease in chlorophyll degradation (Whapham et al. 1993). Yield enhancement effects due to improved chlorophyll content in leaves of various crop plants have been attributed to the betaines present in the seaweed (Genard et al. 1991, Blunden et al. 1997). It has been indicated that betaine may work as a nitrogen source when provided in low concentration and serve as an osmolyte at higher concentrations (Naidu et al. 1987). Betaines have been shown to play a part in successful formation of somatic embryos from cotyledon tissues and mature seeds of tea (Wachira and Ogada 1995, Akula et al. 2000).

3.3 Sterols

As with many other eukaryotic cells, sterols are an essential group of lipids. Generally, a plant cell contains a mixture of sterols, such as β-sitosterol, stigmasterol, 24-methylenecholesterol, and cholesterol (Nabil and Cosson 1996). Brown seaweed chiefly contains fucosterol and fucosterol derivatives, whereas red seaweeds primarily contain cholesterol and cholesterol derivatives. Green seaweed accumulates mainly ergosterol and 24-methylenecholesterol (Govindan et al. 1993, Nabil and Cosson 1996) (Table 1).

Sterols are structural components of cell membrane and regulate membrane fluidity and permeability. They are composed of four rings (A–D) with a hydroxyl group in carbon-3, two methyl groups at C18 and C19 carbons and a side chain at C17 (Fig. 4). The main sterols in macroalgae are cholesterol, fucosterol, isofucosterol, and clionasterol (Kumari et al. 2013).

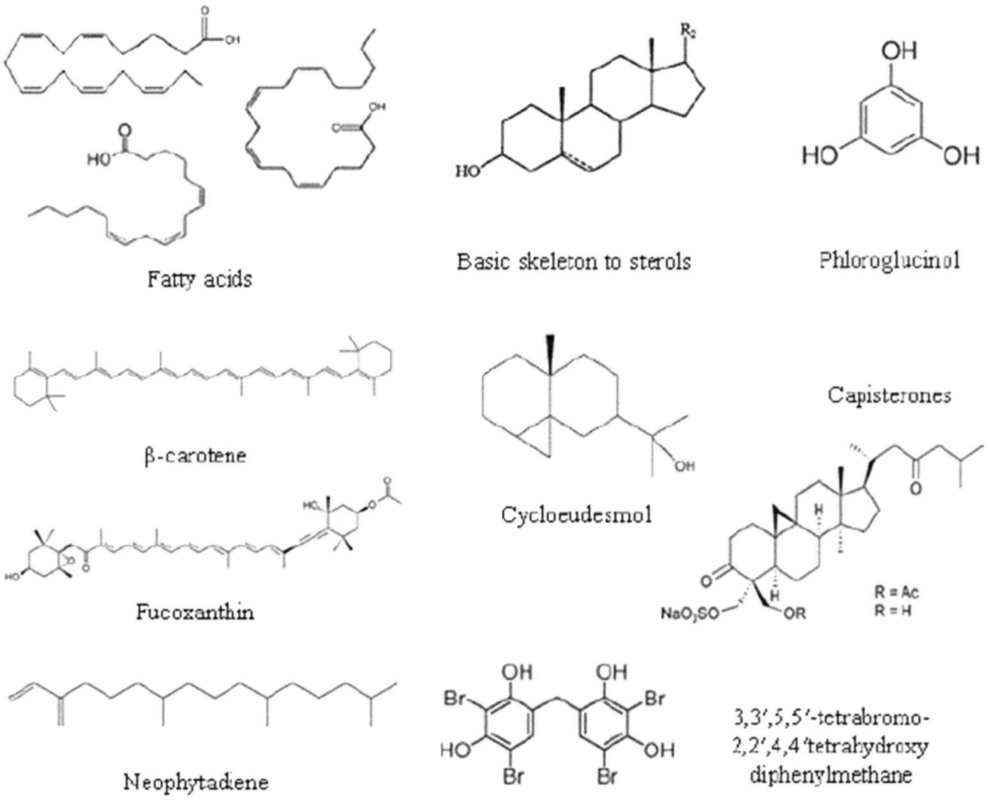

Figure 4. Chemical structures of fatty acids, sterol, phloroglucinol, carotenoids (β-carotene and fucoxanthin), terpenes (neophytadiene and cycloeudesmol) and a brominated compound.

3.4 Lipids and fatty acids

Algal lipids content in seaweed ranges from 0.12% to 6.73% (dry weight). Algal lipids are composed mainly of phospholipids, glycolipids and non-polar glycerolipids (neutral lipids) (Kumari et al. 2013). Phospholipids are located in extra-chloroplast membranes and account for 10% to 20% of

Table 1. Common sterol constituents of green, red, and brown seaweeds

Seaweed	Type of sterol
Chlorophyta (green algae)	22-Dehydrocholesterol
	24-Methylenecholesterol
	24-Methylenecycloartanol
	28-Isofucosterol
	Brassicasterol
	Cholesterol
	Clerosterol
	Clionasterol
	Codisterol
	Cycloartannol
	Cycloartenol
	Decortinol
	Decortinone
	Ergosterol
	Fucosterol
	Isodecortinol
	Ostreasterol
	ß-Stitosterol
	Zymosterol
	Chondrillasterol
	D5-Ergostenol
	D7-Ergostenol
	Poriferastenol
	24-Methylenophenol
Rhodophyta (red algae)	22-Dehydrocholesterol
	24-Methylenecholesterol
	Campesterol
	Cholesterol
	Cycloartenol
	Desmosterol
	Fucosterol
	Stigmasterol
	Brassicasterol
	5-Dihydroergosterol
	D5-Ergostenol
	Obusifoliol
	D4,5-Ketosteroids
Phaeophyceae (brown algae)	22-Dehydrocholesterol
	Cycloartenol
	24-Methylenecycloartenol
	24-Methylenecholesterol
	Fucosterol
	Cholesterol
	Campesterol
	Stigmasterol
	Brassicasterol
	Clionasterol

total lipids in algae. They are characterized by higher contents of n-6 fatty acids, and the major fatty acids present are oleic, palmitic, stearic, arachidonic and eicosapentanoic acids. The most dominant phospholipid in algae is phosphatidylglycerol in green algae, phosphatidylcholine in red algae, and phosphatidylcholine and phosphatidylethanolamine in brown algae. Glycolipids are located in photosynthetic membranes and constitute more than half of the lipids in the main algal groups. They are characterized by high n-3 polyunsaturated fatty acids. Three major types of glycolipids are monogalactosyldiacylglycerides, digalactosyldiacylglycerides, and sulfoquinovosyldiacylglycerides (Plouguerné et al. 2014). Triacylglycerol is the most prevalent neutral lipid, their content ranging from 1% to 97% with a function of storage and energy reservoir.

Fatty acids are carboxylic acids with aliphatic chains and prevalent even carbon numbers (C4-C28) that may be straight or branched, saturated or unsaturated. According to the double bond, fatty acids are classified as monounsaturated (MUFA) or polyunsaturated (PUFA), and the latter can be classified as n-3 or n-6 depending on the position of the first double bond from the methyl end. Green algae are rich in C18 PUFAs, mainly linolenic (C18:3 n-3), stearidonic (C18:4 n-3) and linoleic (C18:2 n-6) acids; red algae are rich in C20 PUFAs, mainly arachidonic (C20:4 n-6) and eicosapentanoic (C20:5 n-3) acids, and brown algae exhibit both. Oxylipins are the oxygenated products of fatty acids and are mainly derived from C16, C18, C20 and C22 PUFAs and confer innate immunity in response to biotic and abiotic stress, such as pathogenic bacteria and herbivores (Kumari et al. 2013).

3.5 Phenolic compounds

Phenolic compounds are secondary metabolites because they are not directly involved in primary processes such as photosynthesis, cell division or reproduction of algae. They are characterized by an aromatic ring with one or more hydroxyl groups and the antimicrobial action is due to the alteration of microbial cell permeability and the loss of internal macromolecules or by interference with the membrane function and loss of cellular integrity and eventual cell death (Abu-Ghannam and Rajauria 2013).

Chemically, structures ranging from simple phenolic molecules to complex polymers with a wide range of molecular sizes (126-650 kDa) have been described (Cardoso et al. 2014). Polyphenols can be divided into phloroglucinols and phlorotannins. Phloroglucinol contains an aromatic phenyl ring with three hydroxyl groups (Fig. 3). Phlorotannins are oligomers or polymers of phloroglucinol with additional halogen or hydroxyl groups; and, according to the inter-linkage, phlorotannins can be subdivided into six specific groups: (1) phlorethols (with aryl-ether linkage); (2) fucols (with aryl-aryl bonds); (3) fucophlorethols (with ether or phenyl linkage); (4) eckols (with dibenzodioxin linkages; Wijesinghe et al. 2011); (5) the less frequent fuhalols (with ortho-/para-arranged ether bridges containing an additional hydroxyl on one unit); and (6) carmalols (with dibenzodioxin moiety) (Singh and Sidana 2013). The presence of simple phenols, such as hydroxycinnamic and benzoic acids and derivates, and flavonoids was reported in the green seaweed (Gupta and Abu-Ghannam 2011), but brown seaweed has higher contents of phenolic compounds than green and red macroalgae. The typical phlorotannin profile from brown algae with antimicrobial activity mainly consists of phloroglucinol, eckol and dieckol (Suleria et al. 2015).

3.6 Pigments

Algae as photosynthetic organisms can synthesize the three basic classes of pigments found in marine algae: chlorophylls, carotenoids and phycobiliproteins, allowing classification of seaweed into Chlorophyta (green algae), Phaeophyceae (brown algae) and Rhodophyta (red algae). The green color is due to the presence of chlorophylls a and b, the greenish-brown color is attributed to the fucoxanthin, chlorophylls a and c, and the red color is attributed to the phycobilins, such as phycoerythrin (Kraan 2013).

The antimicrobial mechanism proposed for carotenoids could lead to the accumulation of lysozyme, an immune enzyme that digests bacterial cell walls (Abu-Ghannam and Rajauria 2013). Carotenoids are present in all algae and are lipid-soluble, natural pigments composed of eight units of five carbons, namely tetraterpenoids, with up to 15 conjugated double bonds. Carotenoids are usually divided into two classes: carotenes (when the chain ends with a cyclic group, containing only carbon and hydrogen atoms) and xanthophylls or oxycarotenoids (which have at least one oxygen atom as a hydroxyl group, as an oxy-group or as a combination of the two). β-Carotene is the most common carotene (Fig. 3), whereas lutein, fucoxanthin (Fig. 3) and violaxanthin belong to the xanthophylls class (Christaki et al. 2013). β-Carotene, lutein, violaxanthin, neoxanthin and zeaxanthin are found in green seaweed species; α- and β-carotene, lutein and zeaxanthin are present in red seaweed; and β-carotene, violaxanthin and fucoxanthin are found in brown algae (Kraan 2013).

3.7 Lectins

Lectins are natural bioactive ubiquitous proteins or glycoproteins of non-immune response that bind reversibly to glycans of glycoproteins, glycolipids and polysaccharides possessing at least one non-catalytic domain causing agglutination. Algal lectins differ from terrestrial lectins because they are monomeric, low molecular weight proteins, exhibiting high content of acidic amino acids, with isoelectric point in the range of 4–6. They do not require metal ions for their biological activities, and most of them show higher specificity for oligosaccharides and/or glycoproteins than for monosaccharides. Based on the binding properties to glycoproteins, algal lectins are divided into three major categories: complex type N-glycan-specific lectins, high mannose (HM) type N-glycan-specific lectins, and lectins with specificity to both the above types of N-glycans (Singh et al. 2015). Lectins from marine organisms are also classified into C-type lectins, F-type lectins, galectins, intelectins, and rhamnose-binding lectins (Cheung et al. 2015).

3.8 Alkaloids

An alkaloid is a compound that has nitrogen atom(s) in a cyclic ring. Numerous biological amines and halogenated cyclic nitrogen-containing substances are included in the term "alkaloid". The latter is specific to marine organisms and marine algae. They have not been found in terrestrial plants. Alkaloids in marine algae were classified in three groups as follows (Güven et al. 2010): (i) phenylethylamine alkaloids; (ii) indole and halogenated indole alkaloids; and (iii) other alkaloids, such as 2,7-naphthyridine derivatives. Alkaloids isolated from marine algae mostly belong to 2-phenylethylamine and indole groups. Halogenated alkaloids are specific for algae, being bromine- and chloride-containing alkaloids particularly dominant in Chlorophyta. Most of the alkaloids of the indole group are concentrated in Rhodophyta (Barbosa et al. 2014).

4 Biostimulation Proficiency on Plant Growth

Marine macroalgae are regarded as valuable resources for plant improvement because of their higher contents of mineral substances, amino acids, vitamins, and plant growth regulators, including auxins, cytokinin and gibberellins (Stirk and Van Staden 1997a, b). Brown algal extracts, as well as algae themselves, are widely used in agriculture. They have been shown to increase the productivity of a variety of agricultural plants, including potato, grasses, citrus plants, tomato, beet and legumes. Application of marine macroalgae in plant biotechnology has been shown to produce healthy plants, in addition to significant increase in the number and weight of fruits. Also, they offer a non-toxic alternative for disease management (Baloch et al. 2013). It also has been reported that aqueous algal extracts obtained by various means (e.g., boiling, autoclaving, or homogenization) showed positive effects on health, growth, and crop yield of many plants.

Plant growth regulators mainly differ from fertilizers in several points: (i) they alter and manage cell division, (ii) they control root and shoot elongation, and (iii) they initiate flowering and other metabolic functions. Fertilizers clearly supply nutrients needed for normal plant growth (Allen et al. 2001). Cytokinin is regarded as the most important plant growth regulator in marine algae, while trace minerals present in marine macroalgal extracts play important roles in plant nutrition and physiology, probably as enzyme activators (Senn 1987). The exogenous application of the *A. nodosum* extract on turf and forage grasses increased the antioxidant metabolites in plants such as α-tocopherol, ascorbic acid and β-carotene in testing plants as well as antioxidant enzyme activities such as superoxide dismutase, GSH reductase and ascorbate peroxidase (Allen et al. 2001). A biostimulant is an organic substance that when applied in small amounts enhances plant growth and development and such response cannot be achieved by application of traditional plant nutrients (EBIC 2012). Macroalgal extracts have been used as agricultural biostimulants (ABs) (EBIC 2012). The use of macroalgal ABs on crop plants can generate numerous benefits with reported effects including enhanced rooting, higher crop and fruit yields, enhanced photosynthetic activity, and resistance to fungi, bacteria and virus (Sharma et al. 2014). ABs include various formulations of compounds, substances and other products, such as microorganisms, trace elements, enzymes, plant growth regulators and macroalgal extracts that are applied to plants or soils to organize and enhance the crop's physiological processes, therefore making them more efficient. ABs act on the physiology of the plant through diverse pathways to improve crop vigor, yields, quality and post-harvest condition (EBIC 2012). Macroalgal ABs have been shown to influence respiration, photosynthesis, nucleic acid synthesis and ion uptake (Khan et al. 2009). Consequently, these products can enhance nutrient availability, improve water-holding capacity, increase antioxidants, enhance metabolism and increase chlorophyll production in plants (Zhang 1997, Khan et al. 2009). In addition, organic fertilizers based on seaweed extract improved the essential oil yield and composition in rosemary (*Rosmarinus officinalis* L.): both the quality and quantity of rosemary essential oil were enhanced in comparison with plants grown with the inorganic fertilizers. The sprayed seaweed fertilizer showed a significantly higher percentage of β-pinene, α-terpinene (monoterpenes), α-phellandrene and 3-methylenecycloheptene than inorganic fertilizer on *R. officinalis*. Italicene, α-thujene, α-bisabolol (sesquiterpenes), and E-isocitral (monoterpenes) occurred in significantly higher percentages for plants watered with the seaweed extract. The levels were significantly different from controls and from plants treated with the inorganic fertilizer. The seaweed treatments caused a significant increase in oil amount and leaf area as compared with both inorganic treatments and the control regardless of application method (Tawfeeq et al. 2016). Moreover, applications of different extract types have been reported to improve plants' tolerance to a wide range of abiotic stresses, i.e., salinity, drought and temperature extremes. Extensive studies on the chemical composition of various extracts made from a diversity of seaweed species revealed that the nutrient content (typically macronutrients including N, P, K) of the extracts was insufficient to elicit physiological responses at the typical concentrations at which the seaweed extracts were applied in the field (Blunden 1991, Khan et al. 2009). Thus, it has long been suggested that the physiological effect of seaweed extracts is largely mediated by growth-promoting compounds and elicitors. Also, seaweed extracts played an important role in enhancing growth and phytochemical composition of medicinal shrubs. They enhanced plant growth and the phytochemical composition and antioxidant capacity of plant leaves of both species during moderate drought conditions (Elansary et al. 2016). Seaweed extracts and 5-aminolevulinic acid had stimulatory synergistic effects on the growth and secondary metabolites of *Ascophyllum nodosum* subjected to saline conditions. Several mechanisms are involved in such effects, including gas exchange control, sugar buildup, increasing non-enzymatic and enzymatic antioxidants control of reactive oxygen species accumulation, as well as transcriptional and metabolic regulation of environmental stress (Al-Ghamdia and Elansary 2018). *Ecklonia maxima* is one of the brown seaweeds widely used commercially as a biostimulant to improve plant growth and crop protection.

Eckol, a phenolic compound isolated from *E. maxima* (Ochrophyta, Phaeophyceae), has shown stimulatory effects in maize, indicating its potential use as a plant biostimulant. Eckol acted as a plant growth stimulant as well as exhibiting insecticidal effect for cabbage plants. Commercially there are very few organic compounds that have dual effects as shown by eckol. Eckol-based biostimulants therefore would be beneficial to the cabbage industry, where it faces consistent losses due to aphid infestation. More importantly, it will also help in reducing the use of chemical fertilizers and pesticides for sustainable agricultural systems (Rengasamy et al. 2016).

In particular, seaweed extracts made from different starting raw materials, and by different procedures, are reported to have a number of beneficial effects such as increased nutrient uptake, biotic and abiotic stress tolerances, and improvement in the quality of products (Fig. 5) (Battacharyya et al. 2015). Therefore, seaweed extracts are recommended especially when there is a stressful condition that affects crop performance. Moreover, they are considered an organic farm input as they are environmentally benign and safe for the health of animals and humans.

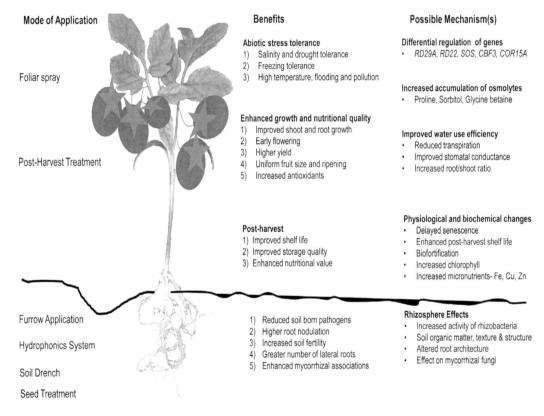

Figure 5. Methods of application of seaweed extracts, and their effects on plant and mechanisms of action.

5 Conclusion

Commercial extracts from raw materials of different seaweeds have received a greater acceptance in plant biostimulants. Extracts of various types are now being used widely and a number of the larger agrochemical companies include seaweed extract and formulations among their offerings.

Acknowledgments

The authors would like to express their appreciation to the Faculty of Agriculture, Cairo University, Department of Biochemistry and the Faculty of Education, Ain Shams University, Department of Biological and Geological Sciences for continuing cooperation to support research and providing the facilities necessary to accomplish the research objectives.

References Cited

Abbassy, M.A., M.A. Marzouk, E.I. Rabea and A.D. Abd-Elnabi. 2014. Insecticidal and fungicidal activity of *Ulva lactuca* Linnaeus (Chlorophyta) extracts and their fractions. Annu. Res. Rev. Biol. 4: 2252–2262.

Abdel Aziz, N.G., M.H. Mahgoub and H.S. Siam. 2011. Growth, flowering and chemical constituents performance of Amaranthus tricolor plants as influenced by seaweed (*Ascophyllum nodosum*) extract application under salt stress conditions. J. Appl. Sci. Res. 7: 1472–1484.

Abu-Ghannam, N. and G. Rajauria. 2013. Antimicrobial activity of compounds isolated from algae. pp. 287–306. *In*: Domínguez, H. (ed.). Functional Ingredients from Algae for Foods and Nutraceuticals. Woodhead Publishing, Cambridge, UK.

Akula. A., C. Akula and M. Bateson. 2000. Betaine: a novel candidate for rapid induction of somatic embryogenesis in tea (*Camellia sinensis* [L.] O. Kuntze). Plant Growth Regul. 30: 241–246.

Alalwani, B.A., M.A. Jebor and T.A.I. Hussain. 2012. Effect of seaweed and drainage water on germination and seedling growth of tomato (*Lycopersicon* spp.). Euphrates J. Agric. Sci. 4: 24–39.

Alam, Z.M., G. Braun, J. Norrie and D.M. Hodges. 2014. Ascophyllum extract application can promote plant growth and root yield in carrot associated with increased root-zone soil microbial activity. Can. J. Plant Sci. 94: 337–348.

Al-Ghamdia, A.A. and H.O. Elansary. 2018. Synergetic effects of 5-aminolevulinic acid and *Ascophyllum nodosum* seaweed extracts on *Asparagus* phenolics and stress related genes under saline irrigation. Plant Physiol. Biochem. 129: 273–284.

Al-Hameedawi, A.M.S. 2016. Effect of hletab, kelpak and paisein on vegetative growth and yield of fig trees (*Ficus carica* L). J. Env. Sci. Pollut. Res. 2: 87–89.

Allen, V.G., K.R. Pond, K.E. Saker, J.P. Fontenot, C.P. Bagley, R.L. Ivy, R.R. Evans, R.E. Schmidt, J.H. Fike, X. Zhang, J.Y. Ayad, C.P. Brown, M.F. Miller, J.L. Montgomery, J. Mahan, D.B. Wester and C. Melton. 2001. Tasco: influence of a brown seaweed on antioxidants in forages and livestock – a review. J. Anim. Sci. 79: 21–31.

Asma, M., S. Muhammad and A.A. Nudrat. 2006. Influence of exogenously applied glycine betaine on growth and gas exchange characteristics of maize (*Zea mays* L.). Pak. J. Agri. Sci. 43: 36–41.

Badar, R., M. Khan, B. Batool and S. Shabbir. 2015. Effects of organic amendments in comparison with chemical fertilizer on cowpea growth. Int. J. Appl. Res. 1: 66–71.

Baloch, G.N., S. Tariq, S. Ehteshamul-Haque, M. Athar, V. Sultana and J. Ara. 2013. Management of root diseases of eggplant and watermelon with the application of asafoetida and seaweeds. J. Appl. Bot. Food Qual. 86: 138–142.

Bandoniene, D. and M. Murkovic. 2002. On-line HPLC–DPPH screening method for evaluation of radical scavenging phenols extracted from apples (*Malus domestica* L.). J. Agric. Food Chem. 50: 2482–2487.

Barahona, T., N.P. Chandıa, M.V. Encinas, B. Matsuhiro and E.A. Zuniga. 2011. Antioxidant capacity of sulfated polysaccharides from seaweeds. A kinetic approach. Food Hydrocol. 25: 529–535.

Barbosa, M., P. Valentão and P.B. Andrade. 2014. Bioactive compounds from macroalgae in the new millennium: implications for neurodegenerative diseases. Mar Drugs. 12: 4934–4972.

Bastos, F.J.C., F.A.L. Soares, C.V. Sousa, C.J. Tavares, M.B. Teixeira and A.E.C. Sousa. 2016. Common bean yield under water suppression and application of osmoprotectants. Rev. Bras. Eng. Agric. Ambient. 20: 697–701.

Battacharyya, D., M.Z. Babgohari, P. Rathor and B. Prithiviraj. 2015. Seaweed extracts as biostimulants in horticulture. Sci. Hortic. 196: 39–48.

Blunden, G., A.L. Cripps, S.M. Gordon, T.G. Mason and C.H. Turner. 1986. The characterisation and quantitative estimation of betaines in commercial seaweed extracts. Bot. Mar. 29: 155–160.

Blunden, G., 1991. Agricultural uses of seaweeds and seaweed extracts. pp. 65–81. *In*: M.D. Guiry et al. (ed.). Seaweed Resources in Europe: Uses and Potential. John Wiley and Sons, Inc., Somerset, New Jersey, USA.

Blunden, G. and S.M. Gordon. 1986. Betaines and their sulphono analogues in marine algae. pp. 39–80. *In*: Round, F.E. and Chapman, D.J. (eds.). Progress in Phycological Research. Biopress Ltd., Bristol.

Blunden, G., T. Jenkins and Y. Liu. 1997. Enhanced leaf chlorophyll levels in plants treated with seaweed extract. J. Appl. Phycol. 8: 535–543.

Bradacova, K., N.F. Weber, N.M. Talab, M. Asim, M. Imran, M. Weinmann and G. Neumann. 2016. Micronutrients (Zn/Mn), seaweed extracts, and plant growth-promoting bacteria as cold-stress protectants in maize. Chem. Biol. Technol. Agric. 3: 1–10.

Cardoso, M.S., G.L. Carvalho, J.P. Silva, S.M. Rodrigues, R.O. Pereira and L. Pereira. 2014. Bioproducts from seaweeds: a review with special focus on the Iberian Peninsula. Curr. Org. Chem. 18: 896–917.

Chattopadhyay, N., T. Ghosh, S. Sinha, K. Chattopadhyay, P. Karmakar and B. Ray. 2010. Polysaccharides from Turbinaria conoides: structural features and antioxidant capacity. Food Chem. 118: 823–829.

Cheung, R.C.F., J.H. Wong, W. Pan, Y.S. Chan, C. Yin, X. Dan and T.B. Ng. 2015. Marine lectins and their medicinal applications. Appl. Microbiol. Biotechnol. 99: 3755–3773.

Christaki, E., E. Bonos, I. Giannenas and P. Florou-Paneri. 2013. Functional properties of carotenoids originating from algae. J. Sci. Food Agric. 93: 5–11.

Costa, L.S., G.P. Fidelis, S.L. Cordeiro, R.M. Oliveira, D.A. Sabry, R.B.G. Camara, L.T.D.B. Nobre, M.S.S.P. Costa, J. Almeida-Lima, E.H.C. Farias, E.L. Leite and H.A.O. Rocha. 2010. Biological activities of sulfated polysaccharides from tropical seaweeds. Biomed. Pharmacother. 64: 21–28.

Davari, M., S.N. Sharma and M. Mirzakhani. 2012. Residual influence of organic material, crop residues, and biofertilizers on performance of succeeding mung bean in an organic rice-based cropping system. J. Recycl. Organic Waste Agricult. 1: 1–14.

del Val, G.A., G. Platas, A. Basilio, A. Cabello, J. Gorrochategui, I. Suay, F. Vicente, E. Portillo, J.M. del Río, G.G. Reina and F. Peláez. 2001. Screening of antimicrobial activities in red, green and brown macroalgae from Gran Canaria (Canary Islands, Spain). Int. Microbiol. 4: 35–40.

Devi, K.P., N. Suganthy, P. Kesika and S.K. Pandian. 2008. Bioprotective properties of seaweeds: in vitro evaluation of antioxidant activity and antimicrobial activity against food borne bacteria in relation to polyphenolic content. BMC Complement Altern. Med. 8: 38.

Díaz-Rubio, M.E., J. Serrano, A.J. Borderías and F. Saura-Calixto. 2011. Technological effect and nutritional value of dietary antioxidant *Fucus* fiber added to minced fish muscle. J. Aquat. Food Prod. Technol. 20: 295–307.

Divya, K., M.N. Roja and S.B. Padal. 2015. Effect of seaweed liquid fertilizer of *Sargassum wightii* on germination, growth and productivity of brinjal. Int. J. Adv. Res. Sci. Eng. Technol. 2: 868–871.

EBIC. 2012. European Biostimulant Industry Council. Available online at: http://www.biostimulants.eu/ (acessed on January 7, 2019)

Elansary, H.O., K. Skalicka-Wozniak and I.W. King. 2016. Enhancing stress growth traits as well as phytochemical and antioxidant contents of *Spiraea* and *Pittosporum* under seaweed extract treatments. Plant Physiol. Biochem. 105: 310–320.

El Shoubaky, G.A. and E.A. Salem. 2016. Effect of abiotic stress on endogenous phytohormones profile in some seaweeds. IJPPR 8: 124–134.

Genard, H., J. Le Saos, J.-P. Billard, A. Tremolieres and J. Boucaud. 1991. Effect of salinity on lipid composition, glycine betaine content and photosynthetic activity in chloroplasts of *Suaeda maritima*. Plant Physiol. Biochem. 29: 421–427.

Govindan, M., J.D. Hodge, K.A. Brown and M. Nunez-Smith. 1993. Distribution of cholesterol in Caribbean marine algae. Steroids 58: 178–180.

Gupta, S. and N. Abu-Ghannam. 2011. Recent developments in the application of seaweeds or seaweed extracts as a means for enhancing the safety and quality attributes of foods. Innov. Food Sci. Emerg. Technol. 12: 600–609.

Güven, K.S., A. Percot and E. Sezik. 2010. Alkaloids in Marine Algae. Mar. Drugs 8: 269–284.

Heltan, M.M., J.G. Wakibia, G.M. Kenji and M.A. Mwasaru. 2015. Chemical composition of common seaweeds from the Kenya Coast. J. Food Res. 4: 28–38.

Humphrey, A.J. and M.H. Beale. 2006. Terpenes. pp. 47–101. *In*: Crozier, A. (ed.). Plant Secondary Metabolites in Diet and Health. Blackwell, Oxford.

Ibtissam, C., R. Hassane, M.L. José, D.S.J. Francisco, G.V.J. Antonio, B. Hassan and K. Mohamed. 2009. Screening of antibacterial activity in marine green and brown macroalgae from the coast of Morocco. Afr. J. Biotechnol. 8(7): 1258–1262.

Ismail, M.M. and S.M. El-Shafay. 2015. Variation in taxonomical position and biofertilizing efficiency of some seaweed on germination of *Vigna unguiculata* (L). IJESE 6: 47–57.

Jiang, Z.D. and W.H. Gerwick. 1991. Novel pyrroles from the Oregon red alga *Gracilariopsis lemaneiformis*. J. Nat. Prod. 54: 403–407.

Khan, W., U.P. Rayirath, S. Subramanian, M.N. Jithesh, P. Rayorath, D.M. Hodges, A.T. Critchley, J.S. Craigie, J. Norrie and B. Prithiviraj. 2009. Seaweed extracts as biostimulants of plant growth and development. J. Plant Growth Regul. 28: 386–399.

Kraan, S. 2013. Pigments and minor compounds in algae. pp. 205–251. *In*: Domínguez, H. (ed.). Functional Ingredients from Algae for Foods and Nutraceuticals. Woodhead Publishing, Cambridge, UK.

Kumar, N.J.L., R.N. Kumar, K. Patel, S. Viyol and R. Bhoi. 2009. Nutrient composition and calorific value of some seaweeds from Bet Dwarka, west coast of Gujarat, India. Our Nat. 7: 18–25.

Kumari, P., M. Kumar, C.R.K. Reddy and B. Jha. 2013. Algal lipids, fatty acids and sterols. pp. 87–134. *In*: Domínguez, H. (ed.), Functional Ingredients from Algae for Foods and Nutraceuticals. Woodhead Publishing, Cambridge, UK.

Latique, S., H. Chernane and M. El Kaoua. 2014. Seaweed liquid fertilizer effect on physiological and biochemical parameters of bean plant (*Phaseolus vulgaris* var. *paulista*) under hydroponic system. Eur. Sci. J. 9: 174–191.

Luo, D., Q. Zhang, H. Wang, Y. Cui, Z. Sun, J. Yang, Y. Zheng, J. Jia, F. Yu, X. Wang and X. Wang. 2009. Fucoidan protects against dopaminergic neuron death *in vivo* and *in vitro*. Eur. J. Pharmacol. 617: 33–40.

Manaf, H.H. 2016. Beneficial effects of exogenous selenium, glycine betaine and seaweed extract on salt stressed cowpea plant. Ann. Agric. Sci. 61: 41–48.

Manivannan, K., G. Karthikai Devi, G. Thirumaran and P. Anantharaman. 2008. Mineral composition of macroalge from Mandapam coastal region, southeast coast of India. Am-Euras. J. Bot. 1: 58–67.

Matsukawa, R., Z. Dubinsky, E. Kishimoto, K. Masaki, Y. Masuda and T. Takeuch. 1997. A comparison of screening methods for antioxidant activity in seaweeds. J. Appl. Phycol. 9: 29–35.

Mendes, G.S., A.R. Soares, F.O. Martins, M.C.M. Albuquerque, S.S. Costa, Y. Yoneshigue-Valentin, L.M.S. Gestinari, N. Santos and M.T.V. Romanos. 2010. Antiviral activity of the green marine alga *Ulva fasciata* on the replication of human metapneumovirus. Rev. Inst. Med. Trop. S. Paulo. 52: 3–10.

Michalak, I., L. Tuhy and K. Chojnacka. 2015. Seaweed extract by microwave assisted extraction as plant growth biostimulant. Open Chem. 13: 1183–1195.

Mirparsa, T., H.R. Ganjali and M. Dahmardeh. 2016. The effect of bio fertilizers on yield and yield components of sunflower oil seed and nut. Inter. J. Agri. Biosci. 5: 46–49.

Mohammadi, M., H. Tajik and P. Hajeb. 2013. Nutritional composition of seaweeds from the Northern Persian Gulf. Iran J. Fish Sci. 12: 232–240.

Nabil, S. and J. Cosson. 1996. Seasonal variations in sterol composition of *Delesseria sanguinea* (Ceramiales, Rhodophyta). Hydrobiologia 326: 511–514.

Nabti, E., M. Sahnoune, M. Ghoul, D. Fischer, A. Hofmann, M. Rothballer, M. Schmid and M. Hartmann. 2010. Restoration of growth of durum wheat (*Triticum durum* var. waha) under saline conditions due to inoculation with the rhizosphere bacterium *Azospirillum brasilense* NH and extracts of the marine alga *Ulva lactuca*. J. Plant Growth Regul. 29: 6–22.

Nahas, R., D. Abatis, M.A. Anagnostopoulou and P. Kefalas. 2007. Radical scavenging activity of Aegean sea marine algae. Food Chem. 102: 577–581.

Naidu, B.P., G.P. Jones, L.G. Paleg and A. Poljakoff-Mayber. 1987. Proline analogues in *Melaleuca* species: response of *Melaleuca lanceolata* and *M. uncinata* to water stress and salinity. Aust. J. Plant Physiol. 14: 669–677.

Pacholczak, A., K. Nowakowska and S. Pietkiewicz. 2016. The effects of synthetic auxin and a seaweed-based biostimulator on physiological aspects of rhizogenesis in ninebark stem cuttings. Not. Bot. Horti. Agrobo. 44: 85–91.

Perez, J., E. Falque and H. Dominguez. 2016. Antimicrobial action of compounds from marine seaweed. Mar. Drugs 14: 1–38.

Pereira, R.C. and V.L. Teixeira. 1999. Sesquiterpenes of the marine algae *Laurencia* lamouroux (Ceramiales, Rhodophyta). 1. Ecological significance. Quim. Nova 22: 369–373.

Pereira, L. 2009. Guia Ilustrado das Macroalgas – conhecer e reconhecer algumas espécies da flora Portuguesa (Illustrated Guide of Macroalgae – to know and recognize some species of the Portuguese flora). IUC Coimbra University Press, Coimbra, Portugal.

Pereira, L. 2018. Biological and therapeutic properties of the seaweed polysaccharides. Int. Biol. Rev. 2(2): 1–50.

Pichereau, V., J.A. Pocard, J. Hamelin, C. Blanco and T. Bernard. 1998. Differential effects of dimethylsulfoniopropionate, dimethylsulfonioacetate and other S-methylated compounds on the growth of *Sinorhizobium meliloti* at low and high osmolarities. Appl. Environ. Microbiol. 64: 1420–1429.

Plouguerne, E., B.A.P. da Gama, R.C. Pereira and E. Barreto-Bergter. 2014. Glycolipids from seaweeds and their potential biotechnological applications. Front. Cell. Infect. Microbiol. 4: 1–5.

Ramarajan, S., J.L. Henry and G.A. Saravana. 2013. Effect of seaweed extracts mediated changes in leaf area and pigment concentration in soybean under salt stress condition. RRJoLS 3: 17–21.

Renn, D.W. 1993. Medical and biotechnological applications of marines natural polysaccharides. pp. 181–196. *In*: Attaway, O.H. and Zabrosky, O.R. (eds.). Marine Biotechnology, Vol. 1. Pharmaceutical and Bioactive Natural Products. Plenum Press, New York, USA.

Rengasamy, K.R.R., M.G. Kulkarni, S.C. Pendota and J. Van Staden. 2016. Enhancing growth, phytochemical constituents and aphid resistance capacity in cabbage with foliar application of eckol – a biologically active phenolic molecule from brown seaweed. New Biotechnol. 33: 273–279.

Rezaei, M.A., B. Kaviani and H. Jahanshahi. 2012. Application of exogenous glycine betaine on some growth traits of soybean (*Glycine max* L.) drought stress conditions. Sci. Res. Essays 7: 432–436.

Rupérez, P., O. Ahrazem and J.A. Leal. 2002. Potential antioxidant capacity of sulphated polysaccharides from the edible marine brown seaweed *Fucus vesiculosus*. J. Agric. Food. Chem. 50: 840–845.

Russo, R.O. and G.P. Berlyn. 1990. The use of organic biostimulants to help low imput sustainable agriculture. J. Sustain. Agric. 1: 19–38.

Selvin, J. and A.P. Lipton. 2004. Biopotentials of *Ulva fasciata* and *Hypnea musciformis* collected from the peninsular coast of India. J. Marine Sci.Technol. 12: 1–6.

Senn, T.L. 1987. Seaweed and Plant Growth. Clemson University, Clemson, SC.

Sethi, P. 2012. Biochemical composition of the marine brown algae *Padina tetrastromatica* Hauck. Int. J. Curr. Pharm. Res. 4: 117–118.

Shahbazi, F., S.M. Nejad, A. Salimi and A. Gilani. 2015. Effect of seaweed extracts on the growth and biochemical constituents of wheat. Int. J. Agric. Crop Sci. 8: 283–287.

Sharma, H.S.S., C. Fleming, C. Selby, J.R. Rao and T. Martin. 2014. Plant biostimulants: a review on the processing of macroalgae and use of extracts for crop management to reduce abiotic and biotic stresses. J. Appl. Phycol. 26: 465–490.

Shevchenko, N.M., S.D. Anastyuk, N.I. Gerasimenko, P.S. Dmitrenok, V.V. Isakov and T.N. Zvyagintseva. 2007. Polysaccharide and lipid composition of the brown seaweed *Laminaria gurjanovae*. Russ. J. Bioorg. Chem. 33: 88–98.

Singh, I.P. and J. Sidana. 2013. Phlorotannins. pp. 181–204. *In*: Domínguez, H. (ed.). Functional Ingredients from Algae for Foods and Nutraceuticals. Woodhead Publishing, Cambridge, UK.

Singh, S., M.K. Singh, S.K. Pal, K. Trivedi, D. Yesuraj, C.S. Singh, V.K.G. Anand, M. Chandramohan, R. Patidar, D. Kubavat, S.T. Zodape and A. Ghosh. 2016. Sustainable enhancement in yield and quality of rain-fed maize through *Gracilaria edulis* and *Kappaphycus alvarezii* seaweed sap. J. Appl. Phycol. 28: 2099–2112.

Singh, R.S., S.R. Thakur and P. Bansal. 2015. Algal lectins as promising biomolecules for biomedical research. Crit. Rev. Microbiol. 41: 77–88.

Stratmann, K., W. Boland and D.G. Muller. 1992. Pheromones of marine brown algae: a new branch of eicosanoid metabolism. Angew. Chem. Int. Ed. 3: 1246–1248.

Stirk, W.A. and J.J. Van Staden. 1997a. Comparison of cytokinin- and auxin-like activity in some commercially used seaweed extracts. Appl. Phycol. 8: 503–508.

Stirk, W.A. and J.J. Van Staden. 1997b. Isolation and identification of cytokinins in a new commercial seaweed product made from *Fucus serratus* L. Appl. Phycol. 9: 327–330.

Suleria, H.A.R., S. Osborne, P. Masci and G. Gobe. 2015. Marine-based nutraceuticals: an innovative trend in the food and supplement industries. Mar. Drugs 13: 6336–6351.

Summers, P.S., K.D. Nolte, A.J.L. Cooper, H. Borgeas, T. Leustek, D. Rhodes and A.D. Hanson. 1998. Identification and stereospecificity of the first three enzymes of 3-dimethylsulfoniopropionate in a chlorophyte alga. Plant Physiol. 116: 369–378.

Tawfeeq, A., A. Culham, F. Davis and M. Reeves. 2016. Does fertilizer type and method of application cause significant differences in essential oil yield and composition in rosemary (*Rosmarinus officinalis* L.)? Ind. Crops Prod. 88: 17–22.

Thinakaran, T. and K. Sivakumar. 2013. Antifungal activity of certain seaweeds from Puthumadam coast. Int. J. Res. Rev. Pharm. Appl. Sci. 3: 341–350.

Tuhy, L., M. Samoraj, S. Basladynska and K. Chojnacka. 2015. New micronutrient fertilizer biocomponents based on seaweed biomass. Pol. J. Environ. Stud. 24: 2213–2221.

Van Alstyne, K.L., K.N. Pelletreau and K. Rosari. 2003. The effects of salinity on dimethylsulfoniopropionate production in the green alga *Ulva fenestrata* Postels and Ruprecht (Chlorophyta). Bot. Mar. 46: 350–356.

Vyomendra, C. and N. Kumar. 2016. Effect of algal bio-fertilizer on the *Vigna radiata*: a critical review. Int. J. Eng. Res. Appl. 6: 85–94.

Wachira, F. and J. Ogada. 1995. *In vitro* regeneration of *Camellia sinensis* (L.) O. Kuntze by somatic embryogenesis. Plant Cell Rep. 14: 463–466.

Whapham, C.A., G. Blunden, T. Jenkins and S.D. Hankins. 1993. Significance of betaines in the increased chlorophyll content of plants treated with seaweed extract. J. Appl. Phycol. 5: 231–234.

Wijesinghe, W.A.J.P., Y. Athukorala and Y.J. Jeon. 2011. Effect of anticoagulative sulfated polysaccharide purified from enzyme-assistant extract of a brown seaweed *Ecklonia cava* on Wistar rats. Carbohydr. Polym. 86: 917–921.

Yoo, J.S. 2003. Biodiversity and community structure of marine benthic organisms in the rocky shore of Dongbaekseom, Bunsan. Algae 18: 225–232.

Yuan, Y.V., D.E. Bone and M.F. Carrington. 2005. Antioxidant activity of dulse (*Palmaria palmata*) extract evaluated *in vitro*. Food Chem. 91: 485–494.

Zbakh, H., H. Chiheb, H. Bouziane, V.M. Sanchez and H. Riadi. 2012. Antibacterial activity of benthic marine algal extracts from the mediterranean coast of Marocco. J. Microbiol. Biotechnol. Food Sci. 2: 219–228.

Zhang, X. 1997. Influence of plant growth regulators on turf grass growth, antioxidant status, and drought tolerance. PhD thesis. Virginia Polytechnic Institute and State University, Blacksburg, Virginia, 144 pp.

5

Seaweeds as Plant Biostimulants

Marcia Eugenia Amaral Carvalho[1]* and Paulo Roberto de Camargo e Castro[2]**

[1] Department of Biological Sciences, Luiz de Queiroz College of Agriculture, University of São Paulo, 13418-900, Piracicaba, SP, Brazil
[2] Full Professor, Department of Biological Sciences, Luiz de Queiroz College of Agriculture, University of São Paulo, 13418-900, Piracicaba, SP, Brazil

1 Introduction

In the face of rapid global environmental changes, our ability to maintain or even increase crop quality and yield is becoming difficult. Abiotic and biotic stresses trigger morphological, physiological, biochemical, anatomical and molecular modifications that can dramatically affect plant development and productivity (Soares et al. 2019, Borges et al. 2019). However, the performance of plants subjected to environmental challenges can be improved by applying seaweed-based products, which are eco-friendly alternatives to the use of fertilizers and synthetic biostimulants (Carvalho and Castro 2014, Shukla et al. 2018, Sharma et al. 2019). The use of seaweed extracts as biostimulant in agriculture has increased significantly in the last few decades (Khan et al. 2009, Craigie 2011, Carvalho and Castro 2014, Hernández-Herrera et al. 2016). Biostimulants are a mixture of hormones with compounds of different chemical nature (e.g., amino acids and minerals), which are able to affect plant development (Machado et al. 2014, Castro et al. 2017, Pacheco et al. 2019). Since algal extracts influence both enzymatic and non-enzymatic system of plants (Elansary et al. 2017, Carvalho et al. 2018, Farid et al. 2018), they can be considered as anti-stress agents that can potentially increase plant tolerance of adverse environmental conditions and enhance plant capacity for stress recovery (Nair et al. 2012, Sharma et al. 2019). There is also evidence that plant development can be indirectly improved by using seaweed extracts that can mediate, for instance, the legume-rhizobia symbiotic relationship (Khan et al. 2012). In certain situations, the yield of plants under stresses was even supported by applying algal extracts, when compared to untreated plants grown in optimal conditions (Yıldıztekın et al. 2018).

Algal extracts are generally made from species that inhabit marine water, and *Ascophyllum nodosum* (L.) Le Jolis (Phaeophyceae) stands out among the seaweeds commonly used for agricultural purposes (Ugarte et al. 2006, Khan et al. 2009, Craigie 2011, Arioli et al. 2015). However, other species from the genera *Durvillaea*, *Ecklonia*, *Fucus*, *Kappaphycus*, *Laminaria*, *Sargassum* (Phaeophyceae), and *Ulva* (Chlorophyta) are also employed for the manufacturing of plant biostimulants that are generally supplied to farmers as liquid concentrate and soluble powder. Even at low concentrations, seaweed extract-based products are able to affect plant development, indicating the presence of bioactive compounds (Khan et al. 2009). Therefore, the following question arises: What are these seaweed extracts made of? Their organic matrix is generally complex, being composed of macro- and micronutrients, amino acids, lipids, carbohydrates and plant (or plant-like)

*Corresponding author: *marcia198811@usp.br; **prcastro@usp.br hormones (Santaniello et al. 2017, Dumale et al. 2018, Farid et al. 2018, Michalak et al. 2018, Torres et al. 2018). However, the composition of these products is heavily dependent on several factors (Fig. 1), such as the seaweed species, their harvest season, the geographical location in which they grow or are farmed, and the manufacturing processes (Hernández-Herrera et al. 2016, Duarte et al. 2018, Di Filippo-Herrera et al. 2018, Michalak et al. 2018).

The organic and inorganic components of a commercial product based on *A. nodosum* are shown in Table 1. The high variation in the composition of seaweed extracts and, consequently, their effects is one of the major challenges for the industry of seaweed-based biostimulants (Goñi et al. 2016). Moreover, the mixture of seaweed extracts with other compounds (such as essential and non-essential elements and amino acids) is a usual procedure that affects the efficiency of these agrochemicals (Paungfoo-Lonhien et al. 2017, Kałużewicz et al. 2017), so that farmers do not know whether the (positive or negative) effects of their application on plants are due to the seaweed-dependent substances, the additional compounds or their synergism. This is the case of the mixture of humate supplemented with red seaweed *Ahnfeltiato buchiensis* (Rhodophyta), which provided increases in the root (36%) and shoot (54%) area, as well as in the root (100%) and shoot (54%) dry weight of maize seedlings (Paungfoo-Lonhienne et al. 2017). In this review we have concentrated on promising and emerging topics related to both the mode of action of seaweed-based products and the plant responses to the application of seaweed-based products. For better readability, the text was organized according to key matters such as the influence of seaweed-based products on plant development and yield, their outcomes on the quality of different plant parts, and their impacts on plant performance under biotic and abiotic stresses.

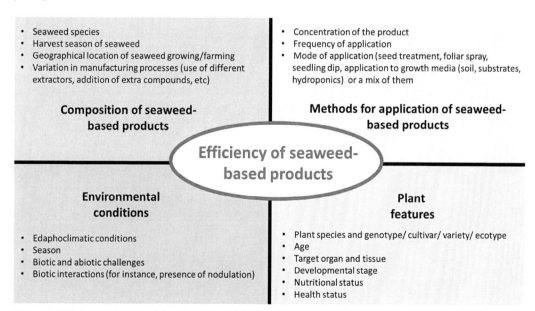

Figure 1. Efficiency of seaweed-based products is highly dependent on several factors, such as their composition and methods of application, as well as environmental conditions and plant features.

2 Plant Development

2.1 Germination

Initial plant development can be positively affected by biostimulants, which are able to increase the percentage of both seed germination and seedling vigor, thus providing better plant establishment and potentially enhancing productivity. Previous studies reported that organic components from *A. nodosum* improved the germination of barley seeds through induction of the activity of gibberellin-

Table 1. Chemical components of the commercial liquid extract from *Ascophyllum nodosum*, according to the label of Acadian® Marine plant extract

Components	
Organic matter	13.00-16.00%
Total nitrogen (N)	0.30-0.60%
Available phosphate (P_2O_5)	< 0.1%
Soluble potassium (K_2O)	5.00-7.00%
Sulfur (S)	0.30-0.60%
Magnesium (Mg)	0.05-0.10%
Calcium (Ca)	0.10-0.20%
Sodium (Na)	1.00-1.50%
Iron (Fe)	30-80 ppm
Copper (Cu)	01-05 ppm
Zinc (Zn)	05-15 ppm
Manganese (Mn)	01-05 ppm
Boron (B)	20-50 ppm
Carbohydrate:	Mannitol, and alginic and laminarian acids
Amino acids (1.01%)	
Alanine	0.08%
Aspartic acid	0.14%
Glutamic acid	0.20%
Glycine	0.06%
Isoleucine	0.07%
Leucine	0.09%
Lysine	0.05%
Methionine	0.03%
Phenylalanine	0.07%
Proline	0.07%
Tyrosine	0.06%
Valine	0.07%
Tryptophan	0.02%

independent amylase, which possibly worked together with the gibberellin-dependent isoform(s) (Rayorath et al. 2008a). Gibberellin promotes the germination of monocotyledon seeds by inducing the synthesis of enzymes, such as amylases, which convert starch to sugars that provide energy from the endosperm to the development of embryonal axis (Taiz et al. 2018). The immersion of bean (*Phaseolus vulgaris* cv. Alvorada) seeds into the solution containing *A. nodosum* extract at 0.8 mL L^{-1} for 5, 10, 15 and 20 min also provided increases in the percentage of seedling emergence in sand on the 6th day after sowing, when compared to the control seeds. Only bean seeds treated for 15 min exhibited superior speed of seedling emergence (Carvalho et al. 2013). However, all treatments presented similar percentage of germinated seeds on the final evaluation (i.e., on the 10th day after sowing) of the germination test in rolled paper towels (Carvalho et al. 2013), as shown in soybean (Gehling et al. 2017).

The extract derived from *Ulva intestinalis* (formerly *Enteromorpha intestinalis*) (Chlorophyta), however, increased soybean germination by 80% over the control seeds, which in comparison to the treated seeds showed lower shoot and root length and reduced content of carbohydrate, protein, amino acids and pigments, for example, chlorophylls and carotenoids (Mathur et al. 2015). Depending on seed lot, treatment with seaweed extract improved the biomass of shoots and roots in soybean seedlings, which also presented increments in the root length (Gehling et al. 2017). There is strong evidence that polysaccharides or their fractions are among the compounds responsible for increments in the seed germination (Goñi et al. 2016; de Borba et al. 2019). For instance, a sulfated polysaccharide named ulvan was associated to the improvements in seed emergence in common bean (de Borba et al. 2019).

In another leguminous species, *Cyamopsis tetragonoloba,* the application of *Ulva lactuca* (Chlorophyta) extract (0.5-5%) increased seed germination by 30% (Balakrishnan et al. 2007). According to Ghaderiardakani et al. (2018), *Ulva* extract is able to alter germination probably by acting on the abscisic acid, ethylene and cytokinin metabolism, as evidenced by different performance of *Arabidopsis thaliana* wild type and its mutants (*abi1, etr1* and *cre1*) after seed treatment. The percentage of germinated seeds of *Vigna radiata* was also enhanced by the application of products based on seaweeds *Sargassum plagiophyllum*, *Turbinaria conoides*, *Padina tetrastromatica*, *Dictyota dichotama* (Phaeophyceae) and *Caulerpa scalpelliformis* (Chlorophyta) (Kavipriya et al. 2011). In general, the most efficient concentrations varied from 0.3 to 0.5%, when treated seeds showed a germination that ranged from 80 to 100%, whereas only 70% of the control seeds were able to germinate (Kavipriya et al. 2011).

2.2 Seedling development

Vigorous seedlings can better withstand environmental challenges and potentially present a higher yield. The action of polysaccharides in seaweed extract was also associated to the growth promotion of seedlings (Hernández-Herrera et al. 2016, Goñi et al. 2016, de Borba et al. 2019). In line with these findings, a study provided evidence that polysaccharide-enriched seaweed extracts act as effective growth-promoting substances in *in vitro* environments, as shown by the faster and greater rooting of mung bean hypocotyl cuttings treated with extracts of *U. lactuca* and *Padina gymnospora* (Phaeophyceae), when compared to the controls (Hernández-Herrera et al. 2016). However, other organic components, such as lipids, are also candidates as plant growth promoters. For instance, the dichloromethane extract from *Crassiphycus caudatus* (formerly *Gracilaria caudata*) (Rhodophyta) improved the root growth in lettuce seedlings in comparison to the untreated ones, and assays indicated that such outcomes are from the action of palmitic acid (Torres et al. 2018). Regardless of the identification of the active substance, several works have shown the positive effects of seaweed extracts or seaweed-based products on seedling development. For instance, increments in the length of main roots and shoots (105% and 106%, respectively), number of lateral roots (123%), and seedling fresh and dry weight (93% and 85%, respectively) were noticed in *V. radiata* seedlings after seed treatment with *S. plagiophyllum*, *T. conoides*, *P. tetrastromatica*, *D. dichotama* and *C. scalpelliformis* (Kavipriya et al. 2011).

The development of *C. tetragonoloba* seedlings was also promoted by seed treatment with distinct doses (from 0.5% to 10%) of algal extracts (*U. lactuca*, *P. tetrastromatica* and *C. caudatus caudatus*), which provided increases in the length of roots and shoots, and their fresh and dry weight (Balakrishnan et al. 2007). In carrot seedlings, increased root length was noticed after seeds were soaked in solution containing seaweed for 12 h (Kanmaz et al. 2018). The extracts from the green seaweed *Ulva intestinalis* below 0.1% stimulated root growth but inhibited lateral root formation in *A. thaliana* seedlings (Ghaderiardakani et al. 2018). These outcomes in roots are similar to that provided by auxin (Taiz et al. 2018), indicating the role of auxin-dependent mechanisms in plants treated with seaweed extracts. This hypothesis is supported by early studies that evidenced the role of auxin or auxin-like compounds as responsible for the root growth in *A. thaliana* treated with *A. nodosum* extract (Rayorath et al. 2008b, Khan et al. 2011).

2.3 Asexual propagation

Seaweed extracts can also improve the asexual propagation of plants, as shown by the increased fresh weight of roots and leaves in sweet potato apical cuttings treated with *A. nodosum* extract, when compared to untreated plants (Neumann et al. 2017). The production of *Anonna glabra* rootstocks also benefited from use of seaweed-based extract, which enhanced stem diameter, number of leaves, and height and dry weight of shoot and roots (da Silva et al. 2016). Micro-propagated eucalyptus cuttings (Euca 103 and Euca 105 from *Eucalyptus urograndis* and clone I 144 from *Eucalyptus urophilla*) cultivated in two different substrates, which were irrigated with seaweed-based solution (0.5, 1.0, 1.5 or 2.0 mL L^{-1}) before and after plant transplantation, exhibited improved performance due to application of this agrochemical (Losi 2010). Depending on the solution concentration, application frequency, substrate composition, and genotypes (clones) from the same plant species, the root growth and biomass were positively or negatively affected by the use of seaweed-dependent products (Losi 2010). In another study, Hungarian researchers evaluated the production of cuttings from stock plants of *Crataegus pinnatifida* (cultivars Da Chang Kao and Liao Hong), *Prunus marianna* cv. GF 8-1 and *Prunus mahaleb* cv. Bogdány after three foliar sprays of seaweed-based products (Szabó and Hrotkó 2009). The number of suitable cuttings depended on the selected species/cultivars, products and their dosage. The results ranged from reductions (50% in *C. pinnatifida* cv. Da Chang Kao) to high increases (587% in *P. marianna* cv. GF 8-1) in the number of shoots. In pine (*Pinus pinea*) seedlings, the foliar spray of seaweed extract provided shoot growth (length and biomass), while soil irrigation induced root growth (Atzmon and Van Staden 1994).

3 Yield

The use of seaweed-based biostimulants is able to improve the yield of reproductive organs (i.e., flowers, fruits and seeds); however, this outcome is more frequently achieved when productivity is based on vegetative organs (i.e., leaves, stems and roots). For instance, the number of spinach leaves was significantly increased after foliar sprays of extracts from *Ecklonia maxima* (Phaeophyceae) (Kulkarni et al. 2019). Under foliar applications of *Kappaphycus alvarezii* (Rhodophyta) extract (KAE–at 5%) in combination with recommended rate of synthetic fertilizers (RRF), enhancements in sugarcane productivity by 12.5% and 8% were observed in plant and ratoon crops, respectively. Moreover, the treatment involving 6.25% KAE + 50% RRF showed yield similar to the control (water + 100% RRF) in ratoon (Singh et al. 2018).

When the *A. nodosum* extract was applied at 3.8 mL L^{-1}, improved number and dry mass of kale (*Brassica oleracea* cv. Couve-Manteiga da Georgia) was noticed (Silva et al. 2012). Furthermore, three-time application of algae extracts (7 mL per plant) on curly lettuce cv. Elba increased its shoot fresh mass (103%) and dry mass (111%) and, consequently, the yield (103%) after either foliar pulverization or soil irrigation at 1.0 g L^{-1} and 0.5 g L^{-1}, respectively (Pinto et al. 2010).

Increments in the number of leaves (17%) and shoot dry and fresh weight (24% and 25%, respectively) were detected in curly lettuce cv. Vera after application of *Sargassum* and *Laminaria* (Phaeophyceae) extracts through foliar pulverization (2 L ha^{-1}) at 14 and 21 days after transplantation (Cecato and Moreira 2013). The weekly or biweekly application of seaweed extract on onion (*Allium cepa* L.) at 3 and 4 mL L^{-1} increased the fresh and dry weight of bulbs, and also decreased the loss of bulb biomass during storage (Bettoni et al. 2010). Plants of 'Ágata' potato, in which 1 L ha^{-1} of seaweed extract was applied at 30, 40 and 50 days after planting (DAP), exhibited increments by 16%, 12% and 36% in the number, fresh weight and diameter of tubers on the onset of their development, i.e., 65 DAP (Bettoni et al. 2008).

Regarding reproductive organs, the application of algae extracts (*Turbinaria murrayana, T. ornata, Sargassum* sp., *S. polycystum, Hydroclathrus* sp. (Phaeophyceae)*, Ulva fasciata, U. fasciculata, Padina* sp. and *Chaetomorpha* sp.) provided increases in the number of tillers and panicules, and in 100-grain weight in rice plants, which exhibited the best performance when *Hydroclathrus* sp. was used as biostimulant (Sunarpi et al. 2010). The application of *A. nodosum* extract also improved the number of spikes in wheat (*Triticum aestivum*) cultivars IAC 364 and BRS Guamirim (from 13% to 20%, respectively), but did not affect the 1000-grain weight (Igna and Marchioro 2010, Carvalho et al. 2014). In winter wheat variety Akteur, beneficial outcomes in the productivity parameters were counterbalanced by disadvantageous effects, so that yield was unaffected (Michalak et al. 2016). For instance, plants treated with Baltic sea algal extract exhibited both higher ear number per m^2 and 1000-grain mass, but lower number of grains than control plants. By contrast, increased number of grains in ears concurrently to decreased grain biomass was noticed after plant treatment with supercritical extract from *Arthrospira platensis* (formerly *Spirulina platensis*) (Cyanobacteria) (Michalak et al. 2016). However, when *Kappaphycus alvarezii* (Rhodophyta) extract was used on wheat cv. GW 322, grain weight was improved up to 34% in addition to increases (37%) in the number of spikes (Zodape et al. 2009).

Coffee trees cv. Catuaí 144 grown in the Brazilian savanna (aka Cerrado) also presented increments in yield (from 37% to 70% in the number of bags) in the first and second harvests due to the application of *Ascophyllum nodosum*, starting from the pre-bloom stage, via soil irrigation and pulverization from the pre-bloom stage (Fernandes and Silva 2011). An Indian study showed that soil irrigation with solution containing *Sargassum swartzii* (formerly *Sargassum wightii*) (Phaeophyceae) or *Ulva lactuca* (Chlorophyta) extracts increased the number (2% to 44%), weight (53% to 61%) and length (38% to 40%) of *Cyamopsis tetragonoloba* pods, as well as the number of grains per pods (up to 27%) as a function of dosage and alga species (Ramya et al. 2010). Another leguminous plant, *Phaseolus radiata* cv. K-851, also exhibited increments in the yield (16% to 30%) due to increases in the number and weight of pods (from 15% to 31%, and 19% to 30%, respectively) after foliar pulverization of solution containing *Kappaphycus alvarezii* extract (Zodape et al. 2010). It was observed that the mode of application and concentration of *Sargassum johnstonii* (Phaeophyceae) extract affected the number, weight and quality of tomato fruits (*Solanum lycopersicum* cv. Pusa Ruby) after 15 applications from the vegetative to the reproductive plant stages (Zodape et al. 2011).

In a Brazilian soybean (*Glycine max* cv. BRS 232) field, the seed treatment with seaweed extract (2 mL kg^{-1}) promoted increases in the 1000-grain weight and improved the crop yield up to 10% (Ferrazza and Simonetti 2010). Enhanced pod production was also observed in peanut (*Arachis hypogaea*) plants treated with *Sargassum swartzii* (formerly *Sargassum wightii*) and *Ulva lactuca* extracts (Sridhar and Rengasamy 2010). Commercial extracts from *Ecklonia maxima* and *Ascophyllum nodosum* augmented strawberry (*Fragaria x ananassa* cv. Elkat) fruit production up to 42% but did not affect yield of the cultivar Salut (Masny et al. 2004). In grape (*Vitis vinifera* cv. Perlette), the use of products containing seaweed extract plus amino acids promoted increases in the number of bunches (61%), rachis length (16%), number of berries per bunch (9%), berry size (6%) and 100-berry weight (17%) (Khan et al. 2012). Moreover, the pulverization of seaweed extract on orange and grapefruit trees for 2 to 3 yr provided increases by 10-25% in the yield over control plants (Koo and Mayo 1994). According to Fornes et al. (1995), the use of seaweed extract for three years in young tangerine and orange plants also improved (30%) fruit production.

4 Quality of Edible Plant Parts

The commercially important characteristics (such as size, color, and nutritional features) of different plant parts can be positively modified by the use of seaweed extracts (Kałużewicz et al. 2017). For instance, an increased content of total chlorophyll and proteins, and improved levels of different phenolic compounds (such as sinapic, vanillic, syringic and salicylic acids) was noticed in spinach (*Spinacea oleracea* L.) leaves after pulverization with a commercial seaweed extract prepared from *Ecklonia maxima* (Kulkarni et al. 2019). In cauliflower (*Brassica oleraceae* cv. Caraflex) inflorescences, the content of phenolic compounds and flavonoids was also increased from 1.3- to 2-fold by applying seaweed-based products (Lola-Luz et al. 2013). The use of solution with *Ascophyllum nodosum* extract improves the concentration of N (9%), P (50%), and K (6%) in leaves of *Amaranthus tricolor*, when compared to untreated plants (Aziz et al. 2011). Grains from wheat plants treated with *Kappaphycus alvarezii* extract exhibited increments in the carbohydrate, protein, and lipid contents, as well as in the nutrient concentrations (P, K, Ca, Fe and Zn, among others) when compared to untreated plants (Zodape et al. 2009). After application of *K. alvarezii* extract at 10%, higher carbohydrate, protein, and total nitrogen contents (5%, 7% and 7%, respectively) were observed in *Phaseolus radiata* seeds, which also had superior concentrations of Na (33%), Mo (53%), Mg (12%) and K (11%) in comparison to the control plants (Zodape et al. 2010).

The pulverization of seaweed extract on orange and grapefruit plants for 2 to 3 yr provided early changes in the color of the fruit peel while preventing premature fruit fall (Koo and Mayo 1994). The use of *Ascophyllum nodosum* extract for 3 yr in orange and tangerine plants promoted increases in the sugar content and reduced acidity of juice from 'Navelina' orange, and 'Satsuma' and 'Clementina' tangerine (Fornes et al. 1995). After a 2 yr pulverization with algal extract (0.2%) on 'Valência' orange trees, fruits from treated plants exhibited higher weight (20-21%) and content of total soluble solids (13-14%), total sugars (11-13%), reducing sugars (32%) and vitamin C (12%), and also decreased acidity (6-7%) in comparison to the fruits from untreated, control plants (Ahmed et al. 2013). In *Calibrachoa* x *hybrida*, an ornamental plant that is also a potential source of antimicrobial compounds, the application of *A. nodosum* extract as soil drench improved leaf number and area, plant height and weight in comparison to the untreated plants (Elansary et al. 2016). Enhancements in the phenolic, flavonol and tannin contents, and antioxidant capacity of *Calibrachoa* leaves, an outcome attributed to the mineral composition of the seaweed extract, were associated to the improved antifungal and antibacterial activities of *Calibrachoa* leaf extract (Elansary et al. 2016). In seeds from common bean var. Aura treated with *Ecklonia maxima*, the content of anthocyanins was increased in relation to seeds from control plants (Kocira et al. 2018). However, the quality of edible plants parts is not always improved by the use of seaweed-based products, as shown by reduction in the Fe and Mn concentration in potato tubers after plant treatment with *Ecklonia maxima* commercial extract (Mystkowska 2018).

5 Anti-Stress Agents

There are several reports about the alleviation of abiotic and biotic stresses by seaweed-based products, and evidence suggests they can act as elicitors in plants (Carvalho et al. 2018). This hypothesis is supported by the fact that application of seaweed extracts improves, for example, the content of compatible osmolytes even before the stress onset, potentially enhancing the plant defense against imminent environmental challenge. For instance, proline accumulation was increased previously to the beginning of drought (Carvalho et al. 2018) and salinity exposure (Yıldıztekın et al. 2018) in plants treated with seaweed-based products. Therefore, the plant defense may become even more responsive to the stresses, sometimes providing results that are similar to or even better than those observed in unstressed plants (Yıldıztekın et al. 2018). The use of *Ulva rigida* (Chlorophyta) extract, for instance, was able to sustain the biomass accumulation in bean plants under moderate water deficit in comparison to untreated plants grown without stress (Mansori et al. 2015). However, this does not mean that such agrochemicals are 100% efficient in protecting plants against such environmental challenges because yield enhancement or maintenance is rarely observed in treated plants under stress. The hypothetical model for the mode of action of seaweed-based products in alleviating abiotic stress in plants is shown in Fig. 2.

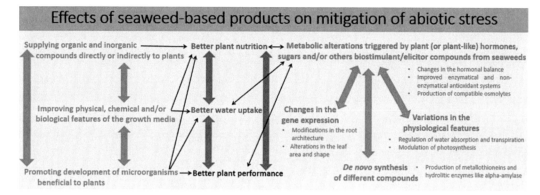

Figure 2. Hypothetical model for the mode of action of seaweed extract-based products in alleviating abiotic stress in plants.

5.1 Abiotic stresses

5.1.1 Drought

Under drought exposure, tolerance of *Arabidopsis thaliana* treated with *Ascophyllum nodosum* extract was associated to a higher leaf water content throughout the dehydration period, being only marginally affected in terms of survival rate when compared to untreated plants (Santaniello et al. 2017). According to the authors, such effects were due to seaweed-extract induction of partial stomatal closure, which reflected in lower stomatal conductance, higher water use efficiency and better photosynthetic performance during the last phase of dehydration period. Several of the above-mentioned physiological effects are similar to those orchestrated by abscisic acid (ABA), as corroborated by detection of modulation in the expression of genes for ABA metabolism (both synthesis and catabolism), the consequent accumulation of this hormone, and the classical outcomes such as increasing closure of stomata in plants treated with seaweed extract (Sharma et al. 2019). In one such study, application of *Gracilaria dura* (Rhodophyta) extract was able to maintain the yield of wheat plants under water stress, in comparison to the control plants, due to the sustenance of the number and length of spikes (Sharma et al. 2019). In addition, an improved recovery of the drought-stressed plants was noticed due to treatment with seaweed extract.

Shukla et al. (2018) also reported higher adaptability in recovering from drought conditions in soybean plants treated with *A. nodosum* extract (ANE), which provided better water management in plants (as measured by differences in the stomatal conductance and leaf water content). According to Martynenko et al. (2016), the improved drought tolerance in soybean treated with ANE is also associated to reductions in the rapid increase in leaf temperature and improved stomatal conductance. In grafted orange plants (*Citrussinensis* x *Poncirus trifoliata* and *P. trifoliata* x *C. paradisi*), soil irrigation or foliar pulverization with solution containing seaweed extract (from 5 to 10 mL L^{-1}) mitigated the side effects of drought stress (Spann and Little 2011). For instance, the scion *Citrus sinensis* cv. Hamlin produced leaves with increased area and biomass (on average, 42% and 25%, respectively), and plants did not exhibit the usual alterations coupled to the drought stress, such as changes in root length and in biomass allocation pattern (root-shoot ratio) after application of algal extract.

The use of seaweed extract with or without amino acids on two gramineous plants (*Poa pratensis* cv. Plush and *Agrostis palustris* cv. Penncross) under water deficit was able to improve their development, as measured by the increased biomass allocated to leaves and roots (47-54% and 29-312%, respectively); changes were also observed in the metabolites related to enhanced tolerance to drought (Zhang and Schmidt 1999, Zhang and Ervin 2004). In the turfgrass *Paspalum vaginatum* cv. Salam subjected to prolonged irrigation intervals (6 d), spraying with a solution of *Ascophyllum nodosum* extract (7 mL L^{-1}) increased turf quality, leaf photochemical efficiency, root length and dry weight, total non-structural carbohydrates, K, Ca, and proline concentrations, and also improved the antioxidant defensive mechanisms by enhancing both enzymatical system (as shown by high catalase (CAT), superoxide dismutase (SOD) and ascorbate peroxidase (APX) activities) and non-enzymatical system (as measured by total and free ascorbate concentrations) that, in turn, were able to provide decreases in the H$_2$O$_2$ production and lipid peroxidation level (Elansary et al. 2017). In maize plants treated with *Kappaphycus alvarezii* extract, the yield advantage under water stress was associated to minimal damage of photosystem, as evidenced by higher pigment content, photosynthetic rate, reduced photoinhibition, and also to the decreased lipid peroxidation due to enhancements in the protection against reactive oxygen species, as indicated by increments in the CAT, APX and glutathione reductase activities, especially under severe water deficit (Trived et al. 2018).

5.1.2 High temperature

High temperatures can trigger plant dehydration, decreasing yield and even causing plant death. However, the pulverization of seaweed-based extract solution on leaves of *Agrostis stolonifera* cv. L-93 subjected to thermal stress (35/25 ºC, day/night) retarded loss of plant quality needed for its commercialization, probably because of the higher quantity of viable roots (from 14% to 57%) in treated plants in comparison to the untreated ones under stress (Zhang and Ervin 2008). These authors provided evidence that such results come from the influence of seaweed-based extracts on the metabolism of hormones such as cytokinin, auxin and gibberellin, and also on the antioxidant system (Zhang and Ervin 2008). In soybean leaves, the application of seaweed extract provided a better regulation of the leaf temperature in stressed plants (Martynenko et al. 2016).

5.1.3 Radiation

The pulverization of algal extract containing humic acid promoted a better photosynthetic efficiency (41% to 50%) until day 12 of continuous UV-B radiation incidence on *Poa pratensis* cv. Georgetown, when compared to the untreated plants (Ervin et al. 2004). The plant quality for commercialization was also superior in plants that received the solution containing the mixture of algal extract and humic acid, even after 42 d of consecutive radiation, probably because of increased pigments chlorophyll, carotenoid and anthocyanin, and improved antioxidant enzymes (Ervin et al. 2004).

5.1.4 Salinity

The area of saline soils has grown worldwide because of unsuitable agricultural practices, so the development of strategies to decrease the impact of soil salinity on crop development and yield is very important.

Mitigation of salinity-induced stress was achieved by spraying *A. nodosum* extract on asparagus plants, which presented less pronounced impacts in their growth and biomass due to better chlorophyll, sugar and proline contents, modifications in the profile of phenolic compounds (such as robinin, rutin, apigein, chlorogenic acid, and caffeic acid), enhancements in the antioxidant enzymes (SOD and CAT), and improved photosynthetic and transpiration rates in treated plants in comparison to untreated plants cultivated under increasing salt levels (Al-Ghamdia and Elansary 2018). According to Latique et al. (2016), the challenges imposed by salt stress were alleviated by applying *Ulva rigida* extract on wheat (*Triticum durum*) plants, which exhibited enhanced vegetative growth, improved leaf chlorophyll content, increased total phenolic content, higher APX activity, and lowered biomass loss in comparison to the untreated plants.

In the turfgrass *Paspalum vaginatum* subjected to salinity (49.7 dS/m), the spray of solution with *Ascophyllum nodosum* extract (7 mL·L^{-1}) increased turf quality, leaf photochemical efficiency, root length and dry weight, total non-structural carbohydrates, K, Ca, and proline concentrations, and also enhanced the antioxidant capacity of the plants while decreasing ROS generation and, consequently, reducing lipid peroxidation level (Elansary et al. 2017). Egyptian researchers reported improvements in the performance of *Amaranthus tricolor* under salinity stress (1000, 2000 and 3000 ppm of NaCl) after application of two doses of *A. nodosum* extract (Aziz et al. 2011). Plants treated with *A. nodosum* extract at 3 mL L^{-1} showed higher fresh and dry biomass of leaves, stems and roots, as well as higher number of leaves in comparison to the untreated plants, regardless of salinity level. Plants that received 2.5 mL L^{-1} of the seaweed extract exhibited enhanced biomass (both fresh and dry) and number of inflorescences when compared to the untreated plants.

5.1.5 Nutrient imbalance

When lettuce was grown in K-deficient hydroponic condition, the application of seaweed extract was able to improve the leaf fluorescence, root length and fresh weight, and plant biomass in comparison to untreated plants cultivated in solution with low level of this macronutrient (Chrysargyris et al. 2018). Yet, the application of brown seaweed-based product also reduced the detrimental effects of K deficiency during storage of fresh cut salad by reducing leaf respiration and whitening rates (Chrysargyris et al. 2018). Increased N levels improved photosynthesis and water use efficiency in optimal water supply conditions in wild rocket plants (*Diplotaxis tenuifolia*), while resulting in harmful outcomes under water stress, but the application of brown seaweed-based product was able to alleviate these side effects (Schiattone et al. 2018). The use of seaweed extracts also provided good results in wheat under suitable N fertilization, stimulating nutrient uptake and increasing (Stamatiadis et al. 2014). However, under N deficiency N fertilization, the application of the seaweed extract presented insignificant effects. In almond (*Prunus dulcis*), increases in the total shoot leaf area as well as alleviation of the negative effects on the trunk diameter and number of branches were noticed in plants grown under low K supply, in comparison to the K-sufficient plants (Saa et al. 2015). A gramineous species (*Agrostis palustris* cv. Southshore) used in golf courses presented, in general, a higher photosynthetic activity, root biomass, and quality score (10%, 15% and 23%, respectively) when treated with seaweed extract mixed with humic acid, regardless of soil fertilization regime (low or high) (Zhang et al. 2002).

5.1.6 Low temperature

Previous studies also showed that application of seaweed extracts can mitigate the effects of low temperatures (Rayirath et al. 2009, Nair et al. 2012, Masondo et al. 2018), which can cause significant losses in crop yield. The use of seaweed-based extracts obtained from hexane, ethyl acetate and chloroform as extractors provided higher tolerance to plants grown at −10 to 0°C, since they presented less tissue damage (as measured by less severe chlorosis and longer duration of green color), lower mortality and higher recovery percentages after the end of the low temperature stress, when compared to the plants that did not receive the seaweed extract fractions (Rayirath et al. 2009). According to these authors, such changes were associated to alterations in the expression of genes for chlorophyll metabolism, and stromal protein with cryoprotective activity (Rayirath et al. 2009). Another study showed that the lipophilic components of the brown seaweed *Ascophyllum nodosum* increased freezing tolerance of *Arabidopsis thaliana* through increments in the concentration of total soluble sugars, alterations in the cellular fatty acid composition, and accumulation of proline, the last by increasing the expression of proline synthesis genes P5CS1 and P5CS2 while marginally reducing the expression of proline catabolism gene (ProDH) (Nair et al. 2012).

In another study, the effects of extracts from *Codium tomentosum* (Chlorophyta), *Gracilaria gracilis* (Rhodophyta) and *Cystoseira barbata* (Phaeophyceae) were tested on the germination of tomato (*Solanum lycopersicum* cv. Rio Grande), pepper (*Capsicum annuum* cv. Demre) and eggplant (*Solanum melongena* cv. Pala) seeds under optimum (25°C) and sub-optimum (15°C) temperatures (Demir et al. 2006). Under ideal temperature, *C. tomentosum* extract improved the germination of pepper and eggplant by 10.4% and 18.7%, respectively, when compared to the untreated seeds. Under cold stress, the extract of this same algae increased the germination of pepper and eggplant by 10.6-48.6% and 7.1-16.7%, respectively. The suspensions made from the other algae species had no effect on the germination potential, regardless of the temperature. In addition, the germination of tomato seeds was not affected by the application of any of these algal extracts (Demir et al. 2006), showing again that plant response depends on plant species.

Low temperature (10°C) completely inhibited *Ceratotheca triloba* seed germination but, when temperature was increased to 15°C, seeds soaked in solution with *E. maxima* extract showed even better germination than the control, untreated seeds (Mansodo et al. 2018).

5.2   Biotic stress

Seaweed-based products can exert a broad spectrum of action on the control of plant diseases and pests, thus potentially improving the crop yield. This capacity is partially linked to the presence of pesticide, fungicide and bactericide compounds, which are also able to combat virus (Manilal et al. 2009, Jiménez et al. 2011, Mathur et al. 2015, Silva et al. 2018, de Borba et al. 2019). In addition, the seaweed extract-induced improvements in the development of microorganisms beneficial to plant growth, such as root nodulation (Raverkar et al. 2018), can potentially enhance plant tolerance to stresses. The literature offers evidence that seaweed-based products have both direct and indirect actions on the mitigation of agents and pests that cause plant disease.

5.2.1   Diseases

In vitro assays showed that ethanolic extracts of *Lessonia trabeculata* (Phaeophyceae) can inhibit by 40% to 60% the growth of *Erwinia carotovora* and *Pseudomonas syringae* (Bacteria) (Jiménez et al. 2011). Furthermore, aqueous and ethanolic extracts from *Macrocystis pyrifera* (formerly *Macrocystis integrifolia*) (Phaeophycae) decreased *Pseudomonas syringae* growth by 50%, while those from *Agarophyton chilense* (formerly *Gracilaria chilensis*) (Rhodophyta) reduced the growth of *Phytophthora cinnamomi* up to 50% (Jiménez et al. 2011). In *in planta* assays, a sulfated hetero-polysaccharide (aka ulvan) extracted from *Ulva fasciata* (Chlorophyta) retarded the occurrence of the typical symptoms of *Fusarium oxysporum* f.sp. *phaseoli* and decreased its severity, probably because of reductions in the number of *Fusarium* propagules (Fungi) (de Borba et al. 2019). In another study, decreases in the number of bacteria (*Pseudomonas syringae* DC3000, *P. aeruginosa* and *Xanthomonas campestris*) found in both MS media and within the seedlings *(P. syringae* DC3000 and *P. aeruginosa)* were noticed after seeds were treated with the *Ascophyllum* extract for 24 h, and these results were probably related to the oxidative burst mediated by the application of the seaweed-based product (Cook et al. 2018). The potential mechanisms by which seaweed extracts mediate the control of plant diseases are shown in Fig. 3.

Brazilian researchers showed that each of the extracts of the following seaweeds— *Stypopodium zonale*, *Ascophyllum nodosum*, *Sargassum muticum*, *Pelvetia canaliculata* and *Fucus spiralis* (Phaeophyceae) and *Laurencia dendroidea* (Rhodophyta)—exhibited an efficient control on *Colletotrichum lagenarium* growth, but none of the 20 tested species was able to inhibit the development of *Aspergillus flavus* (Fungi) (Peres et al. 2012). The effects of seaweed species (*Macrocystis pyrifera* (formerly *Macrocystis integrifolia*), *Lessonia nigrescens*, *Lessonia trabeculata*, *Durvillaea antarctica* (Phaeophyceae), *Gracilaria chilensis*, *Pyropia columbina* (formerly *Porphyra columbina*), *Gigartina skottsbergii* (Rhodophyta), and *Ulva nematoidea* (formerly *Ulva costata*, Chlorophyta), type of extractor (water and ethanol), concentration (100, 1,000, 5,000 and 10,000 ppm), and harvest season (spring, summer and autumn) were evaluated on phytopathogens. It was observed that ethanolic extracts from *Lessonia trabeculata* (5,000 and 10,000 ppm) reduced damage triggered by *Botrytis cinerea* (Fungi) in tomato leaves (from 72% to 95%, depending on dose and harvest season). In this same study, it was verified that both aqueous and ethanolic extract from *Durvillaea antarctica* (5,000 or 10,000 ppm) decreased up to 95% the symptoms of tobacco mosaic virus in tobacco leaves, as measured by reductions in the number and size of leaf lesions (Jiménez et al. 2011).

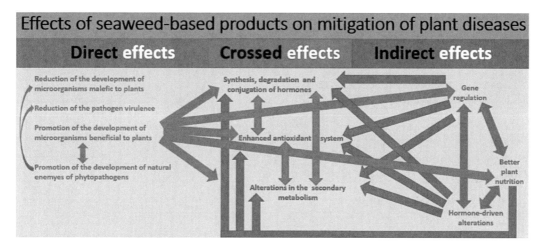

Figure 3. Hypothetical mechanisms of action (both direct and indirect) of seaweed extracts on disease control.

The ethanolic extract from *Cystoseira tamariscifolia* (Phaeophyceae) had a good bactericide effect on the growth of *Agrobacterium tumefaciens*, *Escherichia coli*, *Staphylococcus aureus* and *Pseudomonas aeruginosa* (Bacteria), and it was also an efficient inhibitor of the growth of fungi *Botrytis cinerea*, *Fusarium oxysporum* and *Verticillium albo-atrum* (Fungi) (Bennamara et al. 1999). *Ascophyllum nodosum* extract (aqueous, chloroform and ethyl acetic fractions) also decreased the symptoms from *Pseudomonas syringae* (Bacteria) (14% to 22% in the disease incidence) and *Sclerotinia sclerotiorum* (Fungi) (1.6- to 2.1-fold in the lesion sizes) in *Arabidopsis thaliana* wild types and mutants (Subramanian et al. 2011). Differences in the expression of genes (PDF-1.2, PR-1 and ICS-1) involved in the plant defense against pathogens were also noticed, so that authors suggested seaweed extract acts through a jasmonic acid-dependent mechanism to provide a higher tolerance to these diseases. The foliar spray of the commercial extract from *Ascophyllum nodosum* (0.2%) also decreased the severity of diseases caused by *Alternaria radicina* (4 to 57%) and *Botrytis cinerea* (4% to 53%) (Fungi) in carrot leaves on day 10 and 25 after inoculation, when compared with plants treated with fungicides and water, respectively) (Jayaraj et al. 2008). In addition, increments in the activity of enzymes (chitinase, peroxidase, polyphenol oxidase, phenylalanine ammonia lyase and lipoxygenase) and expression of genes (PR-1, NPR-1, PR-5, among others) related to the plant defense were also reported (Jayaraj et al. 2008).

According to Jayaraman et al. (2010), the use of commercial extract from *Ascophyllum nodosum* (foliar pulverization plus soil irrigation at 0.5%) in cucumber plants decreased the incidence level of diseases caused by *Alternaria cucumerinum* (70%), *Didymella applanata* (47%), *Fusarium oxysporum* (46%) and *Botrytis cinerea* (88%) (Fungi), probably as a result of changes in the enzymatic activities (chitinase, peroxidase, polyphenol oxidase, phenylalanine ammonia lyase and lipoxygenase) and gene expression (LOX, PO and PAL). In addition, French researchers observed that a carbohydrate from *Laminaria digitata* (Phaeophyceae), laminarin, is able to induce the plant defense system in grape, modulating the gene expression (CHIT1b, CHIT4c, GLU1, PIN, LOX, PAL, among others) that is responsible for the synthesis of enzymes chitinase and glucanase, and also for the production of phytoalexin and other compounds, which have antimicrobe properties or participate in the perception and signaling during the infection (Aziz et al. 2007). This extract also mitigated the development of *Botrytis cinerea* and *Plasmopara viticula* in plants already infected by these microorganisms (Aziz et al. 2007). Chilean researchers showed that oligosaccharides from *Lessonia trabeculata*, *Lessonia flavicans* (formerly *Lessonia vadosa*) (Phaeophyceae) and *Schizymenia binderi* (Rhodophyta) promoted increase in both defense enzymes and plant growth in tobacco (Jiménez et al. 2011).

5.2.2 *Pests*

Nematodes

Decreases in both the number of *Meloidogyne javanica* and *Meloidogyne incognita* at the second juvenile stage and the recovery of their eggs were noticed in roots of tomato after soil irrigation with *Ascophyllum nodosum* extract (Wu et al. 1997, Wu et al. 1998). Crouch and Van Staden (1993) also reported reductions in the *Meloidogyne incognita* infestation by applying *Ecklonia maxima* extract in the soil. There is evidence that betaines are the main active compounds responsible for these outcomes (Wu et al. 1997). However, terpenoids were also indicated as the bioactive compounds with nematicide activity in *Stoechospermum polypodioides* (formerly *Stoechospermum marginatum* (Phaeophyceae) extract (Abid et al. 1997). One of the mechanisms by which seaweed extract acts on the control of nematode population is by decreasing female fecundity (Whapham et al. 1994). However, the effects of seaweed-based products on nematodes can also be indirect, either through plant responses (Wu 1996) or through the promotion of antagonist organisms to nematodes (Becker et al. 1988). The extracts from *Dictyota dichotoma* (Phaeophyceae) and *Hypnea pannosa* (Rhodophyta) (at 1.23 and 1.25 mg mL^{-1}, respectively) reduced by 50% the number of nematodes in *in vitro* assay (Manilal et al. 2009). These same authors showed that extracts of *Caulerpa racemosa*, *Chaetomorpha antennina, Valoniopsis pachynema* (Chlorophyta) and *Jania spectabilis* (formerly *Cheilosporum spectabile*) (Rhodophyta), when used in concentrations that ranged from 2 to 4 mg mL^{-1}, were able to decrease the nematode quantity by 31% to 100%.

Insects

Seaweed-based products can be potentially used in pest control. For instance, previous studies showed that *Acrosiphonia orientalis* (Chlorophyta), *Padina tetrastromatica* (Phaeophyceae) and *Centroceras clavulatum* (Rhodophyta) decreased the number of *Culex quinquefasciatus* larvae (Manilal et al. 2009). The ether-petroleum extract from *Acanthophora muscoides* (Rhodophyta) and *Microdictyon pseudohapteron* (Chlorophyta) also presented good results in the control of this same mosquito species (Devi et al. 1997). Yet, ethanolic extract from *Ulva intestinalis* (formerly *Enteromorpha intestinalis*) (Chlorophyta), *Dictyota dichotoma* (Phaeophyceae) and *Acanthopora spicifera* (Rhodophyta) exhibited larvicide activity on *Aedes aegypti*, when saponins and terpenoids were pointed out as the potential bioactive compounds (Ravikumar et al. 2011). Moreover, *Caulerpa racemosa* (Chlorophyta) extract, which contained saponins, flavonoids, alkaloids, protein and sugars, presented larvicide activity on *Anopheles stephensi*, *Aedes aegypti* and *Culex quinquefasciatus* (Ali et al. 2013). The potential mechanisms by which seaweed extracts mediate the control of pests are shown in Fig. 4.

6 Conclusion

The effects of the application of seaweed extracts on plant development and yield are diverse and, frequently, contradictory, possibly because of their dependence on complex and intricate factors (Fig. 1). Moreover, seaweed extracts are frequently mixed with other compounds, and it is difficult to know whether the effects are from the extract, the additional compounds or their synergism. In addition, the results from most studies are relatively limited, since they were carried out under controlled environmental conditions, in potted and small-scale experiments, using model plants (for instance, *Arabidopsis thaliana*) and/or species that are generally from temperate/cold regions (such as *Pinus*, wheat, or barley). Therefore, it is necessary to run field experiments (especially under tropical conditions) in order to obtain more realistic results. In such environments, natural variations of temperature and humidity can affect the physicochemical properties of the seaweed-based products. Furthermore, tropical plants differ in their physiology, anatomy and morphology from plants from temperate/cold environment, so their response to the application of seaweed

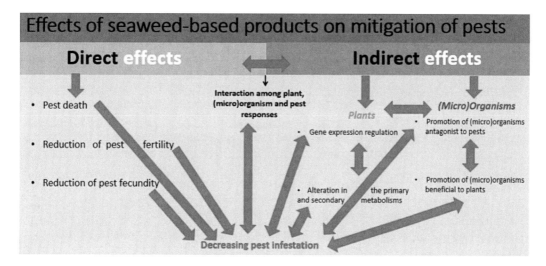

Figure 4. Hypothetical mechanisms of action (both direct and indirect) of seaweed extracts on pest control.

extracts may differ from those observed in previous works. In any case, almost all studies showed slight outcomes from the use of seaweed extract on plants (e.g., enhancements in the secondary metabolites), with little or no influence on plant yield (Carvalho et al. 2018). In this context, the isolation, identification and characterization of the bioactive components from algal extracts are essential for the development of strategies that efficiently improve plant performance while lowering costs. In addition, some reports indicate that algal extracts should be used cautiously, since the inhibitory effects on early development may outweigh any benefits if the concentration of extract is too high (Ghaderiardakani et al. 2018), food contamination by heavy metals from the algae-based products is a potential risk (Michalak et al. 2015), and economic analysis does not always favor the application of seaweed-based products (Singh et al. 2018).

References Cited

Abid, M., V. Sultana, M.J. Zaik and M.A. Maqbool. 1997. Nematicidal properties of *Stoechospermum marginatum*, a seaweed. Pakistan J. Phytopathol. 9: 143–147.

Ahmed, F.F., A.E.M. Mansour, M.A.A. Montasser, M.A. Merwad and E.A.M. Mostafa. 2013. Response of Valencia orange trees to foliar application of roselle, turmeric and seaweed extracts. J. Appl. Sci. Res. 9: 960–964.

Ali, M.Y.S., S. Ravikumar and J.M. Beula. 2013. Mosquito larvicidal activity of seaweeds extracts against *Anopheles stephensi*, *Aedes aegypti* and *Culex quinquefasciatus*. Asian Pac. J. Trop. Dis. 3: 196–201.

Al-Ghamdia, A.A. and H.O. Elansary. 2018. Synergetic effects of 5-aminolevulinic acid and Ascophyllum nodosum seaweed extracts on Asparagus phenolics and stress related genes under saline irrigation. Plant Physiol. Biochem. 129: 273–284.

Arioli, T., S.W. Mattner and P.C. Winberg. 2015. Applications of seaweed extracts in Australian agriculture: past, present and future. J. Appl. Phycol. 27: 2007–2015.

Atzmon, N. and J. Van Staden. 1994. The effect of seaweed concentration on the growth of *Pinus pinea* seedlings. New Forest. 8: 279–288.

Aziz, A., A. Gauthier, A. Bézier, B. Poinssot, J.-M. Joubert, A. Pugin, A. Heyraud and F. Baillieul. 2007. Elicitor and resistance-inducing activities of β-1,4 cellodextrins in grapevine, comparison with β-1,3 glucans and α-1,4 oligogalacturonides. J. Exp. Bot. 58: 1463–1472.

Aziz, N.G.A., M.H. Mahgoub and H.S. Siam. 2011. Growth, flowering and chemical constituents performance of *Amaranthus tricolor* plants as influenced by seaweed (*Ascophyllum nodosum*) extract application under salt stress conditions. J. Appl. Sci. Res. 7: 1472–1484.

Balakrishnan, C.P., V. Kumar, V.R. Mohan and A.T. Athiperumalsami. 2007. Study on the effect of crude seaweed extracts on seedling growth and biochemical parameters in *Cyamopsis tetragonoloba* (L.) Taub. Plant Arch. 7: 563–567.

Becker, J.O., E. Zavaletamejia, S.F. Colbert, M.N. Schroth, A.R. Weinhold, J.G. Hancock and S.D. Van Gundy. 1988. Effects of rhizobacteria on rootknot nematodes and gall formation. Phytopathology 78: 1466–1469.

Bennamara, A., A. Abourrichea, M. Berradaa, M. Charroufa, N. Chaibb, M. Boudoumab and F.X. Garneau. 1999. Methoxybifurcarenone: an antifungal and antibacterial meroditerpenoid from the brown alga *Cystoseira tamariscifolia*. Phytochemistry 52: 37–40.

Bettoni, M.M., R. Koyama, V.C. Pacheco, W.M. Adam and A.F. Mógor. 2010. Produção, classificação e perda de peso durante o armazenamento de cebola orgânica em função da aplicação foliar de extrato de algas. Hortic. Bras. 28: S2880–S2886.

Bettoni, M.M., W.M. Adam and A.F. Mógor. 2008. Tuberização de batata em função da aplicação de extrato de alga e cobre. Hortic. Bras. 26: S5256–S5260.

Borges, K.L.R., F.W.R. Hippler, M.E.A. Carvalho, R.S. Nalin, F.I. Matias and R.A. Azevedo. 2019. Nutritional status and root morphology of tomato under Cd-induced stress: comparing contrasting genotypes for metal-tolerance. Sci. Hortic. 246: 518–527.

Carvalho, M.E.A., P.R.C. Castro, S.A. Gaziola and R.A. Azevedo. 2018. Is seaweed extract an elicitor compound? Changing proline content in drought-stressed bean plants. Comunicata Scientiae 9: 292–297.

Carvalho, M.E.A., F.A. Piotto, M.L. Nogueira, F.G. Gomes-Junior, H.M.C.P. Chamma, D. Pizzaia and R.A. Azevedo. 2018a. Cadmium exposure triggers genotype-dependent changes in seed vigor and germination of tomato offspring. Protoplasma 255: 989–999.

Carvalho, M.E.A., F.A. Piotto, S.A. Gaziola, A.P. Jacomino, M. Jozefczak, A. Cuypers and R.A. Azevedo. 2018b. New insights about cadmium impacts on tomato: plant acclimation, nutritional changes, fruit quality and yield. Food Energy Secur. 7: e00131.

Carvalho, M.E.A., F.A. Piotto, M.R. Franco, K.L.R. Borges, S.A. Gaziola, P.R.C. Castro and R.A. Azevedo. 2018c. Cadmium toxicity degree on tomato development is associated with disbalances in B and Mn status at early stages of plant exposure. Ecotoxicology 27: 1293–1302.

Carvalho, M.E.A., P.R.C. Castro, A.D.L.C. Novembre and H.M.C.P. Chamma. 2013. Seaweed extract improves the vigor and provides the rapid emergence of dry bean seeds. Am. Eurasian J. Agric. Environ. Sci. 13: 1104–1107.

Carvalho, M.E.A. and P.R.C. Castro. 2014. Extratos de algas e suas aplicações na agricultura. Piracicaba: ESALQ – Divisão de Biblioteca.

Carvalho, M.E.A., P.R.C. Castro, L.A. Gallo and M.V.C. Ferraz Junior. 2014. Seaweed extract provides development and production of wheat. Agrarian 7: 1–5.

Castro, P.R.C., M.E.A. Carvalho, A.C.C.M. Mendes and B.G. Angelini. 2017. Manual de estimulantes vegetais – nutrientes, biorreguladores, bioestimulantes, bioativadores, fosfitos e biofertilizantes na agricultura tropical. Editora Agronômica Ceres, Ouro Fino, Minas Gerais.

Cecato, A. and G.C. Moreira. 2013. Aplicação de extrato de algas em alface. Cultivando o Saber, 6: 89–96.

Chrysargyris, A., P. Xylia, M. Anastasiou, I. Pantelides and N. Tzortzakis. 2018. Effects of Ascophyllum nodosum seaweed extracts on lettuce growth, physiology and fresh cut salad storage under potassium deficiency. J. Sci. Food Agric. 98: 5861–5872.

Colapietra, M. and A. Alexander. 2006. Effect of foliar fertilization on yield and quality of table grapes. Acta Hortic. 721: 213–218.

Cook, J., J. Zhang, J. Norrie, B. Blal and Z. Cheng. 2018. Seaweed extract (Stella Maris®) activates innate immune responses in Arabidopsis thaliana and protects host against bacterial pathogens. Mar. Drugs 16: 221.

Craigie, J.S. 2011. Seaweed extract stimuli in plant science and agriculture. J. Appl. Phycol. 23: 371–393.

Craigie, J.S., S.L. Mackinnon and J.A. Walter. 2007. Liquid seaweed extracts identified using ^{1}H NMR profiles. J. Appl. Phycol. 20: 665–671.

Crouch, J. and J. Van Staden. 1993. Effect of seaweed concentrate from *Ecklonia maxima* (Osbeck) Papenfuss on *Meloidogyne incognita* infestation on tomato. J. Appl. Phycol. 5: 37–43.

da Silva, C.C., Í.G. Arrais, J.P.N. de Almeida1, L.L.G.R. Dantas, S.O. Francisco and V. Mendonça. 2016. Seaweed extract of Ascophyllum nodosum (L.) Le Jolis in production of rootstock Anonna glabra L. Rev. Ciê. Agr. 39: 234–241.

de Borba, M.C., M.B. de Freitas and M.J. Stadnik. 2019. Ulvan enhances seedling emergence and controls Fusarium oxysporum f. sp. phaseoli on common bean (Phaseolus vulgaris L.). Crop Protec. 118: 66–71.

Demir, N., B. Dural and K. Yildirim. 2006. Seaweed suspensions on seed germination of tomato, pepper and aubergine. J. Biol. Sci. 6: 1130–1133.

Devi, P., W. Solimabi, L. D'souza and S.Y. Kamat. 1997. Toxic effects of coastal and marine plant extracts on mosquito larvae. Bot. Mar. 40: 533–553.

Di Filippo-Herrera, D.A., M. Muñoz-Ochoa, R.M. Hernández-Herrera and G. Hernández-Carmona. 2018. Biostimulant activity of individual and blended seaweed extracts on the germination and growth of the mung bean. J. Appl. Phycol. 31: 2025–2037.

Duarte, I.J., S.H. Álvarez Hernández, A.L. Ibañez and A.R. Canto. 2018. Macroalgae as soil conditioners or growth promoters of Pisum sativum (L). Annual Res. Rev. Biol. 27: 1–8.

Dumale, J.V., G.R. Gamoso, J.M. Manangkil and C.C. Divina. 2018. Detection and quantification of auxin and gibberellic acid in Caulerpa racemose. Int. J. Agric. Technol. 4: 653–660.

Elansary, H.O., K. Yessoufou, A.M.E. Abdel-Hamid, M.A. El-Esawi, H.M. Ali and M.S. Elshikh. 2017. Seaweed extracts enhance Salam turfgrass performance during prolonged irrigation intervals and saline shock. Front. Plant Sci. 8: 830.

Ervin, E.H., X. Zhang and J.H. Fike. 2004. Ultraviolet-b radiation damage on Kentucky bluegrass II: hormone supplement effects. HortScience 39: 1471–1474.

Farid, R., C. Mutale-joan, B. Redouane, E.L.M. Najib, A. Abderahime, S. Laila and E.L.A. Hicham. 2018. Effect of microalgae polysaccharides on biochemical and metabolomics pathways related to plant defense in Solanum lycopersicum. Appl. Biochem. Biotechnol. doi: 10.1007/s12010-018-2916-y.

Fernandes, A.L.T. and R.O. Silva. 2011. Avaliação do extrato de algas (*Ascophyllum nodosum*) no desenvolvimento vegetativo e produtivo do cafeeiro irrigado por gotejamento e cultivado em condições de cerrado. Enciclopédia Biosfera 7: 147–157.

Ferrazza, D. and A.P.M.M. Simonetti. 2010. Uso de extrato de algas no tratamento de semente e aplicação foliar, na cultura da soja. Cultivando o Saber 3: 48–57.

Fornes, F., M. Sánchez-Perales and J.L. Guardiola. 1995. Effect of a seaweed extract on citrus fruit maturation. Act. Hortic. 379: 75–82.

Ghaderiardakani, F., E. Collas, D.K. Damiano, K. Tagg, N.S. Graham and J. Coates. 2018. Effects of green seaweed extract on *Arabidopsis* early development suggest roles for hormone signalling in plant responses to algal fertilisers. Sci. Rep. 9: 1983.

Goñi, O., A. Fort, P. Quille, P.C. McKeown, C. Spillane and S. O'Connell. 2016. Comparative transcriptome analysis of two Ascophyllum nodosum extract biostimulants: same seaweed but different. J. Agric. Food Chem. 64: 2980-2989.

Hernández-Herrera, R.M., F. Santacruz-Ruvalcaba, J. Zañudo-Hernández and G. Hernández-Carmona. 2016. Activity of seaweed extracts and polysaccharide-enriched extracts from Ulva lactuca and Padina gymnospora as growth promoters of tomato and mung bean plants.

Igna, R.D. and V.S. Marchioro. 2010. Manejo de *Ascophyllum nodosum* na cultura do trigo. Cultivando o Saber 3: 64–71.

Jayaraj, J., A. Wan, M. Rahman and Z.K. Punja. 2008. Seaweed extract reduces foliar fungal diseases on carrot. Crop Prot. 27: 1360–1366.

Jayaraman, J., J. Norrie and Z.K. Punja. 2010. Commercial extract from the brown seaweed *Ascophyllum nodosum* reduces fungal diseases in greenhouse cucumber. J. Appl. Phycol. 23: 353–361.

Jiménez, E., F. Dorta, C. Medina, A. Ramírez, I. Ramírez and H. Peña-Cortés. 2011. Anti-phytopathogenic activities of macro-algae extracts. Marine Drugs 9: 739–756.

Kałużewicz, A., M. Gąsecka and T. Spiżewski. 2017. Influence of biostimulants on phenolic content in broccoli heads directly after harvest and after storage. Folia Hort. 29: 221–230.

Kanmaz, M.G., T. Ozsan, O. Esen and A.N. Onus. 2018. Effects of different organic extracts on seed germination of some carrot (Daucus carota L.) cultivars. Int. J. Agric. Nat. Sci. 1: 06–09.

Kavipriya, R., P.K. Dhanalakshmi, S. Jayashree and N. Thangaraju. 2011. Seaweed extract as a biostimulant for legume crop, green gram. J. Ecobiotechnol. 3: 16–19.

Keser, M., J.T. Swenarton and J.F. Foertch. 2005. Effects of thermal input and climate change on growth of *Ascophyllum nodosum* (Fucales, Phaeophyta) in eastern Long Island Sound (USA). J. Sea Res. 54: 211–220.

Khan, A.S., B. Ahmad, M.J. Jaskani, R. Ahmad and A.U. Malik. 2012. Foliar application of mixture of amino acids and seaweed (*Ascophylum nodosum*) extract improve growth and physicochemical properties of grapes. Int. J. Agric. Biol. 14: 383–388.

Khan, W., D. Hiltz, A.T. Critchley and B. Prithiviraj. 2011. Bioassay to detect *Ascophyllum nodosum* extract-induced cytokinin-like activity in *Arabidopsis thaliana*. J. Appl. Phycol. 23: 409–414.

Khan, W., U.P. Rayirath, S. Subramanian, M.N. Jithesh, P. Rayorath, D.M. Hodges, A.T. Critchley, J.S. Craigie, J. Norrie and B. Prithiviraj. 2009. Seaweed extracts as biostimulants of plant growth and development. J. Plant Growth Regul. 28: 386–399.

Kocira, A., M. Świeca, S. Kocira, U. Złotek and A. Jakubczyk. 2018. Enhancement of yield, nutritional and nutraceutical properties of two common bean cultivars following the application of seaweed extract (Ecklonia maxima). Saudi J. Biol. Scie. 25: 563–571.

Koo, R.C.J. and S. Mayo. 1994. Effects of seaweed sprays in citrus fruit production. Proceed. Florida Stat. Hortic. Sci. 107: 82–85.

Kulkarni, M.G., K.R.R. Rengasamy, S.C. Pendotaa, J. Gruz, L. Plačková, O. Novák, K. Doležal and J. Van Staden. 2019. Bioactive molecules derived from smoke and seaweed Ecklonia maxima showing phytohormone-like activity in Spinacia oleracea L. New Biotechnol. 48: 83–89.

Latique, S., H. Chernane, M. Mansori and M.E. Kaoua. 2016. Biochemical modification and changes in antioxidant enzymes in Triticum durum by seaweed liquid extract of Ulva rigida macroalgae under salt stress condition. pp: 35–54. *In*: J.A. Daniels (ed.). Advances in Environmental Research. Nova Science Publishers, New York.

Limberger, P.A. and J.A. Gheller. 2013. Efeito da aplicação foliar de extrato de algas, aminoácidos e nutrientes via foliar na produtividade e qualidade de alface crespa. Cultivando o Saber 6: 14–21.

Lola-Luz, T., F. Hennequart and M.T. Gaffney. 2013. Enhancement of phenolic and flavonoid compounds in cabbage (*Brassica oleracea*) following application of commercial seaweed extracts of the brown seaweed (*Ascophyllum nodosum*). Agric. Food Sci. 22: 288–295.

Losi, L.C. 2010. Uso de *Ascophyllum nodosum* para o enraizamento de microestacas de eucalipto. M.Sc. thesis, Faculdade de Ciências Agronômicas, Universidade Estadual Paulista "Júlio de Mesquita Filho", Botucatu, Brazil.

Machado, V.P.O., A.C. Pacheco and M.E.A. Carvalho. 2014. Effect of biostimulant application on production and flavonoid content of marigold (Calendula officinalis L.). Ceres 6: 983–988.

Mackinnon, S.L., D. Hiltz, R. Ugarte and C.A. Craft. 2010. Improved methods of analysis for betaines in *Ascophyllum nodosum* and its commercial seaweed extracts. J. Appl. Phycol. 22: 489–494.

Manilal, A., S. Sujith, G.S. Kiran, J. Selvin, C. Shakir, R. Gandhimathi and M.V.N. Panikkar. 2009. Biopotentials of seaweeds collected from southwest coast of India. J. Marine Sci. Technol. 17: 67–73.

Mansori, M., H. Chernane, S. Latique, A. Benaliat, D. Hsissou and M. El Kaoua. 2015. Seaweed extract effect on water deficit and antioxidative mechanisms in bean plants (Phaseolus vulgaris L.). J. Appl. Phycol. 27: 1689–1698.

Martynenko, A., K. Shotton, T. Astatkie, G. Petrash, C. Fowler, W. Neily and A.T. Critchley. 2016. Thermal imaging of soybean response to drought stress: the effect of Ascophyllum nodosum seaweed extract. Springer Plus 5: 1393.

Masny, A., A. Basak and E. Zurawicz. 2004. Effects of foliar applications of Kelpak SL and Goëmar BM 86® preparations on yield and fruit quality in two strawberry cultivars. J. Fruit Ornamental Plant Res. 12: 23–27.

Masondo, N.A., M.G. Kulkarni, J.F. Finnie and J. Van Staden. 2018. Influence of biostimulants-seed-priming on Ceratotheca triloba germination and seedling growth under low temperatures, low osmotic potential and salinity stress. Ecotoxicol. Environ. Saf. 147: 43–48.

Mathur, C., S. Rai, N. Sase, S. Krish and M.A. Jayasri. 2015. Enteromorpha intestinalis derived seaweed liquid fertilizers as prospective biostimulant for Glycine max. Braz. Arch. Biol. Technol. 58: 813–820.

Michalak, I., S. Lewandowska, J. Detyna, S. Olsztyńska-Janus, H. Bujak and P. Pacholska. 2018. The effect of macroalgal extracts and near infrared radiation on germination of soybean seedlings: preliminary research results. Open Chem. 16: 1066–1076.

Mystkowska, I. 2018. The content of iron and manganese in potato tubers treated with biostimulators and their nutritional value. Appl. Ecol. Environ. Res. 16: 6633–6641.

Nabati, D.A., R.E. Scmidt, E.S. Khaleghi and D.S. Parrish. 2008. Assessment of drought stress on physiology growth of *Agrostis palustris* Huds as affected by plant bioregulators and nutrients. Asian J. Plant Sci. 7: 717–723.

Nair, P., S. Kandasamy, J. Zhang, X. Ji, C. Kirby, B. Benkel, M.D. Hodges, A.T. Critchley, D. Hiltz and B. Prithiviraj. 2012. Transcriptional and metabolomic analysis of *Ascophyllum nodosum* mediated freezing tolerance in *Arabidopsis thaliana*. BMC Genomics, 13: 643.

Neumann, E.R., J.T.V. Resende, L.K.P. Camargo, R.R. Chagas and R.B. Lima Filho. 2017. Produção de mudas de batata doce em ambiente protegidocom aplicação de extrato de *Ascophyllum nodosum*. Hortic. Brasil. 35: 490–498.

Pacheco, A.C., L.A. Sobral, P.H. Gorni and M.E.A. Carvalho. 2019. Ascophyllum nodosum extract improves phenolic compound content and antioxidant activity of medicinal and functional food plant Achillea millefolium L. Aust. J. Crop Sci. 13: 418–423.

Pacholczak, A., W. Szydło, K. Zagórska and P. Petelewicz. 2012b. The effect of biopreparations on the rooting of stem cuttings in *Cotinus coggygria* 'Young Lady'. Hort. Land. Architect. 33: 33–41.

Paungfoo-Lonhienne, C., T.G.A. Lonhienne, A. Andreev, L.V. Zhiltsova, S. Kovalev, A. Belov, I. Grebenyuk, N. Kinaev and E. Sagulenko. 2017. Effects of humate supplemented with red seaweed (Ahnfeltia tobuchiensis) on germination and seedling vigour of maize. Aust. J. Crop Sci. 11: 690–693.

Peres, J.C.F., L.R. Carvalho, E. Gonçalez, L.O.S. Berian and J.D. Felicio. 2012. Evaluation of antifungal activity of seaweed extracts. Ciência e Agrotecnologia, 36: 294–299.

Pinto, P.A.C., N.G.N. Santos, G.F.S. Germino, T.D. Deon and A.J. Silva. 2010. Eficiência agronômica de extratos concentrados de algas marinhas na produção da alface em neossolo flúvico. Horticultura Brasileira 28: S3980–S3986.

Ramya, S.S., S. Nagaraj and N. Vijayanand. 2010. Biofertilizing efficiency of brown and green algae on growth, biochemical and yield parameters of *Cyamopsis tetragonolaba* (L.) Taub. Recent Res. Sci. Technol. 2: 45–52.

Ravikumar, S., M.S. Ali and J.M. Beula. 2011. Mosquito larvicidal efficacy of seaweed extracts against dengue vector of *Aedes aegypti*. Asian Pacific J. Tropical Biomed. 1: S143–S146.

Rayirath, P., B. Benkel, D.M. Hodges, P. Allan-Wojtas, S. Mackinnon, A.T. Critchley and B. Prithiviraj. 2009. Lipophilic components of the brown seaweed, *Ascophyllum nodosum*, enhance freezing tolerance in *Arabidopsis thaliana*. Planta 230: 135–147.

Rayorath, P., W. Khan, R. Palanisamy, S.L. Mackinnon, R. Stefanova, S.D. Hankins, A.T. Critchley and B. Prithiviraj. 2008a. Extracts of the brown seaweed *Ascophyllum nodosum* induce gibberellic acid (GA$_3$)-independent amylase activity in barley. J. Plant Growth Regul. 27: 370–379.

Rayorath, P., M.N. Jithesh, A. Farid, W. Khan, R. Palanisamy, S.D. Hankins, A.T. Critchley and B. Prithiviraj. 2008b. Rapid bioassays to evaluate the plant growth promoting activity of *Ascophyllum nodosum* (L.) Le Jol. using a model plant, *Arabidopsis thaliana* (L.) Heynh. J. Appl. Phycol. 20: 423–429.

Saa, S., A. Olivos-Del Rio, S. Castro and P.H. Brown. 2015. Brown Foliar application of microbial and plant based biostimulants increases growth and potassium uptake in almond (*Prunus dulcis* [Mill.] D.A. Webb). Front. Plant Sci. 6: 87.

Santaniello, A., A. Scartazza, F. Gresta, E. Loreti, A. Biasone, D. Di Tommaso, A. Piaggesi and P. Perata. 2017. *Ascophyllum nodosum* seaweed extract alleviates drought stress in *Arabidopsis* by affecting photosynthetic performance and related gene expression. Front. Plant Sci. 8: 1362.

Schiattone, M.I., B. Leoni, V. Cantore, M. Todorovic, M. Perniola and V. Candido. 2018. Effects of irrigation regime, leaf biostimulant application and nitrogen rate on gas exchange parameters of wild rocket. Acta Hortic. doi: 10.17660/ActaHortic.2018.1202.3.

Sharma, C. Chen, K. Khatri, M.S. Rathore and S.P. Pandey. 2019. *Gracilaria dura* extract confers drought tolerance in wheat by modulating abscisic acid homeostasis. Plant Physiol. Biochem. doi: https://doi.org/10.1016/j.plaphy.2019.01.015.

Shukla, P.S., K. Shotton, E. Norman, W. Neily, A.T. Critchley and B. Prithiviraj. 2018. Seaweed extract improve drought tolerance of soybean by regulating stress-response genes. AoB Plant. 10: plx051.

Silva, C.P., K.G.V. Garcia, M.S. Roseano, L.A.A. Oliveira and M.S. Tosta. 2012. Desenvolvimento inicial de mudas de couve-folha em função do uso de extrato de alga (*Ascophyllum nodosum*). Revista Verde de Agroecologia e Desenvolvimento Sustentável 6: 7–11.

Silva, P., C. Fernandes, L. Barros, I.C.F.R. Ferreira, L. Pereira and T. Gonçalves. 2018. The antifungal activity of extracts of Osmundea pinnatifida, an edible seaweed, indicates its usage as a safe environmental fungicide or as a food additive preventing post-harvest fungal food contamination. Food Funct. 9: 6187–6195.

Singh, I., V.A.K. Gopalakrishnan, S. Solomon, S.K. Shukla, R. Rai, S.T. Zodape and A. Ghosh. 2018. Can we not mitigate climate change using seaweed based biostimulant: a case study with sugarcane cultivation in India. J. Clean Prod. 204: 992–1003.

Soares, C., M.E.A. Carvalho, R.A. Azevedo and F. Fidalgo. 2019. Plants facing oxidative challenges—A little help from the antioxidant networks. Environ. Exp. Bot. 161: 4–25.

Spann, T.M. and H.A. Little. 2011. Applications of a commercial extract of the brown seaweed *Ascophyllum nodosum* increases drought tolerance in container-grown 'Hamlin' sweet orange nursery trees. Hortscience 46: 577–582.

Spinelli, F., G. Fiori, M. Noferini, M. Sprocatti and G. Costa. 2009. Perspectives on the use of a seaweed extract to moderate the negative effects of alternate bearing in apple trees. J. Horticult Sci. Biotechnol. Special volume: 131–137.

Sridhar, S. and S. Rengasamy. 2010. Significance of seaweed liquid fertilizers for minimizing chemical fertilizers and improving yield of *Arachis hypogaea* under field trial. Recent Res. Sci. Technol. 2: 73–80.

Stamatiadis, S., L. Evangelou, J.-C. Yvin, C. Tsadilas, J.M.G. Mina and F. Cruz. 2015. Responses of winter wheat to Ascophyllum nodosum (L.) Le Jol. extract application under the effect of N fertilization and water supply. J. Appl. Phycol. 27: 589–600.

Subramanian, S., J.S. Sangha, B.A. Gray, R.P. Singh, D. Hiltz, A.T. Critchley and B. Prithiviraj. 2011. Extracts of the marine brown macroalga, *Ascophyllum nodosum*, induce jasmonic acid dependent systemic resistance in *Arabidopsis thaliana* against *Pseudomonas syringae* pv. *tomato* DC3000 and *Sclerotinia sclerotiorum*. European J. Plant Pathol. 131: 237–248.

Sunarpi, J.A., R. Kurnianingsih, N.I. Julisaniah and A. Nikmatullah. 2010. Effect of seaweed extracts on growth and yield of rice plants. Nusantara Biosci. 2: 73–77.

Szabó, V. and K. Hrotkó. 2009. Preliminary results of biostimulator treatments on *Crataegus* and *Prunus* stock plants. Bull. UASVM Horticult. 66: 223–228.

Taiz, L., E. Zeiger, I.M. Møller and A. Murphy. 2018. Plant physiology and development. Sinauer Associates, Oxford.

Torres, P., P. Novaes, L.G. Ferreira, J.P. Santos, E. Mazepa, M.E.R. Duarte, M.D. Noseda, F. Chow and D.Y.A.C. dos Santos. 2018. Effects of extracts and isolated molecules of two species of Gracilaria (Gracilariales, Rhodophyta) on early growth of lettuce. Algal Res. 32: 142–149.

Trivedi, K., K.G.V. Anand, D. Kubavat, R. Patidar and A. Ghosh. 2018. Drought alleviatory potential of Kappaphycus seaweed extract and the role of the quaternary ammonium compounds as its constituents towards imparting drought tolerance in Zea mays L. J. Appl. Phycol. 30: 2001–2015.

Ugarte, R.A., G. Sharp and B. Moore. 2006. Changes in the brown seaweed *Ascophyllum nodosum* (L.) Le Jol. Plant morphology and biomass produced by cutter rake harvests in southern New Brunswick, Canada. J. Appl. Phycol. 18: 351–359.

Whapham, C.A., T. Jenkins, G. Blunden and S.D. Hankins. 1994. The role of seaweed extracts, *Ascophyllum nodosum*, in the reduction in fecundity of *Meloidogyne javanica*. Fundamen. Appl. Nematol. 17: 181–183.

Wu, Y. 1996. Biologically active compounds in seaweed extracts. 1996. PhD. Thesis. University of Portsmouth, Portsmouth.

Wu, Y., T. Jenkins, G. Blunden, C. Whapham and S.D. Hankins. 1997. The role of betaines in alkaline extracts of *Ascophyllum nodosum* in the reduction of *Meloidogyne javanica* and *M. incognita* infestations of tomato plants. Fundamen. Appl. Nematol. 20: 99–102.

Wu, Y., T. Jenkins, G. Blunden, N. Von Mend and S.D. Hankins. 1998. Suppression of fecundity of the rootknot nematode, *Meloidogyne javanica*, in monoxenic cultures of *Arabidopsis thaliana* treated with an alkaline extract of *Ascophyllum nodosum*. J. Appl. Phycol. 10: 91–94.

Yıldıztekın, M., A.L. Tuna and C. Kaya. 2018. Physiological effects of the brown seaweed (Ascophyllum nodosum) and humic substances on plant growth, enzyme activities of certain pepper plants grown under salt stress. Acta Biol. Hung. 69: 325–335.

Zhang, X. and E.H. Ervin. 2004. Cytokinin-containing seaweed and humic acid extracts associated with creeping bentgrass leaf cytokinins and drought resistance. Crop Sci. 44: 1737–1745.

Zhang, X. and E.H. Ervin. 2008. Impact of seaweed extract-based cytokinins and zeatin riboside on creeping bentgrass heat tolerance. Crop Sci. 48: 364–370.

Zhang, X. and R.E. Schmidt. 1999. Antioxidant response to hormone-containing product in Kentucky bluegrass subjected to drought. Crop Sci. 39: 545–551.

Zhang, X., R.E. Schmidt, H.E. Ervin and S. Doak. 2002. Creeping bentgrass physiological responses to natural plant growth regulators and iron under two regimes. Hortscience 37: 898–902.

Zodape S.T., S. Mukherjee, M.P. Reddy and D.R. Chaudhary. 2009. Effect of *Kappaphycus alvarezii* (Doty) Doty ex Silva. extract on grain quality, yield and some yield components of wheat (*Triticum aestivum* L.). Int. J. Plant Prod. 3: 97–102.

Zodape, S.T., A. Gupta, S.C. Bhandari, U.S. Rawat, D.R. Chaudhary, K. Eswaran and J. Chikara. 2011. Foliar application of seaweed sap as biostimulant for enhancement of yield and quality of tomato (*Lycopersicon esculentum* Mill.). J. Sci. Industrial and Res. India 70: 215–219.

Zodape, S.T., S. Mukhopadhyay, K. Eswaran, M.P. Reddy and J. Chikara. 2010. Enhanced yield and nutritional quality in green gram (*Phaseolus radiata* L.) treated with seaweed (*Kappaphycus alvarezii*) extract. J. Sci. Ind. Res. 69: 468–471.

Seaweeds in the Control of Plant Diseases and Insects

Levi Pompermayer Machado*, Maria Cândida de Godoy Gasparoto, Norival Alves
Santos Filho and Ronaldo Pavarini

Campus Experimental de Registro, UNESP – São Paulo State University, Registro, São Paulo, Brazil

1 Knowledge of Seaweed to Control Plant Diseases and Insects in Agriculture

The world history of seaweed use is mainly related to their direct harvest from wild stocks or their cultivation to obtain biomass. Marine macroalgae have been used in human food and extraction of carrageenans and agar, which together correspond to 90% of the market of macroalgae derivatives (FAO 2018).

Research on the application of extracts of macroalgae in agriculture is nowadays concentrated in three main areas: (1) fertilizer or growth stimulant and development; (2) inducing of resistance in plants to pathogens or stress caused by environmental factors; (3) phytopathogen and phytophage control.

Many microorganisms cause diseases in plants. However, fungi are the major group related to plant diseases on tropical crops. The lack of products with low toxicity in the market has motivated the search for new effective and safe technologies to control these pathogens (Shuping and Eloff 2017). Additionally, technologies to serve as alternative to conventional fungicides (or as a complement of conventional agrochemicals) are now becoming more acceptable in agriculture and can play an important role in an integrated disease management program (Sharma et al. 2014). This interest shown by farmers is related to the rising threat of fungal resistance to fungicides and the environmental appeal of biological products such as seaweed extracts.

Global food security is strongly impacted by fungal losses at different stages of production. The Food and Agriculture Organization of the United Nations (FAO) estimates 30% of food losses globally wasted or lost per year. About 95% of investment of research and development is applied to increase food production, but only 5% is directed towards reducing losses (Gustavsson et al. 2011).

Phytophagous arthropods such as insects and mites are routinely reported as pests, as they can cause a significant reduction in the productivity of agricultural and forest crops. Traditionally, these pests are controlled by successive applications of agricultural pesticides composed of synthetic molecules. Such formulations are costly. Also, their continued and improper use may have undesirable effects, such as the development of resistant populations, the appearance of new pests or the resurgence of others, the occurrence of biological imbalances, and detrimental effects on humans, natural enemies, fish and other non-target organisms.

---

*Corresponding author: levi.p.machado@unesp.br

The damage caused by phytophages, in addition to compromising the quality of food production, impairs the yield, causing producers to suffer high losses (Copatti et al. 2013, Mossi et al. 2011). About 7.7% of Brazilian agricultural production is lost annually because of pest attack, while the world losses due to pest attacks are estimated to reach an astonishing $1.4 trillion, or nearly 5% of world GDP (Gross Domestic Product) (Oliveira et al. 2014).

In a search through scientific databases (Google Scholar, Web of Science and Scielo) for papers published between 2010 and 2018 using the search terms "seaweed or macroalgae", "antifungal" and "control of phytophagous pests", 42 scientific papers were verified, highlighting the interest in what is a critical issue in agriculture and food production. (The studies are summarized in Table 1.)

In spite of the number of studies, specific surveys on the effect of seaweed on pest control are lacking; the subject is not prominent in published reviews of bioactive marine macroalgae or natural extracts. Research on land plant extracts in agriculture was more advanced, including commercial derivative products (Ncama et al. 2019).

2 Assays and Screening of Biotechnological Potential

Assays for investigation of seaweed compounds applied to the control of phytopathogenic fungi or plant insects need multiple and interdisciplinary approaches. The chosen methods should meet the specific objectives of the study. However, a few bioactive compounds from macroalgae have been converted into commercial products. When we approach the area of plant pathogens or agricultural pests, there is a lot of distance between scientific publications and commercial products development.

In our search for studies about bioactive potential of macroalgae in agriculture, we also realized the discrepancy in the number of studies carried out *in vitro* and *in vivo*. It is important to establish correlations between *in vitro* and *in vivo* assays. In this context, it is necessary to know the main methods, their objectives, and the possible results for a standardized and multidisciplinary approach, which contemplates a platform to accelerate the research and development processes of new seaweed-based products in agriculture.

2.1 Antifungal activity

2.1.1 In vitro assays

Thin layer chromatography (TLC)
Thin layer chromatography has the advantages of simplicity, speed, and relatively low cost, the ability to analyze multiple samples on a single plate using small amounts of solvents as the mobile phase, and low susceptibility to contamination. The development of different stationary phases in matrices of layer types allows the amplification spectrum of extracts, polarity and compounds evaluated (Sherma 2017).

This kind of assay can be applied to screen many different seaweed extracts against a broad range of plant pathogens at the same time. These potentials can be accessed for the separation, detection, qualitative and quantitative determination, and preparative isolation.

Techniques of detection (derivatization or visualization) are simple processes used to evaluate a large number of substances, based on physical and chemical methods of detection. These methods use thermochemical reaction and reaction with chemical agents to modify the molecules in the extract, revealing detailed information that determines the identity or classes of the presented compounds (Hahn-Deinstrop 2007).

Another visualization assay is based on biological-physiological methods of detection. Living organisms or enzymes are used to determine biological action of active substances. Bioautography can be used with multiple targets, e.g., antifungal, antibacterial or antioxidative (Kagan and Flythe 2014).

Table 1. Scientific papers published between 2010 and 2018 related to seaweed and antifungal action or control of phytophagous pests

Seaweed (phylum)	Biomass processing	Extraction solvent	Extract and/or bioactive compound	Assays	Host—disease (and/or plant pathogen)	Reference
Osmundea pinnatifida (Rhodophyta)	Freeze-dried	Organic solvents of different polarities— methanol dichloromethane and n-hexane	Extract mainly composed of palmitic acid (hexadecanoic acid, 29.60% ± 0.60), phytol isomer 1 (12.80% ± 0.30), oleic acid (octadecanoic acid, 9.62% ± 0.14), stearic acid (octadecanoid acid, 6.15% ± 0.02) and D-(-)-tagatofuranose (4.11% ± 0.04)	1. *In vitro* (radial growth inhibition); fungal morphology and 2. chitin and β-(1,3)-D-glucan cell wall content)	*Alternaria infectoria* and *Aspergillus fumigatus* (Fungi, Ascomycota)	Silva et al. 2018
Gracilariopsis longissima (formerly *Gracilaria confervoides*) (Rhodophyta)	Air-dried at 45°C for 5 d in a hot-air oven	Chemical analysis— methanol, ethyl acetate, acetone, benzene, and chloroform	Extracts and dry algae powder (chemical analysis of chloroform and ethyl acetate extracts	1. *In vitro* (radial growth inhibition); 2. *In vivo* (algae powder applied on soil—greenhouse)	*Rhizoctonia solani* (Fungi, Ascomycota) *Fusarium solani*, and *Macrophomina phaseolina* (Fungi, Basidiomycota)	Soliman et al. 2018 Melo et al. 2018
Ulva lactuca (Chlorophyta)	Dry biomass	Extracted with hot oxalate solution for 3 h at 90°C followed by acid precipitation using HCl. The precipitate was dissolved in water and lyophilized to obtain the glucuronan polysaccharide fraction	β-Δ-(4,5)-oligoglucuronans	*In vitro* (radial growth inhibition)	*Penicillium expansum* and *Botrytis cinerea* (Fungi, Ascomycota) Apples (*Malus domestica Borkh. cv Golden Delicious*)	Abouraïcha et al. 2017
Sargassum vulgare (Ochrophyta)	Shade-dried for several days at room temperature	Aqueous, methanol, petroleum ether, chloroform and ethyl acetate extractions	Crude extract	1. *In vitro* (poisoned food technique) and 2. *In vivo* (10 tubers per treatment)	Potato tubers cv. Spunta—Pythium leak disease (*Pythium aphanidermatum*) (Chromista, Oomycota)	Ammar et al. 2017

(Contd.)

Seaweed species	Preparation	Extract	Assay	Target	Reference
Cystoseira tamariscifolia and *Bifurcaria bifurcata* (Ochrophyta)	Dried at room temperature	Algal powdered samples extraction with CHCl$_3$/EtOH (5/5) and crude extract extracted with hexane. Aqueous extract *in vivo* (only).	Extracts (hexane and aqueous)	1. *In vitro* (radial growth inhibition) and 2. *In vivo* (post harvest of tomato fruit)	*Botrytis cinerea* (Fungi, Ascomycota)	Bahammou et al. 2017
Laminaria digitata and *Undaria pinnatifida* (Ochrophyta); *Porphyra umbilicalis*, *Eucheuma denticulatum* and *Gelidium pusillum* (Rhodophyta)	Fresh biomass was dried in an industrial dryer	Seaweed extracts used in the experiments were obtained by a SC-CO$_2$ pilot-extractor. Dry extract was suspended in n-hexane, or boiling distilled water, or 80% aqueous methanol (1:5 w/v) for 2 h	Crude and fractionated seaweed extract and chemical analysis of fractioned extract total lipids and fatty acids; water-soluble polysaccharides and phenolic compounds and phlorotannins	1. *In vitro* (radial growth inhibition); 2. Conidial germination assays on 96-microwell plates); and 3. *In vivo* (wounded fruit)	*Botrytis cinerea, Monilinia laxa* and *Penicillium digitatum* (Fungi, Ascomycota) —*In vivo*: grey mold on strawberries, brown rot on peaches, and green mold on lemons	De Corato et al. 2017
Cystoseira humilis var. *myriophylloides* (formerly *Cystoseira myriophylloides*), *Laminaria digitata*, and *Fucus spiralis* (Ochrophyta)	Air-dried and then oven-dried at 65°C for 24 h. Dried seaweeds were ground and sieved into a fine powder	Aqueous seaweed extracts and methanolic seaweed extracts	Extract: methanolic and aqueous	1. *In vitro* (radial growth inhibition) and 2. *In vivo* (greenhouse, spray application)	Tomato plants (*Solanum lycopersicum*) var. 'Pomodoro'—*Verticillium dahliae* (Fungi, Ascomycota)	Esserti et al. 2017
Sargassum latifolium, *Hydroclathrus clathratus* and *Padina gymnospora* (Ochrophyta)	Air-dried under shade for 72 h	Methanol	1. Crude methanolic extracts. 2. Chemical analysis of fatty acids, saccharides and phenolic compounds	*In vitro* (minimal inhibition concentration—MIC)	1. *Fusarium solani, Rhizoctonia solani* (Fungi, Ascomycota); 2. *Solanum melongena* L. (eggplant)	Ibraheem et al. 2017
Ulva lactuca (Chlorophyta), *Sargassum filipendula* (Ochrophyta) and	Air-dried in a fume hood at 26°C for 3–4 d	Dried seaweeds were blended into a fine powder and soaked in 2% potassium hydroxide (KOH) for 48	Alkaline crude extracts	*In vitro* (radial growth inhibition)	Tomato plants (Hybrid 61 variety)—*Alternaria solani* (Fungi, Ascomycota and *Xanthomonas*	Ramkissoon et al. 2017

(Contd.)

Table 1. (*Contd.*)

Seaweed (phylum)	Biomass processing	Extraction solvent	Extract and/or bioactive compound	Assays	Host—disease (and/or plant pathogen)	Reference
Gelidium serrulatum (Rhodophyta)		h, concentrated under reduced pressure at 45°C until dry, and resuspended in sterile distilled water			*campestris* pv. *Vesicatoria* (Bacteria, Proteobacteria)	
Ascophyllum nodosum (Ochrophyta)	Commercial powder extract	Water	Commercial extract	*In vivo* (greenhouse, spray and/or root drenched)	Cucumber (*C. sativus*) plants (cv. Negin F1)— *Phytophthora melonis* (Chromista, Oomycota)	Abkhoo and Sabbagh 2016
Turbinaria conoides (Ochrophyta)	Chopped fine and dried under sun for 4 d, then oven-dried for 24 h at 60°C	1. Acetone. 2. Hydroalcoholic	Crude extract	*In vitro* (radial growth inhibition)	*Fusarium oxysporum* (Fungi, Ascomycota)	Begum et al. 2016
Sargassum muticum, Dictyota bartayresiana, Padina gymnospora, Pseudochnoospora implexa (formerly *Ochnoospora implexa*)*, Sargassum swartzii* (formerly *Sargassum wightii*) (Ochrophyta)	Air-dried under shade and chopped and pulverized after drying	Powdered sample extracted for 7 d in 1:1(v/v) chloroform: methanol	Crude extracts	1. *In vitro* (spore germination, paper disc assay, agar well method); 2. *In vivo* (seed treatment, prophylactic spraying in greenhouse)	Rice plants *Rhizoctonia solani* (Fungi, Basidiomycota)	Raj et al. 2016

(Contd.)

Species	Drying	Solvent	Extract	Assay	Plant/pathogen	Reference
Cymopolia barbata, Chlorella sp. and *Ulva rigida* (Chlorophyta), *Cystoseira* sp., *Ecklonia* sp., *Lobophora variegata; Sargassum* sp. (Ochrophyta), *Corallina* sp., *Crassiphycus corneus* (formerly *Gracilaria cornea*), *Grateloupia* sp., *Halopithys* sp., *Hypnea* sp., *Solieria* sp. (Rhodophyta)	Dried in greenhouse under environmental lighting for 12 h, dried at 60°C	Water	Aqueous extract	*In vivo* (sprays on cotyledons)	Zucchini (*Cucurbita pepo*) cotyledons—Zucchini plants var. Giambo F1 (Semencoops.r.l., Cesena, FC, Italy) / Powdery mildew (*Podosphaera xanthii*) (Fungi, Ascomycota)	Roberti et al. 2016
Caulerpa racemosa (Chlorophyta), *Sargassum polycystum* (formerly *Sargassum myriocystum*) (Ochrophyta) and *Hydropuntia edulis* (formerly *Gracilaria edulis*) (Rhodophyta)	Shade-dried for 2 wk, then oven-dried at 40°C for 24 h and powdered	Ethanol	Crude extract	*In vitro* (radial growth inhibition)	Pigeonpea var. CO (Rg) 7 (*Cajanus cajan* (L.) / *Macrophomina phaseolina* (Fungi, Ascomycota)	Ambika and Sujatha 2015
NI	NI	NI	Laminarin pure and mixed with a microbial extract of *Saccharomyces*	*In vivo* (field)	Strawberry (*Fragaria* x *ananassa*, 'Alba' and 'Romina' cvs.) fruit/ postharvest decay	Feliziani et al. 2015

(Contd.)

Table 1. (*Contd.*)

Seaweed (phylum)	Biomass processing	Extraction solvent	Extract and/or bioactive compound	Assays	Host—disease (and/or plant pathogen)	Reference
Sargassum vulgare, Cystoseira barbata, Dictyopteris polypodioides (formerly *Dictyopteris membranacea*), *Dictyota dichotoma,* and *Colpomenia sinuosa* (Ochrophyta)	Shade-dried, chopped fine, and ground into powder. Extracted successively with different solvents using Soxhlet apparatus for 8 h	Six organic extracts (ethanol, acetone, chloroform, ethyl acetate and cyclohexane)	Crude extracts	*In vitro* (radial growth inhibition)	*Alternaria alternata, Cladosporium cladosporioides, Fusarium oxysporum, Epicoccum nigrum, Aspergillus niger, Aspergillus ochraceus, Aspergillus flavus,* and *Penicillium citrinum* (Fungi, Ascomycota)	Khalil et al. 2015
Ulva lactuca (Chlorophyta)	Shade-dried for 2 wk, partly powdered using domestic blender	Acetone, chloroform, ethanol, methanol and petroleum ether	Extracts fractionated over silica gel column chromatography	*In vitro* (radial growth inhibition)	Three phytopathogenic fungi: *Aspergillus niger, Penicillium digitatum* and *Rhizoctonia solani* (Fungi, Ascomycota)	Abbassy et al. 2014
Caulerpa racemosa (Chlorophyta), *Sargassum polycystum* (formerly *Sargassum myriocystum*) (Rhodophyta)	Shade-dried for 2 wk, then oven-dried at 40°C for 24 h and powdered	Ethanol	Crude extract	*In vitro* (radial growth inhibition)	*Fusarium oxysporum* f. sp. *udum* (Fungi, Ascomycota)	Ambika and Sujatha 2014
Ascophyllum nodosum (Ochrophyta) and cinnamon (*Cinnamomum* sp.—Magnoliophyta) extracts	Commercial product	NI	NI	*In vivo* (spray until runoff)	Strawberry plants cultivar 'Camarosa' from nurseries/ strawberry anthracnose (*Colletotrichum acutatum*) (Fungi, Ascomycota)	Aguado et al. 2014

(Contd.)

Seaweed	State	Solvent	Compound/description	Method	Target/Disease	Reference
Ulva fasciata (Chlorophyta)	Dried	Water and ethanol	Ulvan (an algal polysaccharide known for its ability to induce resistance to plant diseases)	1. *In vitro* (conidial germination and appressoria formation); 2. *In vivo* (apple leaves)	'Gala' apple/Glomerella leaf spot (pathogenic strain MANE147 of *Colletotrichum gloeosporioides*) (Fungi, Ascomycota)	Araújo et al. 2014
Ulva lactuca, Caulerpa sertularioides (Chlorophyta), *Padina gymnospora* and *Sargassum liebmannii* (Ochrophyta)	Oven-dried for 72 h at 60°C	Water and ethanol	Algal cell wall polysaccharide extracts (or polyssaccharide-enriched seaweed extracts)	1. *In vivo* foliar spray with seaweed extract solutions	Tomato plants cv. Rio Fuego	Hernández-Herrera et al. 2014
Leathesia marina (formerly *Leathesia nana*), *Rhodomela confervoides,* and *Rhodomela confervoides* (Rhodophyta)	NI	NI	Bis(2,3-dibromo-4,5-dihydroxybenzyl) ether (BDDE) is a bromophenol isolated from marine algae	1. *In vitro* (inhibition on mycelial growth and spore germination of *Botrytis cinerea* only); and 2. *In vivo* (inhibition of fruit decay)	*B. cinerea, Valsa mal, Fusarium graminearum, Coniothyrium diplodiella, Colletotrichum gloeosporioides, Alternaria mali* and *Alternaria porri* (Fungi, Ascomycota)	Liu et al. 2014
Ochtodes secundiramea (Rhodophyta)	Fresh biomass	Dichloromethane / methanol (2:1)	Crude extract and halogenated monoterpene were the major compound	*In vitro* (agar dilution) inhibition on mycelial growth	Papaya and banana/ Anthracnose (*Colletotrichum gloeosporioides*). In addition, toxicity analyses were carried out (Fungi, Ascomycota)	Machado et al. 2014a

(Contd.)

Table 1. (*Contd.*)

Seaweed (phylum)	Biomass processing	Extraction solvent	Extract and/or bioactive compound	Assays	Host—disease (and/or plant pathogen)	Reference
Ochtodes secundiramea (Rhodophyta)	Biomass fresh, freeze-dried and oven-dried for 72 h at 50°C	Dichloromethane/methanol (2:1)	Halogenated monoterpene	*In vitro*—thin layer bioautography (guided fractionation)	Cucumber anthracnose / *Colletotrichum lagenarium*) and *Cladosporium sphaerospermum* (Fungi, Ascomycota)	Machado et al. 2014b
Ochtodes secundiramea, Palisada flagellifera and *Hypnea pseudomusciformis* (Rhodophyta)	Fresh biomass collected from nature and produce in laboratory culture	Dichloromethane/methanol (2:1)	NI	*In vitro* (agar dilution) inhibition on mycelial growth	Papaya—Anthracnose/ *Colletotrichum gloeosporioides* (Fungi, Ascomycota)	Machado et al. 2014c
Ascophyllum nodosum (Ochrophyta)	NI	Water	Commercial *A. nodosum* extract, Acadian®, which was used in tests, presents 5.3% of K_2O and 6.0% of total	*In vivo* (biofilm evaluated postharvest)	NI	Melo et al. 2018
Alga 600 (Commercial product based on *Sargassum vulgare* (Ochrophyta)	NI	Water	Composition: organic matter (40-50%), alginic acid (12%), total nitrogen (0.6%), phosphorus (6%), potassium (20%), amino acid (4%), mannitol (3%), Mg (0.06%), Ca (0.4-1.6%), Fe (0.15-0.3%), Cu (25-45 ppm), S (1.0-1.5 ppm), I (300-600 ppm), pH (9-10)	*In vivo* (seaweed extract applied on postharvest analyses)	NI	Omar 2014

(*Contd.*)

Seaweed	Form	Extraction	Notes	Method	Target / Disease	Reference
Spatoglossum variabile, Polycladia indica (formerly *Stokeyia indica*) and *Melanothamnus afaqhusainii* (Ochrophyta)	Shade-dried	Seaweed mixed in sandy loam soil	NI	*In vivo* (field)	Eggplant and watermelon / *Fusarium solani* and *Macrophomina phaseolina*	Baloch et al. 2013
Commercial product Alginure® (seaweed extract), *Ascophyllum nodosum* and *Laminaria* sp. (Ochrophyta)	Commercial powder extract	NI	NI	*In vivo* (field)	Strawberry cv. 'Induka' / Grey mold (*Botrytis cinerea*) and leaf spot (*Mycosphaerella fragariae*)	Bocek et al. 2012
Laurencia dendroidea (Rhodophyta) *Stypopodium zonale, Pelvetia canaliculata, Sargassum muticum, Ascophyllum nodosum* and *Fucus spiralis* (Ochrophyta)	Washed, air-dried	Ethanol	Confirmed terpenes phenolic compounds in bioactivity extract	*In vitro* (1. direct bioautography assays and 2. radial growth inhibition)	*Colletotrichum lagenarium* and *Aspergillus flavus* (Fungi, Ascomycota)	Peres et al. 2012
Codium fragile (Chlorophyta) and *Padina gymnospora* (Ochrophyta)	Air-dried	Ethyl acetate and methyl alcohol	Extract	*In vitro* (radial growth inhibition)	*Alternaria alternata, Ulocladium botrytis* and *Botryotricum piluliferum* (Fungi, Ascomycota)	Galal et al. 2011
Ulva rigida (formerly *Ulva armoricana*) (Chlorophyta)	Fresh biomass	Material incubated in water and the homogenate heated for 2 h at 90°C	Crude extract (chemical analyses—ulvans)	*In vivo* (greenhouse-spraying algal extract in detached leaves or entire plants)	Common bean, grapevine and cucumber / *Erysiphe polygoni, E. necator* and *Sphaerotheca fuliginea*	Jaulneau et al. 2011

(Contd.)

Table 1. (*Contd.*)

Seaweed (phylum)	Biomass processing	Extraction solvent	Extract and/or bioactive compound	Assays	Host–disease (and/or plant pathogen)	Reference
Stimplex™, a marine plant extract formulation from ***Ascophyllum nodosum*** (Ochrophyta)	NI	NI	NI	*In vivo* (combined spray and root drenching)	Cucumber (*Cucumis sativus* var. *sativus*)/ (*Alternaria cucumerina*, *Didymella applanata*. *F. oxysporum*, and *Botrytis cinerea* (Fungi, Ascomycota)	Jayaraman et al. 2011
Lessonia trabeculata (Ochrophyta), *Agarophyton chilense* (formerly *Gracilaria chilensis*) (Rhodophyta) and *Durvillaea antarctica* (Ochrophyta)	Biomass frozen in liquid nitrogen and ground	Ethanolic extracts; aqueous and ethanolic extracts; aqueous (more effective and seasonally independent) and ethanolic extracts	Crude extracts	1. *In vitro* (radial growth inhibition) 2. *In vivo* (leaves kept in incubators)	*Botrytis cinerea* and *Pytophthora cinnamomi*	Jiménez et al. 2011
Commercial product Laminarin	NI	NI	Laminarin	*In vivo* (field)	Commercial strawberry plantations / grey mold	Meszka and Bielenin 2011
Ascophyllum nodosum (Ocrophyta)	NI	Extracts of *A. nodosum* [aqueous (ANE), chloroform (C-ANE) and ethylacetate fractions, (E-ANE)]	Extracts	*In vivo* (root treatment)	*Sclerotinia sclerotiorum*	Subramanian et al. 2011
Melanothamnus afaqhusainii (Rhodophyta), *Spatoglossum variabile*	Dry powder of seaweeds mixed with sandy loam soil	NI	Dry powders	*In vivo* (greenhouse and field)	Sunflower and tomato plants/*Fusarium solani*, *Rhizoctonia solani* and	Sultana et al. 2011

(*Contd.*)

Seaweed	Biomass	Solvent	Extract	Method	Target	Reference
(Ocrophyta) and *Halimeda tuna* (Chlorophyta)					*Macrophomina phaseolina*	Machado et al. 2011
Ulva spp., a mixture of several *Ulva* species (mostly *U. rigida*, formerly *U. armoricana*)	Fresh biomass	Material incubated in water and homogenate heated for 2 h at 90°C	A fraction containing most exclusively the sulfated polysaccharide	*In vivo*	*Nicotiana tabacum/ Erysiphe polygoni, E. necator* and *Sphaerotheca fuliginea*	Jaulneau et al. 2010
Kelpak (commercial product), seaweed concentrate from *Ecklonia maxima* (Ocrophyta)	NI	Water	Aqueous extract	*In vivo* (greenhouse)	Pepper plants (cv. Šorok-šari)—Verticillium wilt (*Verticillium dahliae*)	Rekanovic et al. 2010
Caulerpa cupressoides, C. racemosa, Codium isthmocladum, Halimeda cuneata (Chlorophyta), *Hypnea musciformis, Laurencia dendroidea, Ochtodes secundiramea* and *Pterocladiella capillacea* (Rhodophyta)	Fresh biomass	Dichloromethane/ methanol (4:1)	NI	*In vitro* (radial growth inhibition)	Papaya—Anthracnose (*Colletotrichum gloeosporioides*)	Machado et al. 2011
Sargassum swartzii (formerly *Sargassum wightii*) and *Padina pavonica* (Ochrophyta)	Fresh biomass	Hexane/chloroform/ methanol/water	Crude extract	*In vitro* (biological development)	*Dysdercus cingulatus* (Hemiptera: Pyrrhocoridae)	Asharaja and Sahayaraj 2013
Ulva fasciata and *Ulva lactuca* (Chlorophyta)	Fresh biomass	Hexane/chloroform/ methanol/water	Crude extract	*In vitro* (biological development)	*Dysdercus cingulatus* (Hemiptera: Pyrrhocoridae)	Asha et al. 2012

*NI – Not identified.

Bioautography and derivatization assays applied together provide ample information. It is possible to determine the retention factor of bioactive molecules and their chemical identity, which is necessary information for isolation through bioguided fractionation (Weller 2012).

In preliminary screenings, this qualitative test consists in the application of 10 µL aliquots containing 100 µg of extracts of TLC layer forming small spots. The chromatography can be developed in different mobile phases selected as a function of the polarity of the evaluated extract.

The fungal spore suspensions are then extracted in a solution containing glucose and salt (Rahalison et al. 1994) to a final concentration of 10^8 spores.mL^{-1}, which is sprayed on TLC with extract. The TLC plates are incubated for 72 h at 28°C. After incubation, clear inhibition zones appear against the dark backgrounds of the TLC plates. The retention factor is defined as the ratio of the distance from the origin to the location of the solvent front over the distance from the origin to the center of each spot (Machado et al. 2014b).

The effect of different mobile phases in separation and retention factor substances can be noted in a practical example. The macroalgae *Ochtodes secundiramea* (Gigartinales, Rhodophyta) can be observed with different visualization methods. Silica-gel TLC plates are revealed in bioautography with *Cladosporium sphaerospermum* and derivatization with *p*-hydroxybenzaldehyde (reagent with terpenoids affinity) followed by heating (Fig. 1) (Machado 2014, Machado et al. 2014a).

Natural fluorescent compounds can be detected by inspection under 366 nm. Compounds that absorb UV light at or near 254 nm can be detected on commercial layers containing a fluorescent indicator, because of a fluorescence quenching, characterized by dark zones on a fluorescent background. This method has the advantage of not altering the nature of the sample and is a tool for monitoring the extract's components (Rabel and Sherma 2017).

The practical example of this analytical approach is provided in Fig. 2. The antifungal terpenes verified in extract of red algae *Ochtodes secundiramea* by bioautography and derivatization can be monitored by nondestructive observation of TLC under 254 and 366 nm (Machado 2014).

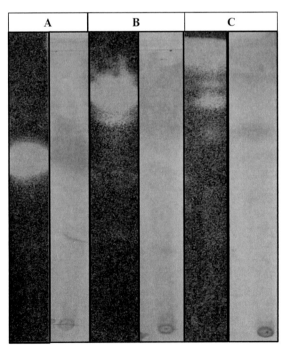

Figure 1. Bioautography (*Cladosporium sphaerospermum*) and derivatization (*p*-hydroxybenzaldehyde) of 100 µg of *Ochtodes secundiramea* extract, in response to different polarity of eluents in silica-gel TLC chromatography. A: chloroform, eluents ethyl acetate (EtOAc), methanol (MeOH) and 70:20:10. B: MeOH and dichloromethane (90:10). C: MeOH and dichloromethane (99:1) (Machado 2014).

The application of different mobile phases integrated with destructive and nondestructive methods of analysis provides better results in the separation of the different bioactive compounds from the extract. This result is the basis for preparative chromatography for bioguided fractionation by TLC approach (Rabel and Sherma 2017).

Quantitative data are accessed by analysis of limit of detection (LD) of activity, which can be assessed by a dose response curve, following the same pattern described for qualitative antifungal TLC. The LD of antifungal activity against *Cladosporium sphaerospermum* and *Colletotrichum lagenarium* (postharvest anthracnose of cucumbers) of *Ochtodes secundiramea* extract were analyzed by application of 10 µL solutions containing 0.5, 1, 5, 10, 25, and 50 µg of extract. Result determined LD of 1 µg for each fungus (Fig. 3) (Machado et al. 2014a, Machado 2014).

These qualitative and quantitative applications have multiple advantages: chemical analysis, nondestructive monitoring, rapid technique and relatively low cost. Bioautography in TLC is therefore a popular method for final clean-up of extractive fractions to obtain pure compounds or for screening of crude extracts for bioactivity and in the bioactivity-guided fractionation (Balouiri et al. 2016).

One limitation of the technique is to standardize the test for phytopathogenic fungi that produce low amounts of spores or that are hyaline. The better conditions for visualization and resolution of results are obtained with dark-colored spores.

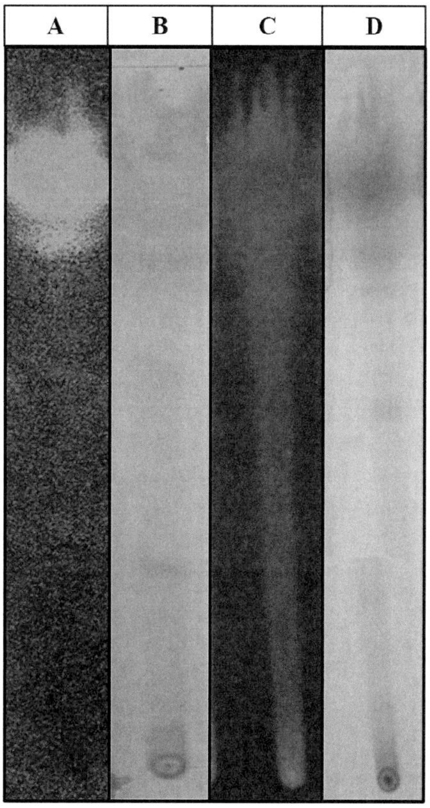

Figure 2. Chromatography in silica gel on TLC plates with fluorescent indicator of *Ochtodes secundiramea* extract eluted at MeOH and dichloromethane (90:10). Destructive analyses. A: Bioautography (*Cladosporium sphaerospermum*). B: Derivatization (*p*-hydroxybenzaldehyde), nondestructive method. C: observation at 366 nm. D: Observation at 254 nm (Machado 2014).

Color version at the end of the book

Diffusion and dilution methods

Many researchers use diffusion and dilution methods to evaluate the antimicrobial activity of seaweed extracts (Peres et al. 2012, Machado et al. 2014b,c, Machado et al. 2011, Ambika and Sujatha 2014, Begum et al. 2016, Bahammou et al. 2017, Reis et al. 2018). This method is well recognized in the field of phytopathology to test fungicides, plant extracts and other antimicrobial products. Diffusion assay, also called poisoned food technique (Ambika and Sujatha 2014, Begum

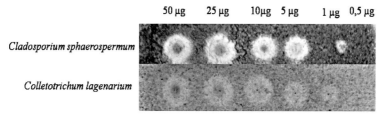

Figure 3. Bioautography of *Ochtodes secundiramea* extract for determination of LD antifungal activity against *Cladosporium sphaerospermum* and *Colletotrichum lagenarium* (Machado 2014, Machado et al. 2014a).

Color version at the end of the book

et al. 2016, Ammar et al. 2017), consists of procedures in which the extracts are added separately to the autoclaved and melted culture medium.

The extracts can be evaluated in different concentrations. When the studies are carried out with fungal pathogens, this medium is usually potato dextrose agar (PDA) that is poured into sterilized plates to solidify. A uniform disc of approximately 5 mm diameter from the test fungal colony is placed in the center of the solid PDA medium and the plates are incubated in a controlled chamber, with temperature and photoperiod regulated according to the fungus. The medium without the tested extracts or products serve as control. During the growth period, colony diameters are measured and the percentage of reduction in mycelial growth of the fungal pathogen is calculated in each treatment and in comparison with the control treatment.

In the scientific literature, we can find some adaptations of this method (Peres et al. 2012, Raj et al. 2016, Ibraheem et al. 2017). One of them is to use filter paper disk containing spore suspension of the fungus being studied. The paper disk is applied on the PDA with the treatments. At the end of the experiment, antifungal bioactivity of the extract is evaluated by measuring the zone of inhibition against the test fungus. Application of paper disk is applied on the PDA for screening of antifungal activity in seaweed extract. Conidial suspension applied to paper disks is shown in Fig. 4 (Machado et al. 2011).

Rather than evaluating the mycelial growth of the tested fungus, some studies evaluate the effect of extracts on its spore germination (Araújo et al. 2014). Studies that evaluate spore germination are more unusual than mycelial growth experiments. Despite this, conidial germination assays should be encouraged in addition to mycelial growth studies. Both of them are important to develop better understanding of the mode of action of the evaluated seaweed extracts. The culture medium used in this assay is water-agar amended with the tested extracts or not (control).

In vitro assay using the diffusion method has the advantage of being fast and easily reproducible. It is therefore useful for initial screening of extracts (Peres et al. 2012, Khalil et al. 2015). However,

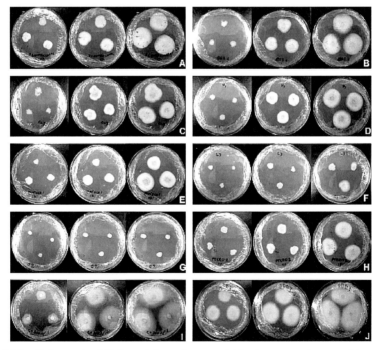

Figure 4. Mycelial growth of *Colletotrichum gloeosporioides* (papaya anthracnose) in response to different seaweed extracts after 2, 4 and 6 days. Conidial suspension was applied on paper disks (Machado et al. 2010). A to H: Crude extracts obtained from different seaweeds. I and J: solvent control and negative control respectively.

it is highly recommended that laboratory results be used to continue the tests in greenhouse and field conditions. In some situations, the results obtained in laboratory conditions are not the same as those obtained in greenhouse and/or field conditions (Esserti et al. 2017, Ramkisoon et al. 2017).

2.1.2 *In vivo assays*

Greenhouse

In the past few years, some researchers have been concerned about the reproducibility of *in vitro* results in greenhouse conditions. Because of this, they have carried out *in vivo* studies to evaluate the bioactivity of seaweed extracts in the control of plant diseases. Most of these studies have been carried out in greenhouse conditions with potted plants (Hernández-Herrera et al. 2014, Esserti et al. 2017, Ibraheem et al. 2017, Soliman et al. 2018).

The importance of working with the same pathogens and extracts *in vivo* that have earlier been tested *in vitro* lies in the fact that, when they are placed in contact with the plants, we may evaluate not only their direct antimicrobial activity. As marine algae extracts trigger plant defenses and enhance disease resistance, we can analyze both gene expression and protein activity to complement results obtained in greenhouse (Jaulneau et al. 2011, Abkhoo and Sabbagh 2016). Therefore, we can observe promising results *in vivo* in the control of pathogens by extracts that we did not find *in vitro* (Esserti et al. 2017). *In vivo* protective effect of seaweed extracts against pathogens is observed when they are applied before challenge inoculation (Esserti et al. 2017).

Severity and incidence of diseases are evaluated in greenhouse conditions. The researchers can evaluate different methods of application of seaweed extracts on plants, if it is in their interest (Jayaraman et al. 2011, Abkhoo and Sabbagh 2016, Raj et al. 2016, Esserti et al. 2017). Results obtained on plants maintained in these conditions are a good indication of how the extracts will behave in the final phase of tests, that is, in the field. However, it is important to note that greenhouse results may not be the same as those in the field (Sultana et al. 2011).

Another evaluation that should be done *in vivo* is the optimal concentration of the extract. In greenhouse condition, it is possible to observe whether the plant will show symptoms of phytotoxicity, and whether the extract could negatively affect its growth, for example (Esserti et al. 2017). Frequency of the extract application can be another parameter to be evaluated in greenhouse experiments.

Field

If the objective of the research is to generate a commercial product from seaweed extracts, it is essential to evaluate their efficacy on plants in the field. Statistical criteria should guide the procedures adopted in the field to test these extracts (Meszka and Bielenin 2011, Sultana et al. 2011, Bocek et al. 2012, Baloch et al. 2013).

It is important to have a well-prepared team to carry out and evaluate the experiments in the field. The team should be prepared to recognize the disease symptoms. In most situations, it is difficult to find appropriate areas (with disease incidence, for example) to install the assays and in which farmers are willing not to use other methods of disease control. In addition, to obtain reliable results it is of paramount importance to include a control plot without any treatment (including the tested extracts). These may be some of the reasons for the low number of studies with seaweed extracts in the field.

Postharvest

The postharvest phase includes harvesting, threshing and cleaning, drying, storage, transportation and commercialization of agricultural products (Kumar and Kalita 2017). Losses associated with this phase are multivariate and difficult to measure accurately, and that represents one of the most important challenges to research and development for global food security (Aulakh et al. 2013).

Postharvest disease represents one aspect of the complex network related to losses in food production, but a series of protocols are applied that can mitigate the losses. These methods, such as washing with chlorine, ethylene inhibitors/growth regulator, calcium application, thermal treatments, fumigation, irradiation, and waxing, are tested by scientific approach and systematically applied in agroindustry (El-Ramady et al. 2015).

Despite the existence of these guidelines of better practices, postharvest loss is still a significant challenge, caused by lack of skilled human capital, and the existing researcher's networks and agroindustry are not fully connected (Alamar et al. 2017). The research innovations to control postharvest diseases need to break the barrier of experimental assays to be converted into suitable concept proofs for scaled-up production and commercialization. These challenges call for multidisciplinary research and effective partnerships to catalyze innovative collaboration in this field (Arias-Bustos and Moors 2018, Alamar et al. 2017).

Application of natural extracts to biological control of postharvest diseases and increased shelf life has been considered an innovation to be combined with conventional treatments. There are review papers highlighting the extensive potential of extracts of plants and oils essential for the management of postharvest diseases (Singh and Sharma 2018, Ribes et al. 2017, Antunes and Cavaco 2010, Sharma and Alemwati 2010, Tripathi and Dubey 2004). In these reviews published since 2008, which focused on postharvest, the potential of macroalgae extracts was not addressed, while the surveys of antifungals and natural products derived from applied macroalgae are scarcely covered in postharvest applications (Hamed et al. 2018, El-Hossary et al. 2017, Arioli et al. 2015, Arunkumar et al. 2010). This shows a gap in the scientific communication of research related to the use of macroalgae in the control of postharvest diseases, especially in the production of specific reviews of the subject, as the results are fragmented in different areas of knowledge.

Research related to use of macroalgae in the control of postharvest diseases is presented in Table 1. The study was divided into three fronts: (1) screening of *in vitro* inhibitory action of postharvest fungi isolated from fruits; (2) application of macroalgae substances or extracts applied directly to fruits and monitoring of postharvest; and (3) integrated application of extracts with conventional methods of reducing disease incidence.

Much scientific knowledge was gathered, especially on the effect of the extract against the phytopathogens or on decay and ripening fruits or on storage and shelf life. However, these approaches are not sufficient, and more research is needed for development of a proof concept and commercial products.

In general, the studies and results are similar and should be changed to focus on innovation and integration of the research and production sectors with real application in the production chain (Henz 2017).

2.2　Insecticidal activity

Insecticides are chemical products commonly used for pest control that have diverse origins, including vegetable. Plants are a great source of studies in the search for new molecules with insecticidal effect and studies have shown that algae are also a promising substrate for obtaining secondary metabolites with insecticidal effect.

However, few studies have attempted to elucidate the use of seaweed extracts in the control of insect pests, whether in agricultural or forestry. Asharaja and Sahayaraj (2013) tested extracts of the seaweeds *Sargassum swartzii* (formerly *Sargassum wightii*) and *Padina pavonica* to control cotton leaf miner *Dysdercus cingulatus* and concluded that these algae have metabolites with potential to control the nymphs of this true bug. Extracts were tested in laboratory conditions to control this important pest of the cotton crop in India. Different concentrations of the extract (100, 200, 300, 400 and 500 ppm) were used against the third nymphal stage of *D. cingulatus*, in an oral toxicity bioassay. Insect nymphs were fed on cotton soaked in seaweed extract solution for 96 h continuously. The mortality of these nymphs was evaluated at the end. The authors concluded that the seaweed extracts negatively affected several biological parameters of the insect and can be a good compound insecticide (Asharaja and Sahayaraj 2013). In relation to sap-sucking insects researches about plant leaf immersion in seaweed extracts aiming to decrease feeding of these insects are rare.

Incorporation of seaweed extracts in the artificial diet of the young phase of insect pests, such as defoliating caterpillars, may be an interesting possibility to test the effect of these bioproducts.

Many studies have tested the effect of marine macroalgae extracts to control insects in the urban environment, mainly against mosquitoes transmitting diseases. Bianco et. al. (2013) evaluated extracts of the algae *Canistrocarpus cervicornis* (Ochrophyta, Phaeophyceae), *Laurencia dendroidea*, *Hypnea musciformis* (Rhodophyta) and *Chaetomorpha antennina* (Chlorophyta) and concluded that concentrations of 300 ppm presented mortality above 50% against larvae of the fourth stage of *Aedes aegypti*.

Cetin et al. (2010) tested acetone extract of *Caulerpa denticulata* (formerly *Caulerpa scalpelliformis* var. *denticulata*) against the 2nd and 3rd instars of *Culex pipiens* in concentrations ranging from 100 to 2000 parts per million (ppm). At 1200 ppm, the extract caused 70% of larval mortality at 24, 48 and 72 h exposure. The LC_{50} (lethal concentration) and LC_{90} values of *C. denticulata* were 338.91 and 1891.31 ppm, respectively. The authors concluded that seaweed species contain components with larvicidal properties against mosquitoes. Other studies have shown that many seaweed extracts have similar potential in insect control (Ali et al. 2013, Ravikumar et al. 2011, Yu et al. 2015).

Despite the great potential of marine macroalgae, there are no commercial seaweed-based products for use in agriculture for the protection of plants against insects, mites or other phytophagous organisms. This fact is explained in part by the lack of studies on this topic. In addition, there is a clear need for advanced laboratory tests on various species of phytophagous organisms that cause significant damage to cultivated plants. These advances in field tests are sought to understand the behavior of seaweed extracts that demonstrated efficiency under laboratory conditions.

It is important to highlight that the application of seaweed extracts under field conditions may not always control the pests, contradicting any promising results obtained under greenhouse or laboratory conditions. Processes of loss of bioactive compounds can occur by degradation and volatilization, reducing the efficiency of field application. Consequently, the potential of certain products for use in agriculture can be underestimated because of the researcher's concern to reduce the losses of the active compounds. Formulations prepared from these products can reduce the losses of important substances and possibly increase the efficiency of the control.

However, there are few studies aimed at the effect of seaweed extracts on arthropods or agricultural pests, especially in tropical regions. Considering that there is a great abundance of seaweed species in this region of the planet, there is much to be explored in searching for new active molecules against phytophagous insects and mites of economic importance.

3   Seaweed Compounds with Bioactivity

3.1   Antifungal activity

The occurrence of plant diseases caused by microorganisms is one of the main reasons for decline in productivity, as well as decline in the quality of the product, and partial or total loss of production. It may even render certain areas unproductive for cultivation. Around the world, losses caused by phytopathogens are estimated at 20% (Rommens and Kishore 2000). In order to control plant diseases and reduce losses of quality and production of agricultural products, different tools are used, such as the use of agrochemicals (Lopes et al. 2005). Nowadays, fungicides are essential for the effective control of plant diseases; however, the high frequency of application of these pesticides has frequently resulted in the emergence of pathogens resistant to the active principle. An alternative for treatment or control of bacteria and fungi in agriculture is the use of seaweeds (or macroalgae) or their isolated macromolecules, since the biotechnological potential of these organisms as a source of bioactive compounds against pathogenic microorganisms has been confirmed in different studies.

During the course of evolution, animals and plants have developed different strategies to survive in many different environments. Marine organisms have produced a diversity of compounds and secretions that allow them to adapt to the biological pressure. Some authors suggest that these bioactive components could be related to ecological competition with fungi, bacteria and other algae, forcing the production of secondary metabolites as a response to competition for space and predation, limiting the development and growth of other competitive microorganisms (Pérez et al. 2016). Furthermore, a number of seaweeds have been studied for their fungal associations (Singh et al. 2015). Additionally, age of the seaweed, climatic changes and environmental pressures such as salinity, light, temperature, and tide changes could contribute to secondary metabolite production (El-Amraoui et al. 2014, Zheng et al. 2005).

Bioactive compounds from marine algae have been studied for applications in agriculture due to their antimicrobial, insecticide, biostimulant and plant protectant activity (Aziz et al. 2003, Delattre et al. 2005, Chandía and Matsuhiro 2008, Paulert et al. 2009). A few studies have focused on the identification and characterization of isolated antifungal biomolecules and mainly the specific mode of action of these molecules (Machado et al. 2014a, b, Perry et al. 1991). In fact, studying the mechanism of action of molecules is an important requirement to make them good models for the development of new products to control plant diseases. The main components from seaweeds with antimicrobial activities studied to date could be characterized as polysaccharides (such as alginates, carrageenans, galactans, fucoidans/fucans, ulvans and others), lipids (fatty acids, sterols, phloroglucinol, beta-carotene and fucoxanthin), terpenes (such as neophytadiene and cycloeudesmol), brominated compounds, phenolic compounds (such as phlorotannins), pigments, proteins (lectins and antimicrobial and cyclic peptides), alkaloids, terpenes and halogenated compounds (Abouraïcha et al. 2017, Araújo et al. 2014, El-Hossary et al. 2017, Feliziani et al. 2015, Liu et al. 2014, Lopes et al. 2013, Machado et al. 2014a, b, Meszka and Bielenin 2011, Mickymaray and Alturaiki 2018, Pérez et al. 2016, Vera et al. 2011, Zerrifi et al. 2018).

Obviously, antifungal properties depend on the solvent and consequently the extract composition (Mubarak et al. 2018, Sampaio et al. 2016). Some researchers have studied the action mechanism of isolated compounds. For example, steroids can kill microorganisms interfering with the formation of fungal spores and mycelium (Subsiha and Subramoniam 2005). Terpenoids act by damaging organelles of the fungus and by inhibiting secretion of enzymes, and also damage its morphology (Martínez et al. 2014). Tannins can inhibit ergosterol activity and cause membrane cell disruption through inhibition of chitin synthesis (Hastuti et al. 2017). Phenolic compounds can interfere in protein synthesis and folding, leading to protein inactivation (Elansary et al. 2016). Furthermore, antimicrobial peptides can act on membrane or in specific cell pathways (Santos-Filho et al. 2015).

To date, the majority of studies have screened crude extracts for antimicrobial activities (Table 1). According to their solubility and polarity, solvents show different antimicrobial activity. The most commonly used solvents have been acetone, aqueous extracts, chloroform, dichloromethane, diethyl ether, methanol and ethanol (Engel et al. 2006, Hellio et al. 2000, Hellio et al. 2001, Machado et al. 2014a and b, Peres et al. 2012, Zerrifi et al. 2018). However, it is important to keep in mind that some factors could interfere with (or change) the antimicrobial response, through change of the secondary metabolism, given by geographical variation, season and other factors, as well as the extraction protocols to recover the active metabolites and assay methods (Salvador et al. 2007).

When working with plant (or seaweed) extracts it is important to consider that depending on their solvent solubility and polarity, the solution could show different biological activity, once solvents drive the polarity of the molecules of interest to be extracted. It is necessary to decide which classes of biomolecules and bioactivities hold interest for the researcher, then select the best extraction solvents for each species of macroalgae in order to optimize extraction of the maximum chemical compounds (Zerrifi et al. 2018). Another characteristic to take into account is the geographical variation and seasonality.

In this context, a deep study and standardization of algae harvesting become necessary to understanding how algae metabolites (and activity) may change around the world and around the year, according to different natural factors such as light, temperature, salinity, life stage, geographical location, seasonality, reproductive state and age of the seaweed. This information could give valuable evidence for harvesting algae for biotechnological applications, once the secondary metabolites have changed according to these factors (Salvador et al. 2007).

3.2 Insecticidal activity

Although only a small minority of insects are classified as pests, they destroy 10-14% of the world food supply (Bende et al. 2015, Boyer et al. 2012) and transmit a diverse range of human and animal diseases (Smith et al. 2013, Tedford et al. 2004, Windley et al. 2012). Despite the introduction of transgenic crops and other biological control methods, chemical insecticides continue to be the dominant approach for combating insects considered to be pests (Smith et al. 2013). Increasing incidence of insecticide-resistant species, together with the limited number of molecular targets in insects, has created greater urgency and driven improvements in pest management (Bates et al. 2005, Tedford et al. 2004).

Another important aspect to consider is that evidence suggests that long-term exposure to certain insecticides can be detrimental to the health of humans and other vertebrates. For example, chronic exposure to organophosphates, which have been widely used until recently, as well as pyrethroids, are often associated with decreased fertility and neurodevelopmental problems in children (Koureas et al. 2012). These concerns have resulted in restrictions on use of a large number of insecticidal compounds.

As rising population applies pressure on the agricultural sector and as the number of insecticides available shrinks, it is necessary to search for new insecticide compounds as well as study new molecular targets in insects. Fortunately, nature has a wide range of insecticidal compounds that have evolved within a huge diversity of seaweeds. Microalgae are a rich source of novel bioactive compounds that may find several applications in agriculture (Aziz et al. 2003, Chandía and Matsuhiro 2008, Delattre et al. 2005). Seaweeds have a high chemical diversity of active substances with protective functions against herbivores (Machado et al. 2014a).

As for antimicrobial activity, in most studies only extracts of algae have been tested for potential use in insect control (Asharaja and Sahayaraj 2013, Asha et al. 2012, Vera et al. 2011, Yu et al. 2015). To date, authors believe that in addition to secondary metabolites showing evidence of antimicrobial activity, sulfated polysaccharides have insecticide properties (Paulert et al. 2009). Elicitors such as laminarin, carrageenans and sulfated fucans also could be alternative tools for pest control in agriculture (Aziz et al. 2003, Cluzet et al. 2004, Klarzynski et al. 2003, Mercier et al. 2001).

With respect to insecticide activity, future studies need to explore, isolate and characterize the new molecules or novel mechanism of action with potential to be used in agriculture against phytophagous insects and mites of economic importance. Two objectives ought to be prioritized: the discovery of new molecules that are model of synthesis or semi-synthesis of new insecticide compounds that are more efficient, less toxic and less persistent in the environment, and the obtaining of plant insecticides for natural use in pest control.

4 Challenges and Opportunities in Development of Seaweed-based Products

The discovery of novel bioactive molecules is only a few decades old. It started in the 1980s and integrated applied research in chemistry and ecology (Hay 2014). Studies with metabolites derived from marine organisms have resulted in the isolation of 15,000 unpublished compounds with numerous applications (Cardozo et al. 2007). Research on natural products remains attractive, even after the development of organic synthesis processes, since they provide the discovery of new carbon skeletons and unusual functional groups (Silva 2009).

Studies on marine macroalgae to control plant pathogens and insects indicate a high number of seaweed species and bioactive extracts obtained for application in agriculture (Hamed et al. 2018), but this information is still presented in a dispersed and poorly standardized way.

The word-clouds obtained from the abstracts of the published articles related to the use of macroalgae to combat phytopathogens and phytophagous species show some remarkable differences (Fig. 5). First, it is clear the articles published to date have prioritized evaluation of the studied plant and the search for reduction of food losses (Fig. 5a). Comparing that with the present chapter, it is clear that we believe that for the emergence of new commercial products derived from macroalgae it is necessary to focus on the extracts and the assays (tests); the processing of the biomass and standardized test conditions must be prioritized in the applied research and development (Fig. 5b).

The results of plant extract in agriculture have been developed in a more organized way in relation to general procedures. Part of this is due to the existence of scientific reviews and integration of different areas of expertise. A list of items should be compiled to contribute to the definition of feasible candidates to develop new commercial seaweed-based products.

The major aspects in the research and development of plant extracts applied to agriculture involve: selection of species; procedures for extraction; mechanisms of fungicidal action; standardized assays for screening; application technology on crops; and health and environmental safety and legislation (Ncama et al. 2019). Seaweed potential is well defined in some scientific areas, especially on standardized assays for screening and application technology, while in the others there are only superficial assessments.

Some recommendations are pointed out to extend the commercial status of antifungal products based on plant extracts to seaweed extracts: innovations in bioactive extraction techniques and measurement techniques. Another relevant aspect is to link the research to the market, especially with regard to consumer perception and commercial scale production (Ncama et al. 2019).

These gaps are still usual for seaweed-based products. Another limitation is to obtain biomass with quality and in a constant frequency, since environmental factors modify the chemical composition and the activity of the extracts (Mohy El-Din and Mohyeldin 2018). Until now, most macroalgae production has been focused on human food, phycocolloids industry and biostimulant (FAO 2018).

Figure 5. Word-clouds elaborated with the summaries of the articles presented in Table 1 (a) and with the present chapter (b) (https://www.wordclouds.com/).

Despite that limitation in research and development, seaweed-based products can advance by applying the concepts of biorefinery, allowing the extraction of multiple products of cultivated macroalgae, and promoting efficient use of resources (Balina et al. 2017). Other aspects to be taken into account for the development of these products are: knowledge about the active concentration against fungi; acceptance by consumers; feasibility techniques for application; minimal interference on product quality; benefit or absence of threat to consumers; abundant source for continued supply; non–technology-dependent or expansive procurement; environmentally friendly process; and compliance with regulations for food additives (Ncama et al. 2019).

This approach needs to be correlated to the framework established in programs such as the Risk Assessment for United States Environmental Protection Agency (EPA) or other recognized international regulatory institutions so that the process can be validated and the research results in a safe, commercially viable product (Lewis et al. 2016).

Table 2. Challenges and opportunities to develop seaweed-based products for plant disease control

Challenges	*Opportunities*
Standardization of biomass achieved	Alternative or complement to conventional agrochemicals
Standardization of biomass processing	Discovery of new molecules with different mechanisms of action
Standardization of extraction	Environmentally friendly insecticides and fungicide
Standardization of assays	Integrating different sectors of the bio-economy (e.g., agriculture, fisheries, pharmaceutical)
Toxicity aspects for consumer safety	Seaweed-based products as stimuli of organic agriculture
Understanding of side effects	Startup and partnership development with companies for on-demand research and development
Assurance of reproducibility of laboratory and greenhouse results versus field conditions	Formulation or adoption of a standardized guideline and legislation for research and development of seaweed-based products (government regulatory agencies)
Diffuse scientific knowledge	Biodegradable products and circular methodology
Development of technology and strategies for application	Social innovation through capacity-building
Long-term stability of products in environmental conditions	

Innovations to use seaweed to control plant diseases and pests in agriculture are being developed. An edible polysaccharide-based biofilm (Cian et al. 2014) can be applied in the postharvest of fruits and naturally presents beneficial properties such as antioxidant action and also antimicrobial activity. Seaweed extracts can be enriched with other bioactive molecules. Another promising area is related to the application of alginate extracted from macroalgae for use as a vehicle for encapsulating agents for the biocontrol of pests or crop diseases (De Corato et al. 2018).

5 Conclusion

The evolution of knowledge related to seaweed applied to control plant pathogens and crop pests needs to integrate these aspects and redefine the processes of research and development to create products and practices that promote the fulfillment of real demands of different agricultural systems, sustainability and food security.

References Cited

Abbassy, M.A., M.A. Marzouk, E.I. Rabea and A.D. Abd-Elnabi. 2014. Insecticidal and fungicidal activity of *Ulva lactuca* Linnaeus (Chlorophyta) extracts and their fractions. Annu. Res. Rev. Biol. 4: 2252–2262.

Abkhoo, J. and S.K. Sabbagh. 2016. Control of *Phytophthora melonis* damping-off, induction of defense responses, and gene expression of cucumber treated with commercial extract from *Ascophyllum nodosum*. J. Appl. Phycol. 28: 1333–1342.

Abouraïcha, E.F., Z. El-Alaoui-Talibi, A. Tadlaoui-Ouafi, R. El-Boutachfaiti, E. Petit, A. Douira and C. El-Modafar. 2017. Glucuronan and oligoglucuronans isolated from green algae activate natural defense responses in apple fruit and reduce postharvest blue and gray mold decay. J. Appl. Phycol. 29: 471–480.

Aguado, A., A.M. Pastrana, B. Los Santos, F. Romero, M.C. Sanchéz and N. Capote. 2014. Efficiency of natural products in the control of *Colletotrichum acutatum* monitored by real-time PCR. Acta Hortic. 1049: 329–334.

Ali, M.Y.S., S. Ravikumar and J.M. Beula. 2013. Mosquito larvicidal activity of seaweeds extracts against *Anopheles stephensi*, *Aedes aegypti* and *Culex quinquefasciatus*. Asian Pac. J. Trop. Dis. 3: 196–201.

Ambika, S. and K. Sujatha. 2014. Comparative studies on brown, red and green alga seaweed extracts for their antifugal activity against *Fusarium oxysporum* f. sp. *udum* in Pigeon pea var. CO (Rg)7 (*Cajanuscajan* (L.) Mills.). Journal of Biopesticides 7: 167–176.

Ambika, S. and Sujatha, K. 2015. Antifungal activity of aqueous and ethanol extracts of seaweeds against sugarcane red rot pathogen (*Colletotrichum falcatum*). Sci. Res. Essays 10(6): 232–235.

Ammar, N., H. Jabnoun-Khiareddine, B. Mejdoub-Trabelsi, A. Nefzi, M.A. Mahjoub and M. Daami-Remadi. 2017. Pythium leak control in potato using aqueous and organic extracts from the brown alga *Sargassum vulgare* (C. Agardh, 1820). Postharvest Biol. Technol. 130: 81–93.

Antunes, M.D.C. and A.M. Cavaco. 2010. The use of essential oils for postharvest decay control. A review. Flavour Fragr. J25: 351–366.

Araújo, L., A.E. Gonçalves and M.J. Stadnik. 2014. Ulvan effect on conidial germination and appressoria formation of *Colletotrichum gloeosporioides*. Phytoparasitica 42: 631–640.

Arias-Bustos, C. and E.H.M. Moors. 2018. Reducing post-harvest food losses through innovative collaboration: insights from the Colombian and Mexican avocado supply chains. J. Clean Prod. 199: 1020–1034.

Arioli, T., S.W. Mattner and P.C. Winberg. 2015. Applications of seaweed extracts in Australian agriculture: past, present and future. J. Appl. Phycol. 27: 2007–2015.

Arunkumar, K.S.R., S.R. Sivakumar and R. Rengasamy. 2010. A review on bioactive potential in seaweeds (marine macroalgae): a special emphasis on bioactive of seaweeds against plant pathogens. Asian J. Plant Sci., 9: 227–240.

Asharaja, A. and K. Sahayaraj. 2013. Screening of insecticidal activity of brown macroalgal extracts against *Dysdercus cingulatus* (Fab.) (Hemiptera: Pyrrhocoridae). J. Biopesticides. 6: 193–203.

Asha, A., J.M. Rathi, D. Patric Raja and K. Sahayaraj. 2012. Biocidal activity of two marine green algal extracts against third instar nymph of *Dysdercus cingulatus* (Fab.) (Hemiptera: Pyrrhocoridae) J. Biopest. 5: 129–134.

Aulakh, J., A. Regmi, J.R. Fulton and C. Alexander. 2013. Estimating post-harvest food losses: developing a consistent global estimation framework. *In*: Proceedings of the Agricultural & Applied Economics Association's AAEA & CAES Joint Annual Meeting, August 4–6, Washington, DC, USA.

Alamar, M.D.C., N. Falagán, E. Aktas and L.A. Terry. 2017. Minimising food waste: a call for multidisciplinary research. J. Sci. Food Agric. 98: 8–11.

Aziz, A., B. Poinssot, X. Daire, M. Adrian, A. Bézier, B. Lambert, J.M. Joubert and A. Pugin. 2003. Laminarin elicits defense responses in grapevine and induces protection against *Botrytis cinerea* and *Plasmopara viticola*. Mol. Plant-Microbe Interact. 16: 1118–1128.

Bahammou, N., O. Cherifi, H. Bouamama, K. Cherifi, T. Moubchir and M. Bertrand. 2017. Postharvest control of gray mold of tomato using seaweed extracts. Journal of Materials and Environmental Sciences 8: 831–836.

Balina, K., F. Romagnoli and D. Blumberga. 2017. Seaweed biorefinery concept for sustainable use of marine resources. Energy Procedia 128: 504–511.

Balouiri, M., M. Sadiki and S.K. Ibnsouda. 2016. Methods for *in vitro* evaluating antimicrobial activity: a review. J. Pharm. Biomed. Anal. 6: 71–79.

Baloch, G.N., S. Tariq, S. Ehteshamul-Haque, M. Athar, V. Sultana and J. Ara. 2013. Management of root diseases of eggplant and watermelon with the application of asafetida and seaweeds. J. Appl. Bot. Food Qual. 86: 138–142.

Bates, S.L., J.Z. Zhao, R.T. Roush and A.M. Shelton. 2005. Insect resistance management in GM crops: past, present and future. Nat. Biotechnol. 23: 57–62.

Begum, A.J., P. Selvaraju and A. Vijayakumar. 2016. Evaluation of antifungal activity of seaweed extract (*Turbinaria conoides*) against *Fusarium oxysporum*. J. Appl. & Nat. Sci. 8: 60–62.

Bende, N.S., S. Dziemborowicz, V. Herzig, V. Ramanujam, G.W. Brown, F. Bosmans, G.M. Nicholson, G.F. Kinga and M. Mobli. 2015. The insecticidal spider toxin SFI1 is a knottin peptide that blocks the pore of insect voltage-gated sodium channels via a large β-hairpin loop. FEBS J. 282: 904–920.

Bianco, E.M., L. Pires, G.K.N. Santos, K.A. Dutra, T.N.V. Reis, E.R.T.P.P. Vasconcelos, A.L.M. Concentino and D.M.A.F. Navarro. 2013. Larvicidal activity of seaweeds from northeastern Brazil and of a halogenated sesquiterpene against the dengue mosquito (*Aedes aegypti*). Ind. Crops Prod. 43: 270–275.

Bocek, S., P. Salas, H. Saasková and J. Mokricková. 2012. Effect of alginure® (seaweed extract), mycosin® (sulfuric clay) and polyversum® (*Pythium oligandrum* Drenchs.) on yield and disease control in organic strawberries. Acta Univ. Agric. Silvic. Mendel. Brun. 8: 19–28.

Boyer, S., H. Zhang and G. Lempérière. 2012. A review of control methods and resistance mechanisms in stored-product insects. Bull. Entomol. Res. 102: 213–229.

Cardozo, K.H.M., T. Guaratini, M.P. Barros, V.R. Falcão, A.P. Tonon, N.P. Lopes, S. Campos, M.A. Torres, A.O. Souza, P. Colepicolo and E. Pinto. 2007. Metabolites from algae with economic impact. Comp. Biochem. Physiol. C. Toxicol. Pharmacol. 146: 60–78.

Cetin, H., M. Gokoglu and E. Oz. 2010. Larvicidal activity of the extract of seaweed, *Caulerpa scalpelliformis*, against *Culex pipiens*. J. Am. Mosq. Control Assoc. 26: 433–435.

Chandía, N.P. and B. Matsuhiro. 2008, Characterization of a fucoidan from *Lessonia vadosa* (Phaeophyta) and its anticoagulant and elicitor properties. Int. J. Biol. Macromol. 42: 235–240.

Cian, R.E., P.R. Salgado, S.R. Drago, R.J. González and A.N. Mauri. 2014. Development of naturally activated edible films with antioxidant properties prepared from red seaweed *Porphyra columbina* biopolymers. Food Chem. 146: 6–14.

Cluzet, S., C. Torregrossa, C. Jacquet, C. Lafitte, J. Fournier, L. Mercier, S. Salamagne, X. Briand, M.T. Esquerré-Tugayé and B. Dumas. 2004. Gene expression profiling and protection of *Medicago truncatula* against a fungal infection in response to an elicitor from green algae *Ulva* spp. Plant Cell Environ. 27: 917–928.

Copatti, C.E., R.K. Marcon and M.B. Machado 2013. Avaliação de dano de Sitophilus zeamais, Oryzaephilus surinamensis e Laemophloeus minutus em grãos de arroz armazenados. Rev. Bras. Eng. Agri. Ambi. 17(8): 855–860.

De Corato, U., R. Salimbeni, A. De Pretis, N. Avella and G. Patruno. 2017. Antifungal activity of crude extracts from brown and red seaweeds by a supercritical carbon dioxide technique against fruit postharvest fungal diseases. Postharvest Biol. Technol. 131: 16–30.

De Corato, U., R. Salimbeni, A. De Pretis, N. Avella and G. Patruno. 2018. Use of alginate for extending shelf life in a lyophilized yeast-based formulate in controlling green mould disease on citrus fruit under postharvest condition. J. Food Pack. Shelf Life 15: 76–86.

Delattre, C., P. Michaud, B. Courtois and J. Courtois. 2005. Oligosaccharides engineering from plants and algae applications in biotechnology and therapeutics. Minerva Biotechnol. 17: 107–117.

El-Amraoui, B., M. El-Amraoui, N. Cohen and A. Fassouane. 2014, Anti-Candida and anti-Cryptococcus antifungal produced by marine microorganisms. J. Mycol. Med. 24: 149–153.

El-Hossary, E.M., C. Cheng, M.M. Hamed, A.N. El-Sayed Hamed, K. Ohlsen, U. Hentschel and U.R. Abdelmohsen. 2017. Antifungal potential of marine natural products. Eur. J. Med. Chem. 126: 631–651.

El-Ramady, H.R., E. Domokos-Szabolcsy, N.A. Abdalla, H.S. Taha and M. Fári. 2015. Postharvest management of fruits and vegetables storage. pp. 65–152. *In*: Lichtfouse, E. (ed.). Sustainable Agriculture Reviews, Volume 15. Springer International Publishing, New York.

Elansary, H.O., J. Norrie, H.M. Ali, M.Z. Salem, E.A. Mahmoud and K. Yessoufou. 2016. Enhancement of *Calibrachoa* growth, secondary metabolites and bioactivity using seaweed extracts. BMC Complement. Altern. Med. 16: 341.

Engel, S., M.P. Puglisi, P.R. Jensen and W. Fenical. 2006. Antimicrobial activities of extracts from tropical Atlantic marine plants against marine pathogens and saprophytes. Mar. Biol. 149: 991–1002.

Esserti, S., A. Smaili, L.A. Rifai, T. Koussa, K. Makroum, M. Belfaiza, E.M. Kabil, L. Faize, L. Burgos, N. Albuquerque and M. Faize. 2017. Protective effect of three brown seaweed extracts against fungal and bacterial diseases of tomato. J. Appl. Phycol. 29: 1081–1093.

FAO. 2018. The state of world fisheries and aquaculture. Food and Agriculture Organization of the United Nations, Rome, 227 pp.

Feliziani, E., L. Landi and G. Romanazzi. 2015. Preharvest treatments with chitosan and other alternatives to conventional fungicides to control postharvest decay of strawberry. Carbohydr. Polym. 132: 111–117.

Galal, H.R.M., W.M. Salem and F.N. El-Deen. 2011. Biological control of some pathogenic fungi using marine algae extracts. Research Journal of Microbiology 6: 645–657.

Gustavsson, J., C. Cederberg, U. Sonesson, R. van Otterdijk and A. Meybeck. 2011. Global Food Losses and Food Waste. Food and Agriculture Organization of the United Nations, Rome, Italy.

Hahn-Deinstrop, E. 2007. Applied Thin-Layer Chromatography: Best Practice and Avoidance of Mistakes, 2nd Ed. Wiley-VHC, Weinheim, Germany, 330 pp.

Hamed, S.M., A.A. Abd El-Rhman, N. Abdel-Raouf and I.B.M. Ibraheem. 2018. Role of marine macroalgae in plant protection and improvement for sustainable agriculture technology. Beni-Suef Univ. J. Appl. Sci. 7: 104–110.

Hastuti, S.U., P.I.Y. Ummah and N.H. Khasanah. 2017. Antifungal activity of *Piper aduncum* and *Peperomia pellucida* leaf ethanol extract against *Candida albicans*. AIP Conf. Proc. 1844: 020006-(1-4).

Hay, M.E. 2014. Challenges and opportunities in marine chemical ecology. J. Chem. Ecol. 40: 216–217.

Hellio, C., G. Bremer, A.M. Pons, Y. Le Gal and N. Bourgougnon. 2000. Inhibition of the development of microorganisms (bacteria and fungi) by extracts of marine algae from Brittany, France. Appl. Microbiol. Biotechnol. 54: 543–549.

Hellio, C., D. De La Broise, L. Dufossé, Y. Le Gal and N. Bourgougnon. 2001. Inhibition of marine bacteria by extracts of macroalgae: potential use for environmentally friendly antifouling paints. Mar. Environ. Res. 52: 231–247.

Henz, G.P. 2017. Postharvest losses of perishables in Brazil: what do we know so far? Hortic. Bras. 35: 6–13.

Hernández-Herrera, R.M., G. Virgen-Calleros, M. Ruiz-López, J. Zañudo-Hernández, J.P. Délano-Frier and C. Sánchez-Hernández. 2014. Extracts from green and brown seaweeds protect tomato (*Solanum lycopersicum*) against the necrotrophic fungus *Alternaria solani*. J. Appl. Phycol. 26: 1607–1614.

Ibraheem, B.M., S.M. Hamed, A.A.A. Elrhman, F.M. Farag and N. Abdel-Raouf. 2017. Antimicrobial activities of some brown macroalgae against some soil borne plant pathogens and *in vivo* management of *Solanum melongena* root diseases. Aust. J. Basic Appl. Sci. 11: 157–168.

Jayaraman, J., J. Norrie and Z.K. Punja. 2011. Commercial extract from the brown seaweed *Ascophyllum nodosum* reduces fungal diseases in greenhouse cucumber. J. Appl. Phycol. 23: 353–361.

Jaulneau, V., C. Lafitte, C. Jacquet, S. Fournier, S. Salamagne, X. Briand, M.T. Esquerré-Tugayé and B. Dumas. 2010. Ulvan, a sulfated polysaccharide from green algae, activates plant immunity through the jasmonic acid signaling pathway. J. Biomed. Biotech. 525291.

Jaulneau, V., C. Lafitte, M.F. Corio-Costet, M.J. Stadnik, S. Salamagne, X. Briand, M.T. Esquerré-Tugayé and B. Dumas. 2011. An *Ulva armoricana* extract protects plants against three powdery mildew pathogens. Eur. J. Plant Pathol. 131: 393–401.

Jiménez, E., F. Dorta, C. Medina, A. Ramírez, I. Ramírez and H. Peña-Cortés. 2011. Anti-phytopathogenic activities of macro-algae extracts. Mar. Drugs 9: 739–756.

Kagan, I.A. and M.D. Flythe. 2014. Thin-layer chromatographic (TLC) separations and bioassays of plant extracts to identify antimicrobial compounds. J. Vis. Exp. 85: 1–8.

Khalil, A.M., I.M. Daghman and A.A. Fady. 2015. Antifungal potential in crude extracts of five selected brown seaweeds collected from the Western Libya Coast. J. Microbiol. Modern. Tech. 1: 1–8.

Klarzynski, O., V. Descamps, B. Plesse, J.C. Yvin, B. Kloareg and B. Fritig. 2003. Sulphated fucan oligosaccharides elicit defense responses in tobacco and local and systemic resistance against tobacco mosaic virus. Mol. Plant Microbe Interact. 16: 1156–1122.

Koureas, M., A. Tsakalof, A. Tsatsakis and C. Hadjichristodoulou. 2012. Systematic review of biomonitoring studies to determine the association between exposure to organophosphorus and pyrethroid insecticides and human health outcomes. Toxicol. Lett. 210: 155–168.

Kumar, D. and P. Kalita. 2017. Reducing postharvest losses during storage of grain crops to strengthen food security in developing countries. Foods 6(8): 1–22.

Lewis, K.A., J. Tzilivakis, D.J. Warner and A. Green. 2016. An international database for pesticide risk assessments and management. Hum. Ecol. Risk. Assess. 22: 1050–1064.

Liu, M., G. Wang, L. Xiao, X. Xu, X. Liu, P. Xu and X. Lin. 2014. Bis (2,3-dibromo-4,5-dihydroxybenzil) ether, a marine algae derived bromophenol, inhibits the growth of *Botrytis cinerea* and interacts with DNA molecules. Mar. Drugs 12: 3838–3851.

Lopes, G., E. Pinto, P.B. Andrade and P. Valentão. 2013. Antifungal activity of phlorotannins against dermatophytes and yeasts: approaches to the mechanism of action and influence on *Candida albicans* virulence factor. PLoS One 8: e72203.

Lopes, I., D.J. Baird and R. Ribeiro. 2005. Genetically determined resistance to lethal levels of copper by *Daphnia longispina*: association with sublethal response and multiple/coresistance. Environ. Toxicol. Chem. 24: 1414–1419.

Machado, L.P. 2014. Determinação dos metabólitos com atividade antifúngica e cultivo em biorreatores para obtenção de bioativos de *Ochtodes secundiramea* e de linhagens pigmentares de *Hypnea musciformis* (Rhodophyta). PhD Thesis, Instituto de Botânica da Secretaria de Estado do Meio Ambiente, São Paulo, 203 pp.

Machado, L.P., W.M.S. Bispo, S.T. Matsumoto, R.B. Santos, F.O. Reis and L.F.G. Oliveira Jr. 2011. Triagem de macroalgas com potencial antifúngico no controle in vitro da antracnose do mamoeiro (*Carica papaya* L.). Curr. Agri.l Sci. Technol. 17: 463–467.

Machado, L.P., S.T. Matsumoto, C.M. Jamal, M.B. Silva, D.C. Cruz, P. Colepicolo, L.R. Carvalho and N.S. Yokoya. 2014a. Chemical analysis and toxicity of seaweed extracts with inhibitory activity against tropical fruit anthracnose fungi. J. Sci. Food Agr. 94: 1739–1744.

Machado, L.P., L.R. Carvalho, M.C.M. Young, L. Zambotti-Villela, P. Colepicolo, D.X. Andreguetti and N.S. Yokoya. 2014b. Comparative chemical analysis and antifungal activity of *Ochtodes secundiramea* (Rhodophyta) extracts obtained using different biomass processing methods. J. Appl. Phycol. 26: 2029–2035.

Machado, L.P., S.T. Matsumoto, G.R.F. Cuzzuol and L.F.G. Oliveira Jr. 2014c. Influence of laboratory cultivation on species of Rhodophyta physiological evaluations and antifungal activity against phytopathogens. Rev. Ciênc. Agron. 45: 52–61.

Martínez, A., N. Rojas, L. García, F. González, M. Domínguez and A. Catalán. 2014. *In vitro* activity of terpenes against *Candida albicans* and ultrastructural alterations. Oral Surg. Oral Med. Oral Pathol. Oral Radiol. 118: 553–559.

Melo, T.A., A.A. Sousa, T.Y.O. Sousa, I.M.R.S. Serra and S.F. Pascholati. 2018. Effect of *Ascophyllum nodosum* seaweed extract on post-harvest 'Tommy Atkins' mangoes. Rev. Bras. Frut. 40: e-621.

Mercier, L., C. Lafitte, G. Borderies, X. Briand, M.T. Esquerré-Tugayé and J. Fournier. 2001. The algal polysaccharide carrageenans can act as an elicitor of plant defense. New Phytol. 149: 43–51.

Meszka, B. and A. Bielenin. Activity of laminarin in control of strawberry diseases. 2011. Phytopathologia 62: 15–23.

Mickymaray, S. and W. Alturaiki. 2018. Antifungal efficacy of marine macroalgae against fungal isolates from bronchial asthmatic cases. Molecules 23: e3032.

Mohy El-Din, S.M. and M.M. Mohyeldin. 2018. Component analysis and antifungal activity of the compounds extracted from four brown seaweeds with different solvents at different seasons. J. Ocean U. China 17: 1178–1188.

Mossi, A.J., V. Astolfi, G. Kubiak, L. Lindomar Lerin, C. Zanella, G. Toniazzo, D. Oliveira, H. Treichel, I.A. Devilla, R. Cansian and R. Restello. 2011. Insecticidal and repellency activity of essential oil of Eucalyptus sp. against Sitophilus zeamais Motschulsky (Coleoptera, Curculionidae). J. Sci. Food. Agri. 91(2): 273–277.

Mubarak, Z., A. Humaira, B.A. Gani and Z.A. Muchlisin. 2018. Preliminary study on the inhibitory effect of seaweed *Gracilaria verrucosa* extract on biofilm formation of *Candida albicans* cultured from the saliva of a smoker. F1000 Research 7: 684.

Ncama, K., A. Mditshwa, S.Z. Tesfay, N.C. Mbili and L.S. Magwaza. 2019. Topical procedures adopted in testing and application of plant-based extracts as bio-fungicides in controlling postharvest decay of fresh produce. Crop Protec. 115: 142–151.

Oliveira, C.M., A.M. Auad, S.M. Mendes and M.R. Frizzas. 2014. Crop losses and economic impact of insect pests on Brazilian agriculture. Crop Protec. 56: 50–54.

Omar, A.El-Din.K., 2014. Use of seaweed extract as a promising post-harvest treatment on Washington Navel orange (*Citrus sinensis* Osbeck). Biol. Agric. Horticult. 30(3): 198–210.

Paulert, R., V. Talamini, J.E.F. Cassolato, M.E.R. Duarte, M.D. Noseda, A. Smania Jr. and M.J. Stadnik. 2009. Effects of sulfated polysaccharide and alcoholic extracts from green seaweed *Ulva fasciata* on anthracnose severity and growth of common bean (*Phaseolus vul*garis L.). J. Plant Dis. Protect. 116: 263–270.

Peres, J.C.F., L.R. Carvalho, E. Gonçalez, L.O.S. Berian and J.D. Felicio. 2012. Evaluation of antifungal activity of seaweed extracts. Ciênc. Agrotec. 3: 294–299.

Pérez, M.J., E. Falqué and H. Domínguez. 2016. Antimicrobial action of compounds from marine seaweed. Mar. Drugs 14: e52.

Perry, N.B., J.W. Blunt and M.H. Munro. 1991. A cytotoxic and antifungal 1,4-naphthoquinone and related compounds from a New Zealand brown algae, *Landsburgia quercifolia*. J. Nat. Prod. 54: 978–985.

Rabel, F. and J. Sherma. 2017. Review of the state of the art of preparative thin-layer chromatography. J. Liq. Chromatogr. Relat. Technol. 40: 165–176.

Rahalison, L., M. Hamburger, M. Monod, E. Frenk and K. Hostettmann. 1994. Antifungal tests in phytochemical investigations comparison of bioautographic methods using phytopathogenic and human pathogenic fungi. Planta Med. 60: 41–44.

Raj, T.S., K.H. Graf and H.A. Suji. 2016. Bio chemical characterization of a brown seaweed algae and its efficacy on control of rice sheath blight caused by *Rhizoctonia solani* Kuhn. Int. J. Trop. Agric. 34: 429–439.

Ramkisoon, A., A. Ramsubhag and J. Jayaraman. 2017. Phytoelicitor activity of three Caribbean seaweed species on suppression of pathogenic infections in tomato plants. J. Appl. Phycol. 29: 3235–3244.

Ravikumar, S., M.Y.S. Ali and J.M. Beula. 2011. Mosquito larvicidal efficacy of seaweed extracts against dengue vector of *Aedes aegypti*. Asian Pac. J. Trop. Biomed. 1: 143–146.

Reis, R.P., A.A. Carvalho Junior, A.P. Facchinei, A.C.S. Calheiros and B. Castelar. 2018. Direct effects of ulvan and a flour produced from the green alga *Ulva fasciata* Delile on the fungus *Stemphylium solani* Weber. Algal Res. 30: 23–27.

Rekanovic, E., I. Potocnik, S. Milijasevic-Marcicm, M. Stepanovic, B. Todorovic and M. Mihajlovic. 2010. Efficacy of seaweed concentrate from *Ecklonia maxima* (Osbeck) and conventional fungicides in the control of Verticillium wilt of pepper. Pestic. Phytomed. 25: 319–324.

Ribes, S., A. Fuentes, P. Talens and J.M. Barat. 2017. Prevention of fungal spoilage in food products using natural compounds: a review. CRC Crit. Rev. Food. Sci. Nutr. 58: 2002–2016.

Roberti, R., H. Righini and C.P. Reyes. 2016. Activity of seaweed and cyanobacteria water extracts against *Podosphaera xanthii* on zucchini. Ital. J. Mycol. 45: 66–77.

Rommens, C.M. and G.M. Kishore. 2000. Exploiting the full potential of disease-resistance genes for agricultural use. Curr. Opin. Biotechnol. 11: 120–125.

Salvador, N., A.G. Garreta, L. Lavelli and M.A. Ribera. 2007. Antimicrobial activity of Iberian macroalgae. Sci. Mar. 71: 101–113.

Sampaio, B.L., R. Edrada-Ebel and F.B. Da Costal. 2016. Effect of the environment on the secondary metabolic profile of *Tithonia diversifolia*: a model for environmental metabolomics of plants. Sci. Rep. 6: 29265.

Santos-Filho, N.A., E.N. Lorenzon, M.A. Ramos, C.T. Santos, J.P. Piccoli, T.M. Bauab, A.M. Fusco-Almeida and E.M. Cilli. 2015. Synthesis and characterization of an antibacterial and non-toxic dimeric peptide derived from the C-terminal region of Bothropstoxin-I. Toxicon. 103: 160–168.

Sharma, R.R. and P. Alemwati. 2010. Natural products for postharvest decay control in horticultural produce: a review. Stewart Postharvest Rev. 4: 1–9.

Sharma, R.R., S. Reddy and M. Jhalegar. 2014. Pre-harvest fruit bagging: a useful approach for plant protection and improved post-harvest fruit quality – a review. J. Hortic. Sci. Biotechnol. 89: 101–113.

Sherma, J. 2017. Review of thin-layer chromatography in pesticide analysis: 2014–2016. J. Liq. Chromatogr. Relat. Technol. 40: 226–238.

Shuping, D. and J.N. Eloff. 2017. The use of plants to protect plants and food against fungal pathogens: a review. Afr. J. Tradit. Complement. Altern. Med. 14: 120–127.

Soliman, A.S., A.Y. Ahmed, S.E. Abdel-Ghafour, M.M. El-Sheekh and H.M. Sobhy. 2018. Antifungal bio-efficacy of the red algae *Gracilaria confervoides* extracts against three pathogenic fungi of cucumber plant. Middle East J. Appl. Sci. 8: 727–735.

Subramanian, S., J.S. Sangha, B.A. Gray, R.P. Singh, D. Hiltz, A.T. Critchley and B. Prithiviraj. 2011. Extracts of the marine brown macroalga, *Ascophyllum nodosum*, induce jasmonic acid dependent systemic resistance in *Arabidopsis thaliana* against *Pseudomonas syringae* pv. *tomato* DC3000 and *Sclerotinia sclerotiorum*. Eur. J. Plant Pathol. 131: 237–248.

Sultana, V., G.N. Baloch, J. Ara, S. Ehteshamul-Haque, R.M. Tariq and M. Athar. 2011. Seaweeds as an alternative to chemical pesticides for the management of root diseases of sunflower and tomato. J. Appl. Food Qual. 84: 162–168.

Silva, P., F. Chantal, L. Barros, I.C.F.R. Ferreira, L. Pereira and T. Gonçalves. 2018. The antifungal activity of extracts of *Osmundea pinnatifida*, an edible seaweed, indicates its usage as a safe environmental fungicide or as food additive preventing post-harvest fungal food contamination. Food Funct. 9: 6188–6196.

Silva, P.M. 2009. Atividades biológicas de extratos de algas marinhas brasileiras. PhD Thesis in Biochemistry. Universidade de São Paulo, USP, São Paulo, 91 pp.

Singh, D. and R.R. Sharma. 2018. Postharvest disinfection of fruits and vegetable: an overview. pp. 1–52. *In*: Siddiqui, M.W. (ed.). Postharvest Disinfection of Fruits and Vegetables. Academic Press. An Imprint of Elsevier, London.

Singh, R.P., P. Kumari and C.R. Reddy. 2015. Antimicrobial compounds from seaweeds-associated bacteria and fungi. Appl. Microbiol. Biotechnol. 99: 1571–1586.

Smith, J.J., V. Herzig, G.F. King and P.F. Alewood. 2013. The insecticidal potential of venom peptides. Cell. Mol. Life Sci. 70: 3665–3693.

Subsiha, S. and A. Subramoniam. 2005. Antifungal activities of steroid from *Pallavicinia lyellii*, a liverwort. Indian J. Pharmacol. 37: 304–308.

Tedford, H.W., B.L. Sollod, F. Maggio and G.F. King. 2004. Australian funnel-web spiders: master insecticide chemists. Toxicon. 43: 601–618.

Tripathi, P. and N.K. Dubey. 2004. Exploitation of natural products as an alternative strategy to control postharvest fungal rotting of fruit and vegetables. Review. Postharvest Biol. Technol. 32: 235–245.

Vera, J., J. Castro, A. González and A. Moenne. 2011. Seaweed polysaccharides and derived oligosaccharides stimulate defense responses and protection against pathogens in plants. Mar. Drugs 9: 2514–2525.

Weller, M.G. 2012. A unifying review of bioassay-guided fractionation, effect-directed analysis and related techniques. Sensors 12: 9181–9209.

Windley, M.J., V. Herzig, S.A. Dziemborowicz, M.C. Hardy, G.F. King and G.M. Nicholson. 2012. Spider-venom peptides as bioinsecticides. Toxins 4: 191–227.

Yu, K.X., C.L. Wong, R. Ahmad and I. Jantan. 2015. Mosquitocidal and oviposition repellent activities of the extracts of seaweed *Bryopsis pennata* on *Aedes aegypti* and *Aedes albopictus*. Molecules 20: 14082–14102.

Zerrifi, S.E.A., F. El Khalloufi, B. Oudra and V. Vasconcelos. 2018. Seaweed bioactive compounds against pathogens and microalgae: potential uses on pharmacology and harmful algae bloom control. Mar. Drugs 16: e55.

Zheng, L., X. Han, H. Chen, W. Lin and X. Yan. 2005. Marine bacteria associated with marine macroorganisms: the potential antimicrobial resources. Ann. Microbiol. 55: 119–124.

Algae as a Promising Feed Additive for Horses

Izabela Michalak[1*] and Krzysztof Marycz[2, 3]

[1] Department of Advanced Material Technologies, Faculty of Chemistry, Wrocław University of Science and Technology, Smoluchowskiego 25, 50-372 Wrocław, Poland
[2] Department of Experimental Biology, Wrocław University of Environmental and Life Sciences, Norwida 27b, 50-375, Wrocław, Poland
[3] Faculty of Veterinary Medicine, Equine Clinic-Equine Surgery, Justus-Liebig-University, Giessen 35392, Germany

1 Introduction

Macroalgae (also called seaweeds) and microalgae have been used in both human and animal nutrition since ancient times (Evans and Critchley 2014). However, information about the use of algal biomass in horse feeding is limited. This chapter aims to synthesize the literature on the subject and provide a perspective on the use of algae in horse feeding.

Seaweeds are usually divided into three main groups, according to their pigments: green (Chlorophyta), red (Rhodophyta) and brown seaweeds (Ochrophyta, Phaeophyceae). They are known as a rich source of non-starch polysaccharides, minerals (over 60 microelements), vitamins, pigments and proteins (Evans and Critchley 2014, Michalak and Chojnacka 2015, Cabrita et al. 2016, Makkar et al. 2016, Marinho et al. 2017). However, their composition can vary depending on species, time and location of collection and external conditions such as light intensity, water temperature, and nutrient concentration in water (Makkar et al. 2016, Marinho et al. 2017). Very often, macroalgae grow in unfavourable and rigorous climatic conditions that result in the synthesis of highly bioactive secondary metabolites with therapeutic properties that can be applied in several areas, including pharmaceuticals and medicine (Marinho et al. 2017).

Microalgae are classified into several groups: diatoms (Bacillariophyta), green algae (Chlorophyta), blue-green algae (Cyanobacteria), and golden algae (Ochrophyta, Synurophyceae). The most important species belong to the genera *Arthrospira* (Cyanobacteria), *Dunaliella, Chlorella,* and *Haematococcus* (Chlorophyta) (Madeira et al. 2017). Among the listed microalgae, *Arthrospira* (formerly *Spirulina*), a blue-green microalga, is recognized as a natural feed for horses that has high-quality vitamins (including B complex vitamins and vitamin E), minerals (e.g., manganese, zinc, copper, iron, selenium), carotenoids (including β-carotene), proteins and essential fatty acids that support optimal digestive health and boost the immune system of animals (Belay et al. 1996, Holman and Malau-Aduli 2013, Yaakob et al. 2014, Pearson 2015, Farag et al. 2016, Ememe and Ememe 2017, Madeira et al. 2017). *Arthrospira* possesses some promising biological activities such

---

*Corresponding author: izabela.michalak@pwr.edu.pl

as antimicrobial, antitumour, antiviral, anti-inflammatory, and hypocholesterolaemic effects, and it shows radio-protective and metallo-protective effects (Belay et al. 1996, Farag et al. 2016). This microalga is also known to improve animal growth and fertility, as well as nutritional quality of product (Holman and Malau-Aduli 2013). The same is true of *Chlorella* (Chlorophyta), a rich source of antioxidants, pigments, provitamins and vitamins that can stimulate or enhance the immune system, increase feed intake and utilization, and promote reproduction (Kotrbáček et al. 2015).

The application of algae as a feed ingredient is a very promising alternative to conventional terrestrial animal feed resources, for example, soybean and corn, because it can mitigate the current competition among food, feed and biofuel industries (Madeira et al. 2017). Moreover, algae can be harvested without degrading land, consuming water, or harming the natural environment (Madeira et al. 2017). Algae are characterized by a high growth rate. They can be cultivated in salt water without occupation of arable land (Øverland et al. 2018).

It should also be remembered that the addition of algae to the animal diet can have some disadvantages. Excessive algal supplementation can negatively impact palatability and feed intake (Altomonte et al. 2018). The quantity of algae included in the feed should vary according to both algae and animal species. Special attention should be also paid to water contamination with some algae, especially microalgae. Some species of blue-green algae that are found growing in ponds and lakes (on the surface of the water or below) can be poisonous and should be avoided (Novak and Shoveller 2008).

Table 1 presents a list of review articles that were dedicated to the application of algae in animal feeding and their potential effect on animals. From this table, it can be seen that there are no review papers in the literature on the use of algae in equine nutrition. In this chapter we summarize data concerning supplementation of horse diet with algae and their effect on growth, health and performance of horses.

2 Nutrients in Equine Nutrition

The basic classes of nutrients essential in equine nutrition are energy, protein, minerals, vitamins and water.

2.1 Energy

Horses are non-ruminant herbivores that get energy for growth, exercise and body maintenance mostly from forage and concentrate. The adequate portion of energy that is supplied every day plays a fundamental role in horse welfare and might lead to the development of endocrine disorders due to insulin dysregulation (Anderson et al. 1983). Dietary energy (DE) is usually expressed in terms of kilocalories (kcal) or megacalories (Mcal) of digestible energy, which is calculated on the basis of the horse's maintenance DE requirement, plus the additional energy expended during exercise. The DE can be provided by most common energy sources, i.e., starch, fat, protein and fibre (Knox et al. 1970).

The most common suppliers of energy in equine nutrition are carbohydrates, delivered most commonly with cereals. There are two general types of carbohydrates, i.e., fibrous and non-fibrous, based on their chemical structure, which in turn affects how these carbohydrates are digested by horses. Because horses are not able to produce enzymes for breaking down fibrous carbohydrates, bacterial fermentation plays a crucial role in their digestion. The microbial fermentation of dietary fibre in hindgut supplies daily energy that is covered by the absorption of volatile fatty acids. The main sources of fibrous carbohydrates in equine nutrition are hemicellulose and cellulose fermented in hindgut. Fibre digestibility decreases simultaneously with increasing impact of cellulose as compared to hemicellulose and other smaller fibrous compounds. The non-fibrous carbohydrates are mainly starches and sugars coming from cereals. Different types of plant are characterized by various quantities of nonstructural carbohydrates that depend *inter alia* on their growing phase and

Table 1. Review papers on the application of algae in animal feeding

Algae	Animal	Composition/effect studied	Reference
Seaweeds	Animals generally	Chelated minerals; prebiotic actions	Evans and Critchley 2014
Seaweeds (green, red, brown)	Livestock (ruminant, pig, poultry, rabbit)	Iodine source; prebiotic and health effects	Makkar et al. 2016
Marine macroalgae	Monogastric animals	Sources of protein and bioactive compounds	Øverland et al. 2018
Seaweeds	Poultry, pig, cattle	Improvement of immune function; pigmentation of animal products	Kaladharan 2006
Microalgae (*Chlorella vulgaris*, *Arthrospira* sp.—formerly *Spirulina* sp.)	Livestock including poultry and aquaculture (with emphasis in shrimp farming)	Biomolecules: astaxanthin, lutein, β-carotene, chlorophyll, phycobiliprotein, polyunsaturated fatty acids, β-1,3-glucan	Yaakob et al. 2014
Microalgae	Livestock (ruminant, pig, poultry, rabbit)	Effects on growth performance; meat quality	Madeira et al. 2017
Microalgae (*Arthrospira* sp.)	Poultry	Antioxidant function; hepatoprotective, nephroprotective, neuroprotective, antitumour, anti-inflammatory, hypoglycaemic and hypolipidaemic, antigenotoxic, immunomodulatory effect; Effects on productive and reproductive performance	Farag et al. 2016
Microalgae (*Arthrospira* sp.)	Fishes, poultry (chicken, quail, turkey), pig	Effects on growth and survival: on tissue (carcass) and other qualities; immunomodulatory effect	Belay et al. 1996
Microalgae (*Arthrospira* sp.)	Chicken, pig, ruminant (cattle, sheep), rabbit	Effects on growth, health, productivity, product quality	Holman and Malau-Aduli 2013
Microalgae (*Chlorella*)	Different animal species	Effects on somatic growth, the immune system; a source of pigments; a carrier of selenium and iodine	Kotrbáček et al. 2015
Microalgae	Ruminant	Effects of microalgae on dry matter intake, on milk yield, on milk composition (protein and lactose, fat, fat globules, fatty acid profile, especially DHA and EPA)	Altomonte et al. 2018

maturation. Cereals are characterized mostly by a large amount of nonstructural carbohydrates when compared with stems or leaves. In turn, corn and wheat will have more starches and sugars when compared to oat (Medina et al. 2002).

Another source of energy is fat, which is supplied with various types of oils or seeds rich in fatty acids. Fat can be oxidized only aerobically to produce energy or to be stored in the body as a body fat. Fatty acids cannot be converted to glucose or used to synthesize glycogen, which is believed to have a beneficial effect for muscle glycogen in endurance-trained horses. It was shown that a high-fat diet coming from 15% quantity of soya oil compared to cereals and protein-based diet in horses exercised over long distances resulted in a greater mobilization and utilization of fat. However, in those horses, lower muscle and liver glycogen storage was observed than in the control group, which indicates a need to use an adequate amount of starch in the diet of racehorses (Riberio 2004).

2.2 Protein and amino acids

Proteins are a key component in the diet for horses of any age. Proteins added to daily feed are broken down to single amino acids that in turn serve as a crucial component for equine metabolism. In equine nutrition, amino acids have been classified into two groups: essential and non-essential. The first group includes histidine, isoleucine, leucine, lysine, methionine, phenylalanine, threonine, tryptophan, and valine. This list does not contain arginine, which might be partially synthesized by horses, but arginine plays an unquestionable role in illnesses, because of its crucial role in the function of the immune system and in the disposal of excess nitrogenous wastes (Ball et al. 2007). Other crucial amino acids include tyrosine and cysteine, which can be synthesized from phenylalanine and methionine respectively if their proper quantity is calculated in the diet. The next two essential amino acids, i.e., glycine and proline, are synthesized by horses to a limited extent and are crucial in growing horses; thus, proper quantities must be included in the daily diet. Glutamine has been reported to play a crucial role during stress or illness.

The second group of amino acids important in equine nutrition includes alanine, asparagine, aspartate, glutamate, serine and selenocysteine. Selenocysteine is a critical component of selenoproteins and they directly depend on selenium quantity in the daily diet. Selenocysteine has been shown to not exist as a free amino acid and it is synthesized directly onto the tRNA molecule (Allmang et al. 2009). Amino acids are a product of particular gene expression that takes place in nuclei of every cell of the body. These products of gene expression often are important factors affecting cell metabolism and fate and in consequence may determine the health status of horses. It has to be underlined that protein constitutes 15-18% of the total body mass. Proteins provide a key function for the organism on various levels, including delivery of muscle components (myosin or collagen), nutrient transport in the bloodstream (i.e., haemoglobin, albumin), buffer to minimize fluctuations in body pH as well as nutrient transport across cell membranes. Moreover, one of their most important functions is the regulation of metabolic function resulting from enzyme activity, hormones or peptides. Finally, they have an impact on immune system functionality (immunoglobins).

2.3 Macro- and trace elements

Trace and macroelements are essential for the proper development of every horse and they play a fundamental role in the regulation of particular molecular pathways that affect health status. Recommended daily allowances of trace and macroelements in equine nutrition indicate the amounts that can be safely used. However, trace elements and macroelements are often oversupplied and those excess levels may have a negative effect on equine welfare. In the case of overfeeding with some elements, the organism reaction may be observed in a short time, as with hypocalcaemia. Another issue is the acute reaction for over-supplementation of selenium, which has serious clinical consequences and leads to systemic toxicosis. However, the presence of minerals, i.e., as organic/

chelates, or in inorganic form in equine nutrition seems to be fundamental since more and more data are known in the field of equine metabolomics and on a particular effect of ions on the condition of bacteria in both small and large intestine. It is well known that mineral intake through plant products does not cover horse's requirement for minerals and therefore additional supplementation is required. The macroelements commonly used in equine nutrition are oxides, chlorides or sulfates. Some of them, when in direct contact with mucosal membrane, may cause its damage.

The most important macroelement in equine nutrition is calcium (Ca), of which 99% is deposited in bones. The requirement for calcium basically is covered by hay, legumes and some herbs that are available in the pasture. That source provides around 6 g Ca/kg of dry mass, which in most horses is a physiological level. However, lactating and pregnant mares will require additional calcium supplementation. In contrast, a diet based on grain will provide a lower calcium level with increased phosphorus, which requires supply of external calcium. Most of the calcium in the diet is absorbed in small intestine, while the quantity not absorbed is excreted by the kidney. The main physiological role of calcium is maintaining biomechanical properties of bones. However, supplementation with calcium over requirements does not improve mechanical properties of bones. In sedentary horses, decreased level of calcium in the bloodstream is common; therefore in such horses calcium supplementation must be carefully calculated. In sport horses, the increased calcium concentration may be reached by the combination of physical activity and elevated intake of calcium.

Another crucial component of bone is phosphorus (P), total body content of which is around 8.6 g/kg. Bone ash is about 150-200 g of phosphorus per kg (Vervuert et al. 2010). Phosphorus is secreted pre-caecally into the digestive tract via saliva, pancreatic juice and other intestinal secretions. Phosphorus is required for myriad body functions including bone formation and together with calcium regulates bone mechanical properties. Moreover, phosphorus is a key constituent of ATP and becomes a crucial component in the formation of RNA/DNA. In the peripheral tissues or organs, phosphorus is characterized by a ratio to calcium that is the inverse of the ratio found in the bone (Grace et al. 1999b). Phosphorus in ileocaecal flux enhances the microbial community in the hindgut and thus possesses prebiotic properties. The primary place where phosphorus is absorbed is hindgut and any quantity above the norm is excreted from the urine.

Magnesium (Mg) is a third component of bones. It is estimated that more than 60% of magnesium is stored in the skeleton, 32% in muscle tissue (Grace et al. 1999a), and the rest of it in the external tissues. Magnesium is also an important component of mitochondria or substrates such as ATP, RNA or DNA, playing an important role in metabolic processes including oxidative phosphorylation and protein synthesis. Elevated magnesium concentration together with other minerals and herbs has been shown to be beneficial in horses with equine metabolic syndrome (EMS) (Marycz et al. 2014). It is worth underlining that magnesium, together with manganese (Mn) and chromium (Cr), has been shown to possess the ability to be passively absorbed by brown and green algae. Therefore, the enriched biomass might become an innovative source of elements with potential medical applications (Michalak et al. 2018).

Other crucial components of the mineral part of the equine diet are electrolytes. Electrolytes such as sodium (Na), potassium (K) and chloride (Cl) are responsible for the regulation of osmolality of the extracellular water space and blood plasma. The forage-based diet is usually enough to cover the requirements for electrolytes. As the most abundant cations and anions, they are major contributors to the acid-base balance. Variation between the quantity of Na, K and Cl will induce changes in the systemic pH and bicarbonate concentrations, as well as urinary acidity (Baker et al. 1993, Baker et al. 1998). The loss of water by sweating due to physical activity or higher temperature is a major reason for loss of electrolytes and additional supplementation is required.

The understanding of the role of trace elements in equine metabolism is seriously limited and requires deeper molecular-based research to explain their health benefits and mode of action. All trace elements have a toxic potential. Iron is associated with oxygen transport proteins haemoglobin and myoglobin as well as storing structures ferritin and haemosiderin (Crichton and Ward 2003). A huge quantity of iron constitutes an immune defence mechanism. Iron is usually covered by forage-based diet and does not require additional supplementation under normal conditions; however, in most commercially available products, iron is present. Liver is the most important storage for copper. Copper is involved in multiple cellular functions, e.g., haemoglobin formation, bone formation, keratin synthesis and myelination of neurons. The small intestine is the major site of copper absorption and, because of the low concentration of copper in forage, additional supplementation is required. The liver is also the site of storage for zinc, which is a constituent of many enzymes including superoxide dismutase, liver dehydrogenase and alkaline phosphatase. Zinc is involved in many cellular processes including the replication of DNA and RNA. Moreover, zinc is critical for cell division and the conservation of genetic information. Deficiency of zinc is associated with mental disorders, but excessive zinc is neurotoxic. The increased level of zinc in daily ration improves the keratinization process. Zinc absorption occurs in the gut but also in the small intestine. Zinc intake from forage does not cover the daily requirement, so additional supplementation is required. It was shown that zinc improves the keratinization process in young horses and improves biomechanical properties of hair (Kania et al. 2009).

In the case of manganese (Mn), 90% of its total body concentration is found in muscle, bone and skin. Absorption efficiency is ~30% of intake. Manganese is absorbed in the small intestine (Ducharme et al. 1987) and the quantity not absorbed is excreted in the urine. Manganese becomes a part of important enzymes or activates other enzymes, i.e., manganese-superoxide-dismutase, which is located in the mitochondria or liver, and ensures protection of DNA against superoxide (Ishida et al. 1999). Moreover, Mn-chloride systemic administration increases the superoxide scavenging ability of equine blood (Singh et al. 1992, Singh et al. 1999). Selenium is also characterized by strong antioxidative properties and is an important component of glutathione peroxide, which is one of the most important antioxidant compounds in the body (Rotruck et al. 1973). Selenium is absorbed along the entire gastrointestinal tract and excreted via the kidney. The quantity of selenium in forage does not meet the daily requirement and therefore selenium should be added to the diet. However, overdose of selenium has been shown to be toxic. Its toxicity can occur in any age, breed or sex of animal and is most commonly seen globally in horses on pasture (Raisbeck et al. 1993).

3 Algae as a Valuable Component of Horse Feed

In most animal production systems, domestic animals have little opportunity to feed naturally on seaweeds (Evans and Critchley 2014). However, macroalgae have a long history of use as livestock feed, especially in coastal regions (Makkar et al. 2016). In Iceland, for example, sheep used to graze on seaweeds on the beaches when the grassland was covered with snow or not available. Many farmers sometimes used seaweeds to feed livestock in order to save hay. The sheep were fed entirely on seaweeds for 6-8 weeks and sometimes up to 18 weeks a year. In Iceland, brown and red seaweeds were used to feed sheep and to a lesser extent to feed horses and cattle. In the case of horses, it was found that they picked only the youngest part of the fronds of brown macroalga *Saccharina latissima* (Phaeophyceae) (Hallsson 1964). Other examples of the inclusion of seaweeds in the diet of livestock (ruminants, pigs, poultry and rabbits) are presented in the extensive reviews by Makkar et al. (2016), Madeira et al. (2017) and García-Vaquero (2019) as well as Yaakob et al. (2014) for microalgae (Table 1).

Nowadays, there is a trend to use algae not as a direct feed or feed additive but as a source of a wide spectrum of bioactive compounds (e.g., algal extracts) that can be used in the treatment of many diseases. An exemplary chemical composition of different seaweed species (considering their potential use in horse nutrition) is presented by Makkar et al. (2016) (brown seaweeds *Ascophyllum nodosum*, *Macrocystis pyrifera*, *Laminaria* and *Saccharina* sp., and *Sargassum* sp.; red seaweed *Palmaria palmata*; and green seaweed *Ulva* sp.), as well as by Øverland et al. (2018) for green, red and brown macroalgae. Madeira et al. (2017) summarized the chemical composition (crude protein and amino acids, crude carbohydrates, crude fat and polyunsaturated fatty acids, ash, micro- and macroelements carotenoids, vitamins) of several species of microalgae used as an animal feed: *Arthrospira platensis* (Cyanobacteria), *Chlorella* sp. (Chlorophyta), *Isochrysis* sp. (Haptophyta), *Porphyridium* sp. (Rhodophyta), and *Schizochytrium* sp. (Labyrinthulomycetes). Holman and Malau-Aduli (2013) presented the chemical and nutritional composition of *Arthrospira*. Some of the algal compounds and their role in horse feeding are described below.

3.1 Minerals

Algae, especially marine algae, are known as a rich source of minerals. Seaweeds are known to concentrate minerals from seawater and contain 10-20 times the minerals of land plants (Makkar et al. 2016). Algae contain various micro- and macroelements that are in a natural form—the same as in all terrestrial plant and feedstuffs (Moir et al. 2016). Therefore, seaweeds have a great potential to be used as a source of minerals in animal feeding. However, a great variability among species should be taken into account (Cabrita et al. 2016). Seaweeds are known to be rich in calcium, magnesium, iron, iodine, copper, manganese and selenium but are poor sources of phosphorus and zinc (Cabrita et al. 2016).

Calcium is an important macroelement that participates in buffering excess stomach acid in mammals, including horses, and prevents development of ulcers within the stomach lining (Moir et al. 2016). Nielsen et al. (2011) found that the application of a commercial supplement from a calcified seaweed (AquaCid™) high in calcium, magnesium and other bone-promoting minerals such as silicon and boron increased bone turnover of one-year-old horses as compared with the supplementation of limestone that provided an equivalent amount of calcium, which is recommended for the nutrient requirements of horses according to the National Research Council (NRC).

Iodine is an essential nutrient for reproduction and normal physiological function in horses. Iodine plays an important role as a constituent of thyroid hormones thyroxine (T_4) and triiodothyronine (T_3), which have a powerful effect on the overall health of horses; they influence, for example, heat regulation, feed utilization, proper bone growth and maturation (Huntington et al. 2007). A high content of iodine in seaweeds can limit their use in diets for horses (Cabrita et al. 2016). According to the nutrient requirements of horses (NRC 2007), a dietary content of 0.35 mg of iodine/kg of dry mass (DM) is used (except for the broodmare in late gestation). In the daily rations of mares, 50 mg of dietary iodine is sufficient to produce incidence of goitre in their foals (Huntington et al. 2007). In the work of Cabrita et al. (2016) it was shown that brown macroalga *Saccharina latissima* presented the highest iodine content, 958 mg/kg DM, among the species studied, brown and green. The maximum content of iodine in brown seaweed *Ascophyllum nodosum* was equal to 1000 mg/kg (Evans and Critchley 2014). The iodine content of seaweeds should be examined prior to feeding (especially with pregnant mares) and intake must be carefully controlled (Huntington et al. 2007). Mochizuki et al. (2016) showed also that the concentration of iodine in the serum of horses can result not only from the application of sea algae supplement to the horse diet, but also from the geological differences between areas where horses are living (e.g., there is a high content of iodine in marine environments, a natural aqueous gas resource). Additionally, Mochizuki et al. (2016) stated that equine serum can be a useful sample for iodine monitoring.

Although algae are a natural rich source of micro- and macroelements, it is possible to increase their content (especially microelements) using the biosorption process. Enriched algae can be used as a therapeutic feed additive supplying mineral in a form highly bioavailable to animals (including horses) (Michalak et al. 2011, Michalak et al. 2014, Marycz et al. 2017, Godlewska et al. 2018, Marycz et al. 2018, Michalak et al. 2018). For example, in the work of Godlewska et al. (2018), biomass of freshwater macroalga *Cladophora glomerata* (Chlorophyta) was enriched with Cr(III) ions since it is required for the maintenance of normal glucose and fat metabolism, whereas in the work of Michalak et al. (2018), the same biomass was enriched separately with Cr(III), Mn(II) and Mg(II) ions, because they seem to be crucial in equine clinical nutrition.

Special attention should be paid to the content of toxic metals in algal biomass (especially in algae collected from the natural environment), including cadmium, mercury, lead, nickel, aluminium, and the metalloid arsenic, which may make it unsuitable for use in animal nutrition (Evans and Critchley 2014, Makkar et al. 2016).

3.2 Proteins

Algae are thought to be too expensive to be used widely as a protein source in animal feed. However, it was found that even a very low and economically acceptable addition of *Chlorella* to the animal feed can positively influence growth and performance (Kotrbáček et al. 2015). Single-cell organisms such as algae (e.g., *Spirulina/Arthrospira*) that contain large amounts of proteins can be a source of single-cell proteins. However, some researchers showed neither benefit nor harm from feeding single-cell proteins to horses (Novak and Shoveller 2008).

In blue-green algae, C-phycocyanin, a protein-bound pigment, is also found. It is available as a diet supplement in horses because of its anti-inflammatory and antioxidant properties and is an effective oxygen radical scavenger (Taintor et al. 2014). C-phycocyanin has been found to inhibit cyclooxygenase COX-2, tumour necrosis factor α (cytokine), and oxygen free radicals, which should ameliorate the signs of pain and decrease the inflammatory response associated with degenerative joint disease (Shih et al. 2009).

Another group are lectins (commonly known haemagglutinins or agglutinins) which are carbohydrate binding proteins that are found in many marine macroalgae. More than 250 algal species have been reported to contain haemagglutinins. They cause red blood cells to agglutinate: it was shown that macroalgal extracts agglutinated horse erythrocytes (Hung et al. 2012). Lectins can be used for the detection of carbohydrate markers and are candidates of virucidal and anti-tumour agents (Sharon and Lis 2003). Macroalgal lectins are known to possess *in vitro* or *in vivo* immunomodulatory, anti-tumour and anti-cancer activities (Sugahara et al. 2001, Pinto et al. 2009, Hung et al. 2012). Hung et al. (2012) examined aqueous extracts from 42 species of Vietnamese marine macroalgae (17 Chlorophyta, 22 Rhodophyta and 3 Phaeophyceae) for haemagglutination activity using different native and enzyme-treated horse erythrocytes. It was shown that extracts agglutinated the tested horse erythrocytes. Also, Kakita et al. (1999) isolated and characterized the marine algal haemagglutinins or lectins extracted from red alga *Gracilariopsis longissima* (formerly *Gracilaria verrucosa*) as essential for their potential application as specific carbohydrate affinity ligands.

3.3 Polyunsaturated fatty acids (PUFA)

Seaweeds are known to contain only small amounts of lipids (1-5%), but most of these lipids are polyunsaturated n-3 and n-6 fatty acids (Makkar et al. 2016). The biomass of microalgae can be a source of lipids in animal diet (Holman and Malau-Aduli 2013, Madeira et al. 2017). Hess et al. (2012) found that the incorporation of n-3 fatty acids into blood and muscle of horses depended directly on dietary supply of specific fatty acids. The application of algae increased the content of DHA and EPA in blood and muscle when compared with flaxseed but lowered the content of LA and ALA .

Supplementation with n-3 fatty acids has been suggested as a therapeutic for humans with metabolic dysfunction since they improve insulin sensitivity and reduce inflammation in these individuals (Elzinga 2017). In the work of Hess et al. (2012) it was suggested that the supplementation of n-3 fatty acids in the form of algae and fish oil in the diet of horses can help reduce problems associated with insulin resistance. Parsons (2011) hypothesized that exercising horses supplied with n-3 fatty acids from an algal source, when compared to other sources such as soybean oil and flaxseed, will show the greater increase in plasma DHA concentrations, as well as increased stride lengths and decreased heart rates during exercise, as a result of a decrease of pro-inflammatory cytokines and eicosanoids as measured in the plasma and serum. In the work of Elzinga (2017) it was shown that supplementation of horse diet with DHA-rich microalgae is able to affect circulating fatty acids, reduce inflammation and modulate metabolic parameters in horses with diagnosed EMS.

It should be also noted that horses are herbivores and adding a marine source (e.g., from algae) of n-3 unsaturated fatty acids can lead to palatability problems and refusals (Hess et al. 2012).

3.4 Polysaccharides

Seaweeds also contain a number of complex carbohydrates and polysaccharides. Green algae (Chlorophyta) contain xylans, ulvans, and sulphated galactans. Brown algae (Phaeophyceae) contain alginates, fucoidans and laminarin. Red algae (Rhodophyta) contain agars, carrageenans, xylans, sulphated galactans, and porphyrans (Michalak and Chojnacka 2015, Makkar et al. 2016). The seaweed polysaccharides exhibit prebiotic actions that can enhance health and productivity of animals (Evans and Critchley 2014). Algal mannitol and other chelating agents can improve nutrient utilization (Kaladharan 2006).

4 Algae in the Treatment of Horse Diseases

Powdered algal biomass or algal extracts can be used in animal feed. Compounds extracted from seaweeds present *in vitro* and/or *in vivo* activity. For example, Marinho et al. (2017) showed that 7-keto-stigmasterol from the crude extract of the green Antarctic alga *Prasiola crispa* (Chlorophyta) inactivated equine herpes virus 1 (EHV-1) by direct interaction. The examined sterol probably interferes with EHV-1 attachment to cells with irreversible inactivation of virus infectivity. The maximum level of dietary inclusion of algal biomass to horse diet should strongly depend on their chemical composition (e.g., mineral profile) (Cabrita et al. 2016). According to the published data, algae and their biologically active compounds can be used in the treatment of several diseases in horses. Examples are presented in Table 2.

In feeding experiments on horses, algae-containing commercial products are very often tested, for example, Phycox (Taintor et al. 2014), AquaCid™ (Nielsen et al. 2011), Maxia Complete® (Moir et al. 2016), GNF™ (Hatton et al. 2007), Tasco (Williams et al. 2017), or Glucogard (Marycz et al. 2014). Algae have been tested for their effects on several diseases in horses.

Nielsen et al. (2011) found that the addition of AquaCid™ (a source of calcium) in the diet of young horses resulted in altered bone metabolism when compared with the group of horses fed diet supplemented with limestone with equivalent amount of calcium. AquaCid™ did not alter bone mass but increased bone turnover and may aid in repairing damaged bone and preventing injuries. Another bothersome ailment in horses is joint inflammation, which can lead to reduced performance and mobility. Brennan et al. (2017) examined the effect of microalgae rich in DHA on lameness, an abnormal stance or gait caused by a structural or functional disorder of the locomotor system. Authors examined also whole blood cytokine gene expression following a lipopolysaccharide challenge in mature horses. It was shown that dietary microalgae can mitigate increases in lameness scores, heart rate and cytokine gene expression. Taintor et al. (2014) found that oral daily supplementation of C-phycocyanin (a blue-green algae extract) in equine athletes can also be beneficial in the case of lameness. The application of a commercial product Phycox resulted in improvement in lameness and a decrease in frequency of administration of intra-articular corticosteroids. The use of these products is one of many therapies, among non-steroidal anti-inflammatory drugs, physiological modifiers and biological therapies, that are available for the treatment of horses with degenerative joint disease.

Polyunsaturated fatty acids are another group of algal bioactive compounds that can not only reduce pain and inflammation in patients with rheumatoid arthritis, but also improve inflammatory status and prevent cardiovascular diseases (Hess et al. 2012). Hess et al. (2012) studied the effect of different sources of omega-3 fatty acids, added to a basal diet consisting of hay and barley, on plasma, red blood cell and skeletal muscle fatty acid compositions. Two commercial products were

Table 2. Algae and algal compounds in the treatment of horse diseases

Algae/algal compound	Disease/ailment	Reference
C-phycocyanin (Phycox, a blue-green algae extract)	Degenerative joint disease or osteoarthritis	Taintor et al. 2014
DHA-rich microalgae	Joint inflammation (reduced performance and mobility)	Brennan et al. 2017
Algae (supplement AquaCid™)	Skeletal injuries (influence on skeletal integrity)	Nielsen et al. 2011
n-3 polyunsaturated fatty acids from algae	Chronic lower airway inflammatory diseases: recurrent airway obstruction and inflammatory airway disease	Nogradi et al. 2015
	Cardiovascular disease; arthritis; inflammation	Hess et al. 2012
Macroelement—calcium (algae-based Maxia Complete®)	Gastric ulceration	Moir et al. 2016
Seaweed extract (supplement GNF™ from *Laminaria hyperborea*)	Equine gastric ulcer syndrome	Hatton et al. 2007
Cladophora glomerata enriched with Cr(III) ions	Equine metabolic syndrome	Marycz et al. 2017
Algae—magnesium, manganese, chromium	Equine metabolic syndrome	Marycz et al. 2018
Arthrospira platensis (formerly *Spirulina platensis*)	Equine metabolic syndrome	Nawrocka et al. 2017
Algal DHA	Equine metabolic syndrome	Elzinga 2017
Microalgae	Insulin sensitivity horses	Brennan et al. 2015
Haemagglutinins (lectins)—proteins	Haemagglutination activity of horse erythrocytes	Hung et al. 2012
7-keto-stigmasterol (from *Prasiola crispa*)	Antiviral properties—equine herpes virus 1, which can cause abortion in pregnant mares, respiratory disease, Myeloencephalopathy and perinatal foal mortality	Marinho et al. 2017
Tasco—extract from brown *Ascopyllum nodosum*	Low semen rate	Williams et al. 2017

compared: pellets containing algae and fish oil and a ground flaxseed. Algal product contained ALA, DHA, EPA and DPA, whereas a flaxseed meal contained ALA. It was found that plasma LA and ALA in the group with algae was lower than in the group with flaxseed and in the control group (no fatty acid supplement); EPA and DHA were detected in plasma and red cells only in the group with algae. Authors showed that dietary fatty acid supplementation affected muscle fatty acid composition in horses. Skeletal muscle LA and ALA were lower in the group with algae than in the control and flaxseed group, whereas EPA and DHA were higher. Nogradi et al. (2015) focused on the supplementation of algal n-3 PUFA in the diet of horses with recurrent airway obstruction and inflammatory airway disease. It was shown that n-3 PUFAs increased the content of DHA in plasma, improved clinical signs (cough score), lung function (decrease of a respiratory effort), bronchoalveolar lavage fluid (decrease of neutrophils), and airway inflammation (Nogradi et al. 2015).

Algae are also used in the treatment of equine gastric ulceration, which occurs in both racing and non-racing horses (Hatton et al. 2007) but is mainly found in racehorses and sport horses (eventing, dressage, show-jumping) (Moir et al. 2016). Clinical signs of this disease include weight loss, decreased appetite, diarrhoea, decreased performance, behavioural changes and colic (Hatton et al. 2007). In the work of Moir et al. (2016) it was demonstrated that feeding horses with an organic form of calcium from algae (commercial product Maxia Complete®) reduced ulceration. All horses examined in this study showed a significant improvement in ulcer score—seven horses had a score of zero (fully healed, no evidence of further ulceration) and two had a score of one (some residual inflammation or keratinosis in areas of healed ulcers). Hatton et al. (2007) tested a commercially available supplement GNF™ that contains extract from a brown seaweed *Laminaria hyperborea* (10 mg/100 g), a rich source of vitamins, minerals and proteins (higher than conventional vegetable sources). Other ingredients are calcium carbonate (calcium), magnesium hydroxide (magnesium), fructo-oligosaccharides (prebiotics), glutamine (synthesis of proteins, a mediator in the development of intestinal epithelial cells), threonine (essential amino acid), excipients and binders (full fat soya, kaolin). This supplement assisted in maintaining optimum gut health and function and allowed a maximum utilization of feed. In this study it was also shown that 73% of the supplemented horses showed an overall decrease in ulcer severity.

Equine metabolic syndrome is a steadily growing life-threatening endocrine disorder in horses that is associated with obesity, hyperinsulinaemia, hyperlipidaemia, insulin resistance, systemic inflammation and oxidative stress (Nawrocka et al. 2017, Marycz et al. 2017, 2018). It was shown that EMS causes molecular deterioration of progenitor cells due to elevated oxidative stress, apoptosis and endoplasmic reticulum stress, which intensify insulin resistance (Marycz et al. 2016a, b, Kornicka et al. 2018). The effect of algae on the treatment of this syndrome was examined by several authors. Hess et al. (2013) tested the effect of n-3 fatty acid supplementation (fish and algae supplement) on insulin sensitivity in horses. No statistically significant effect of this supplement was detected, but the group of horses with fish and algae supplement and flaxseed meal had higher insulin sensitivity than the control group (fed no supplemental fatty acid). It is important to add that horses remained healthy throughout the feeding experiments, without adverse effects observed in the case of other treatments. Elzinga (2017) showed that microalgae rich in DHA significantly increased the concentration of many circulating fatty acids, in particular DHA, when compared to the control. Treated horses also had lower serum triglycerides and there was a trend toward reduction in the amount of tumour necrosis factor α produced per lymphocyte in the treated horses. Interestingly, it was also noted that post supplementation, horses given DHA-rich microalgae did not have the same rise in insulin concentrations 60 minutes post oral sugar administration as seen in controls (Elzinga 2017). *In vivo* studies of Nawrocka et al. (2017) showed that horses fed with a diet based on *Arthrospira platensis* lost weight and improved their insulin sensitivity. The application of microalga contributed also to the restoration of adipose-derived mesenchymal stromal cells (ASC) and intestinal epithelial cells (IEC) morphology and function through the reduction of cellular oxidative stress and inflammation. ASC and IEC isolated from EMS-affected individuals exposed to *A. platensis* extract were also characterized by enhanced viability, suppressed senescence and improved proliferation. Moreover, algal extract showed a protective effect against mitochondrial dysfunction and degeneration and effectively suppressed lipopolysaccharide-induced inflammatory responses in macrophages. It is worth adding that *Spirulina* is considered an excellent supplement for horses under immune stress, such as show horses and those exercising at high intensity. It increases the activity of the body's natural antioxidant enzymes (e.g., superoxide dismutase and catalase) and may help to regulate immune and inflammatory processes (Pearson 2015). In the work of Marycz et al. (2017) it was shown that the extract from freshwater green macroalga *Cladophora glomerata*, enriched with Cr(III) ions using biosorption process, reduced apoptosis and inflammation in ASCs of EMS horses through the improvement of mitochondrial dynamics, decreasing of metabolic related genes such as PDK4 expression, and reduction of endoplasmic reticulum stress. Moreover, *C.*

glomerata enriched with Cr(III) induced antioxidative protection coming from enhanced superoxide dismutase activity.

Another type of research was carried out by Jacobs et al. (2018), who studied the supplementation of a commercially available algae-derived omega-3 fatty acid (DHA) administered to mares during the preconceptual and periconceptual period on the endometrial and subsequent embryonic gene expression in horses. This study showed the alteration of inflammatory and prostaglandin signalling in the endometrial tissue obtained from mares given a diet supplemented with DHA. Algae-derived omega-3 fatty acid also effected changes in uterine composition, subsequently altering the uterine environment and impacting conceptus development. Williams et al. (2017) tested a preparation called Tasco, which is a seaweed extract from brown *Ascopyllum nodosum*, that can increase reproductive traits. It was found that the feed intake was not influenced by the Tasco supplement. Body weight and condition were similar among all tested stallions before the study and throughout the trial period. The feed trial of Tasco also had no effect on sperm motility or concentration.

5 Conclusion

Algae can constitute an innovative source of a broad range of bioactive substances, protein, amino acids and lipids and thus can become an excellent feed ingredient that might be successfully used in advanced equine nutrition. Many research groups all over the world are working on the discovery of new active substances or on the potential to improve algal bioactivity in order to create advanced feed additives. Algae per se might cover requirements not only for protein or amino acids, when used as a feed, but also for active ingredients that can be successfully used as an activator of particular mechanisms and pathways. However, there are several challenges in the near future with the intense application of algae in animal feeding, including cost-effective production (if not harvested from the natural environment) and stable chemical composition (if algae are not cultivated and harvested from the natural environment).

Acknowledgments

This project is financed within the framework of a grant entitled "The effect of bioactive algae enriched by biosorption on certain minerals such as Cr(III), Mg(II) and Mn(II) on the status of glucose in the course of metabolic syndrome horses. Evaluation *in vitro* and *in vivo*" (2015/18/E/NZ9/00607) awarded by The National Science Centre in Poland.

References Cited

Altomonte, I., F. Salari, R. Licitra and M. Martini. 2018. Use of microalgae in ruminant nutrition and implications on milk quality – a review. Livest. Sci. 214: 25–35.

Allmang, C., L. Wurth and A. Krol. 2009. The selenium to selenoprotein pathway in eukaryotes: more molecular partners than anticipated. Biochim. Biophys. Acta. 1790: 1415–1423.

Anderson, C.E., G.D. Potter, J.L. Kreider and C.C. Courtney. 1983. Digestible energy requirements for exercising horses. J. Anim. Sci. 56(1): 91–95.

Ball, R.O., K.L. Urschel and P.B. Pencharz. 2007. Nutritional consequences of interspecies differences in arginine and lysine metabolism. J. Nutr. 137: 1626S–1641S.

Baker, L.A., D.L. Wall, D.R. Topliff, D.W. Freeman, R.G. Teeter, J.E. Breazile and D.G. Wagner. 1993. The effect of dietary cation-anion balance on mineral balance in the anaerobically exercised and sedentary horses. J. Equine Vet. Sci. 13(10): 557–561.

Baker, L.A., D.R. Topliff, D.W. Freeman, R.G. Teeter and B. Stoecker. 1998. The comparison of two forms of sodium and potassium and chloride versus sulfur in the dietary cation-anion difference equation: effects on acid-base status and mineral balance in sedentary horses. J. Equine Vet. Sci. 18: 389–395.

Belay, A., T. Kato and Y. Ota. 1996. *Spirulina* (*Arthrospira*): potential application as an animal feed supplement. J. Appl. Phycol. 8: 303–311.

Brennan, K.M., D.E. Graugnard, M.L. Spry, T. Brewster-Barnes, A.C. Smith, R.E. Schaeffer and K.L. Urschel. 2015. Effects of a docosahexaenoic acid-rich microalgae nutritional product on insulin sensitivity after prolonged dexamethasone treatment in healthy mature horses. Am. J. Vet. Res. 76(10): 889–896.

Brennan, K.M., C. Whorf, L.E. Harris and E. Adam. 2017. The effect of dietary microalgae on American Association of Equine Practitioners lameness scores and whole blood cytokine gene expression following a lipopolysaccharide challenge in mature horses. J. Anim. Sci. 95: 166–167.

Cabrita, A.R.J., M.R.G. Maia, H.M. Oliveira, I. Sousa-Pinto, A.A. Almeida, E. Pinto and A.J.M. Fonseca. 2016. Tracing seaweeds as mineral sources for farm-animals. J. Appl. Phycol. 28: 3135–3150.

Crichton, R.R. and R.J. Ward. 2003. An overview of iron metabolism: molecular and cellular criteria for the selection of iron chelators. Curr. Med. Chem. 10: 997–1004.

Ducharme, N.G., J.H. Burton, A.A. Van Dreumel, F.D. Horney, J.D. Baird and M. Arighi. 1987. Extensive large colon resection in the pony. II. Digestibility studies and postmortem findings. Can. J. Vet. Res. 51: 76–82.

Elzinga, S.E. 2017. Inflammation and insulin dysregulation in the horse. Theses and Dissertations. Veterinary Science 31. Available from: https://uknowledge.uky.edu/gluck_etds/31

Ememe, M. and C. Ememe. 2017. Benefits of super food and functional food for companion animals. pp. 309–323. *In*: Waisundara, V. and Shiomi, N. (eds.). Superfood and Functional Food – An Overview of Their Processing and Utilization. IntechOpen, London, U.K.

Evans, F.D. and A.T. Critchley. 2014. Seaweeds for animal production use. J. Appl. Phycol. 26: 891–899.

Farag, M.R., M. Alagawany, M.E.A. El-Hack and K. Dhama. 2016. Nutritional and healthical aspects of *Spirulina* (*Arthrospira*) for poultry, animals and human. Int. J. Pharmacol. 12(1): 36–51.

García-Vaquero, M. 2019. Seaweed proteins and applications in animal feed. pp. 139–161. *In*: Hayes, M. (ed.). Novel Proteins for Food, Pharmaceuticals and Agriculture: Sources, Applications and Advances. Wiley Blackwell, Hoboken, N.J., USA.

Godlewska, K., K. Marycz and I. Michalak. 2018. Freshwater green macroalgae as a biosorbent of Cr(III) ions. Open Chem. 16: 689–701.

Grace, N.D., S.G. Pearce, E.C. Firth and P.F. Fennessy. 1999a. Concentrations of macro- and micro-elements in the milk of pasture-fed thoroughbred mares. Aust. Vet. J. 77: 177–180.

Grace, N.D., S.G. Pearce, E.C. Firth and P.F. Fennessy. 1999b. Content and distribution of macro- and micro-elements in the body of pasture-fed young horses. Aust. Vet. J. 77: 172–176.

Hallsson, S.V. 1964. The uses of seaweed in Iceland. pp. 398–405. *In*: Proceedings of the 4th International Seaweed Symposium, Biarritz, France. Macmillan, New York.

Hatton, E., C.E. Hale and A.J. Hemmings. 2007. An investigation into the efficacy of a commercially available gastric supplement for the treatment and prevention of Equine Gastric Ulcer Syndrome (EGUS). pp. 199–205. *In*: Lindner, A. (ed.). Applied Equine Nutrition and Training: Equine Nutrition Conference (ENUCO). Wageningen Academic Publishers, The Netherlands.

Hess, T.M., J.K. Rexford, D.K. Hansen, M. Harris, N. Schauermann, T. Ross, T.E. Engle, K.G.D. Allen and C.M. Mulligan. 2012. Effects of two different dietary sources of long chain omega-3, highly unsaturated fatty acids on incorporation into the plasma, red blood cell, and skeletal muscle in horses. J. Anim. Sci. 90: 3023–3031.

Hess, T.M., J. Rexford, D.K. Hansen, N.S. Ahrens, M. Harris, T. Engle, T. Ross and K.G. Allen. 2013. Effects of Ω-3 (n-3) fatty acid supplementation on insulin sensitivity in horses. J. Equine Vet. Sci. 33: 446–453.

Holman, B.W.B. and A.E.O. Malau-Aduli. 2013. *Spirulina* as a livestock supplement and animal feed. J. Anim. Physiol. Anim. Nutr. (Berl.) 97: 615–623.

Hung, L.D., B.M. Ly, T.D.T. Vo, T.D.N. Ngo, L.T. Hoa and T.H.T. Phan. 2012. A new screening for hemagglutinins from Vietnamese marine macroalgae. J. Appl. Phycol. 24: 227–235.

Huntington, P.J., E. Owens, K. Crandell and J.D. Pagan. 2007. Nutritional management of mares – the foundation of a strong skeleton. New Zealand Equine Vet. Pract. 12: 14–35.

Ishida, N., Y. Katayama, F. Sato, T. Hasegawa and H. Mukoyama. 1999. The cDNA sequences of equine antioxidative enzyme genes Cu/Zn-SOD and Mn-SOD, and these expressions in equine tissues. J. Vet. Med. Sci. 61: 291–294.

Jacobs, R.D., A.D. Ealy, P.M. Pennington, B. Pukazhenthi, L.K. Warren, A.L. Wagner, A.K. Johnson, T.M. Hess, Ja.W. Knight and R.K. Splan. 2018. Dietary supplementation of algae-derived omega-3 fatty acids influences endometrial and conceptus transcript profiles in mares. J. Equine Vet. Sci. 62: 66–75.

Kakita, H., S. Fukuoka, H. Obika and H. Kamishima. 1999. Isolation and characterisation of a fourth hemagglutinin from the red alga, *Gracilaria verrucosa*, from Japan. J. Appl. Phycol. 11: 49–56.

Kaladharan, P.E. 2006. Animal feed from seaweeds. pp. 83–90. *In*: National Training Workshop on Seaweed Farming and Processing for Food. Kilakarai, India.

Kania, M., D. Mikołajewska, K. Marycz and M. Kobielarz. 2009. Effect of diet on mechanical properties of horse's hair. Acta Bioeng. Biomech. 11(3): 53–57.

Kornicka, K., A. Śmieszek, J. Szłapka-Kosarzewska, J.M. Irwin Houston, M. Roecken and K. Marycz. 2018. Characterization of apoptosis, autophagy and oxidative stress in pancreatic islets cells and intestinal epithelial cells isolated from Equine Metabolic Syndrome (EMS) horses. Int. J. Mol. Sci. 19(10): 1–20.

Kotrbáček, V., J. Doubek and J. Doucha. 2015. The chlorococcalean alga *Chlorella* in animal nutrition: a review. J. Appl. Phycol. 27: 2173–2180.

Knox, K.L., J.C. Crownover and G.R. Wooden. 1970. Maintenance energy requirements for mature idle horse. pp. 186–190. *In*: Schurch, A. and Wenk, C. (eds.). Proceedings of the 7th Equine Nutrition and Physiology Symposium, Warenton, VA.

Madeira, M.S., C. Cardoso, P.A. Lopes, D. Coelho, C. Afonso, N.M. Bandarra and J.A.M. Prates. 2017. Microalgae as feed ingredients for livestock production and meat quality: a review. Livest. Sci. 205: 111–121.

Makkar, H.P.S., G. Tran, V. Heuzé, S. Giger-Reverdin, M. Lessire, F. Lebas and P. Ankers. 2016. Seaweeds for livestock diets: a review. Anim. Feed Sci. Technol. 212: 1–17.

Marinho, R.S.S., C.J.B. Ramos, J.P.G. Leite, V.L. Teixeira, I.C.N.P. Paixão, C.A.D. Belo, A.B. Pereira and A.M.V. Pinto. 2017. Antiviral activity of 7-keto-stigmasterol obtained from green Antarctic algae *Prasiola crispa* against equine herpes virus 1. J. Appl. Phycol. 29: 555–562.

Marycz, K., E. Moll and J. Grzesiak. 2014. Influence of functional nutrients on insulin resistance in horses with equine metabolic syndrome. Pak. Vet. J. 34(2): 189–192.

Marycz, K., K. Kornicka, M. Marędziak, P. Golonka and J. Nicpoń. 2016a. Equine metabolic syndrome impairs adipose stem cells osteogenic differentiation by predominance of autophagy over selective mitophagy. J. Cell. Mol. Med. 20(12): 2384–2404.

Marycz, K., K. Kornicka, K. Basinska and A. Czyrek. 2016b. Equine metabolic syndrome affects viability, senescence, and stress factors of equine adipose-derived mesenchymal stromal stem cells: new insight into EqASCs isolated from EMS horses in the context of their aging. Oxid. Med. Cell. Longev. Vol. 2016: Article ID 4710326, 17 pp.

Marycz, K., I. Michalak, I. Kocherova, M. Marędziak and C. Weiss. 2017. The *Cladophora glomerata* enriched by biosorption process in Cr(III) improves viability, and reduces oxidative stress and apoptosis in equine metabolic syndrome derived Adipose Mesenchymal Stromal Stem Cells (ASCs) and their Extracellular Vesicles (MV's). Mar. Drugs 15(385): 1–20.

Marycz, K., I. Michalak and K. Kornicka. 2018. Advanced nutritional and stem cells approaches to prevent equine metabolic syndrome. Res. Vet. Sci. 118: 115–125.

Medina, B., I.D. Girard, E. Jacotot and V. Julliand. 2002. Effect of a preparation of *Saccharomyces cerevisiae* on microbial profiles and fermentation patterns in the large intestine of horses fed a high fiber or high starch diet. J. Anim. Sci. 80: 2600–2609.

Michalak, I., K. Chojnacka and K. Marycz. 2011. Using ICP-OES and SEM-EDX in biosorption studies. Microchim. Acta 172(1-2): 65–74.

Michalak, I., K. Marycz, K. Basinska and K. Chojnacka. 2014. Using SEM-EDX and ICP-OES to investigate the elemental composition of green macroalga *Vaucheria sessilis*. The Scientific World Journal Vol. 2014: Article ID 891928, 8 pp.

Michalak, I. and K. Chojnacka. 2015. Algae as production systems of bioactive compounds. Eng. Life Sci. 15: 160–176.

Michalak, I., M. Mironiuk and K. Marycz. 2018. A comprehensive analysis of biosorption of metal ions by macroalgae using ICP-OES, SEM-EDX and FTIR techniques. PLoS ONE 13(10): 1–20.

Mochizuki, M., N. Hayakawa, F. Minowa, A. Saito, K. Ishioka, F. Ueda, K. Okubo and H. Tazaki. 2016. The concentration of iodine in horse serumand its relationship with thyroxin concentration by geological difference. Environ. Monit. Assess. 188(226): 1–6.

Moir, T., J. O'Brien, S.R. Hill and L.A. Waldron. 2016. The influence of feeding a high calcium, algae supplement on gastric ulceration in adult horses. J. Appl. Anim. Nutr. 4: 1–3.

Nawrocka, D., K. Kornicka, A. Śmieszek and K. Marycz. 2017. *Spirulina platensis* improves mitochondrial function impaired by elevated oxidative stress in Adipose-Derived Mesenchymal Stromal Cells (ASCs) and Intestinal Epithelial Cells (IECs), and enhances insulin sensitivity in Equine Metabolic Syndrome (EMS) horses. Mar. Drugs 15: 237.

Nielsen, B.D., E.C. Ryan and C.I. O'Connor-Robison. 2011. A marine mineral supplement Aquacid™. pp. 233–234. *In*: Lindner (ed.). Applied Equine Nutrition and Training. A. Wageningen Academic Publishers, The Netherlands.

Nogradi, N., L.L. Couetil, J. Messick, M.A. Stochelski and J.R. Burgess. 2015. Omega-3 fatty acid supplementation provides an additional benefit to a low-dust diet in the management of horses with chronic lower airway inflammatory disease. J. Vet. Intern. Med. 29: 299–306.

Novak, S. and A.K. Shoveller. 2008. Nutrition and Feeding Management for Horse Owners. Alberta Agriculture and Rural Development, Information Packaging Centre, Edmonton, Alberta, Canada, 123 pp.

NRC. 2007. Nutrient Requirements of Horses, 6th Revised Ed. National Academy Press, Washington, 360 pp.

Øverland, M., L.T. Mydland and A. Skrede. 2018. Marine macroalgae as sources of protein and bioactive compounds in feed for monogastric animals. J. Sci. Food. Agric. 99(1): 13–24.

Parsons, A. 2011. Effect of type and amount of omega-3 fatty acids in the diets of exercising horses. Master Thesis, Michigan State University, East Lansing, Michigan, USA, 100 pp.

Pearson, W. 2015. *Spirulina* for horses: a mighty immune modifier. Can. Horse J. Available on: https://www.horsejournals.com/horse-care/feed-nutrition/spirulina-horses.

Pinto, V.P.T., H. Debray, D. Dus, E.H. Teixeira, T.M. Oliveira, V.A. Carneiro, A.H. Teixeira, G.C. Filho, C.S. Nagano, K.S. Nascimento, A.H. Sampaio and B.H. Cavada. 2009. Lectins from the red marine algal species *Bryothamnion seaforthii* and *Bryothamnion triquetrum* as tools to differentiate human colon carcinoma cells. Advances in Pharmacological Sciences, Vol. 2009: Article ID 862162, 6 pp.

Ribeiro, W., S.J. Valbery, J.D. Pagan and B.E. Gustavsson. 2004. The effect of varying dietary starch and fat content on creatine kinase activity and substrate availability in equine polysaccharide storage myopathy. J. Vet. Intern. Med. 18(6): 887–894.

Sharon, N. and H. Lis. 2003. Lectins, 2nd ed. Springer, The Netherlands, 454 pp.

Shih, C.M., S.N. Cheng, C.S. Wong, Y.L. Kuo and T.C. Chou. 2009. Antiinflammatory and antihyperalgesic activity of C-phycocyanin. Anesth. Analg. 108: 1303–1310.

Singh, R.K., K.M. Kooreman, C.F. Babbs, J.F. Fessler, S.C. Salaris and J. Pham. 1992. Potential use of simple manganese salts as antioxidant drugs in horses. Am. J. Vet. Res. 53: 1822–1829.

Singh, A.K., K. Dobashi, M.P. Gupta, K. Asayama, I. Singh and J.K. Orak. 1999. Manganese superoxide dismutase in rat liver peroxisomes: biochemical and immunochemical evidence. Mol. Cell. Biochem. 197: 7–12.

Sugahara, T., Y. Ohama, A. Fukuda, M. Hayashi and K. Kato. 2001. The cytotoxic effect of *Eucheuma serra* agglutinin (ESA) on cancer cells and its application to molecular probe for drug delivery system using lipid vesicles. Cytotechnology 36: 93–99.

Raisbeck, M.F., E.R. Dahl, D.A. Sanchez, E.L. Belden and D. O'Toole. 1993. Naturally occurring selenosis in Wyoming. J. Vet. Diagn. Invest. 5: 84–87.

Rotruck, J.T., A.L. Pope, H.E. Ganther, A.B. Swanson, D.G. Hafeman and W.G. Hoekstra. 1973. Selenium: biochemical role as a component of glutathione peroxidase. Science 179: 588–590.

Taintor, J.S., J. Wright, F. Caldwell, B. Dymond and J. Schumacher. 2014. Efficacy of an extract of blue-green algae in amelioration of lameness caused by degenerative joint disease in the horse. J. Equine Vet. Sci. 34: 1197–1200.

Vervuert, I., S. Klein and M. Coenen. 2010. Short-term effects of a moderate fish oil or soybean oil supplementation on postprandial glucose and insulin responses in healthy horses. Vet. J. 184: 162–166.

Williams, S.J., T.N. Jones, B. Lambert and, R. Harp. 2017. Effects of Tasco supplementation on stallion semen characteristics: a pilot study. J. Equine Vet. Sci. 52: 91.

Yaakob, Z., E. Ali, A. Zainal, M. Mohamad and M.S. Takriff. 2014. An overview: biomolecules from microalgae for animal feed and aquaculture. J. Biol. Res.-Thessalon. 21(6): 1–10.

Bioactive Algae and Cell Therapies – An Irreversible Perspective in Clinical Nutrition of Horses with Endocrine Disorders

Katarzyna Kornicka[1] and **Krzysztof Marycz**[1,2*]

[1] Department of Experimental Biology, Wrocław University of Environmental and Life Sciences, Norwida 27b, 50-375, Wrocław, Poland

[2] Faculty of Veterinary Medicine, Equine Clinic – Equine Surgery, Justus-Liebig University, Gießen, Germany

1 Equine Metabolic Syndrome

Endocrine disorders have become a serious issue both in human and in veterinary medicine. According to the literature, excessive obesity affects approximately 45% of the worldwide horse population (Johnson et al. 2009). Current breeding methods lead to overfeeding of animals and glucose metabolism dysfunction. Diet overload with non-structural carbohydrates and reduction of physical activity in equids usually contributes to the development of equine metabolic syndrome (EMS). In turn, affected individuals become obese and insulin resistant. EMS is clinically characterized by (i) pathological obesity with characteristic signs of local adiposity (around eyes, on the base of tail, and mane—"cresty neck"); (ii) hyperinsulinemia, hyperglycemia, hyperleptinemia; and (iii) past or chronic laminitis. EMS-related laminitis is a systemic disease that manifests with sudden lamellae inflammation causing coffin bone rotation. In the case of EMS, laminitis highly correlates with systemic elevation of insulin levels (above 1,000-1,100 mU/mL) and simultaneously decreased glucose levels, which is recognized as an agent that impairs further hoof structure development. Finally, laminitis often requires euthanasia and is statistically the second most common reason for death, after colic. Thus, development of effective EMS therapy has become a real challenge.

Severe obesity of EMS horses is one of the most characteristic clinical features and entails a broad spectrum of physiological consequences, since adipose tissue is shown to be a key player in the development of systemic inflammation. Insulin-resistant adipose tissue of EMS horses is highly infiltrated with macrophages, T and B cells that are responsible for the induction of local inflammatory state. It was shown that in EMS-affected Welsh ponies, adipose tissue secretes an abundant quantity of pro-inflammatory cytokines, e.g., interleukin 1 (IL-1), interleukin 6 (IL-6) and tumour necrosis factor α (TNF-α), while the amount of tumour growth factor β (TGF-β) is decreased (Basinska et al. 2015). That unfriendly microenvironment of adipose tissue negatively

*Corresponding author: krzysztof.marycz@upwr.edu.pl

affects adipose-derived stem cells (ASC) within it, leading to their cytophysiological impairment. Moreover, EMS horses are characterized by decreased immune protection, which is strongly connected with reduced activity of systemic regulatory T-lymphocytes (Kornicka et al. 2018a). In consequence of this, "inflammaging" in EMS horses seems to be the dominant pathological feature, which suggests a direction for future clinical therapeutic strategies in EMS horses.

2 Mesenchymal Stem Cells

Adipose-derived stem cells belong to the family of mesenchymal stem cells (MSC). These cells have been used in clinical medicine for more than 10 years and afforded promise in the treatment of numerous diseases including tissue damage and immune disorders. MSC possess the ability to self-renew and differentiate into multiple lineages, not only of mesoderm origin. They can give rise to chondrocytes, adipocytes, and osteocytes as well as neurons and beta cells. They are characterized by the expression of CD90, CD44, CD73 and CD105 surface antigens, while they lack the expression of hematopoietic CD45 and CD34 markers. MSC reside in almost all tissues but they are mostly isolated from bone marrow, adipose tissue, and umbilical cord and can be easily expanded *in vitro*. MSC are believed to be responsible for wound healing, growth and replacement of aged, apoptotic cells. Because of their unique properties they have been shown to be effective in the treatment of degenerative and metabolic disorders. Engraftment of bone marrow–derived MSC (BMSC) improved liver function in cases of liver failure caused by hepatitis B (Peng et al. 2011). MSC also have been proved beneficial in the treatment of musculoskeletal disorders. They enhanced regeneration of cells in diabetic critical limb ischemia and bone damage (Lu et al. 2011, Wang et al. 2013). In the context of diabetes research, MSC have been used to generate insulin-producing cells, counteract autoimmunity, enhance islet engraftment and insulin sensitivity (Pileggi 2012). Because of easy access, abundance and need for only local anesthesia, adipose tissue is now one of the major sources of MSC; ASC have been widely used not only in human but also in veterinary medicine. Previous research has proved their utility in the treatment of horses suffering from bone spavin (Nicpoń et al. 2013), phalanx digitalis distalis fracture (Marycz et al. 2012), and tendon disorders and in dogs suffering from degenerative elbow joint disease.

3 Deterioration of Stem Cells During EMS

Although ASC have been proved to be safe and effective, there are still challenges that need to be tackled before their effective clinical application. It has been shown in multiple studies that age and health condition of patients strongly affect ASC cytophysiological properties, diminishing their therapeutic usefulness. Hyperglycemia that occurs during metabolic syndrome affects the stemness of stem cells and undermines stem cell–based therapies. What is more, overproduction of reactive oxygen species (ROS) negatively affects energy homeostasis in cells of metabolic syndrome individuals. It was proved that ASC isolated from EMS horses suffer cytophysiological impairment. They are characterized by decreased proliferation rate, increased apoptosis, endoplasmic reticulum stress and excessive accumulation of ROS (Marycz et al. 2016a). What is more, antioxidative capacities of those cells are strongly limited. Excessive ROS accumulation positively correlates with mitochondria deterioration. Mitochondria from EMS-derived ASC displayed disarrayed cristae, vacuole formation and membrane rupture (Marycz et al. 2016a). Both width and length of those mitochondria were decreased. In EMS-derived ASC, there is significant enhancement of autophagy, a mechanism allowing for selective removal of dysfunctional organelles and proteins. That autophagic flux may be a protective mechanism that helps deteriorated cells to remove dysfunctional mitochondria and maintain proper energy balance (Marycz et al. 2016b, c).

   EMS was also proved to affect different cells and tissues in horse body. It was shown that adipose tissue and liver of EMS horses are characterized by increased mitochondrial damage and mitophagy followed by activation of apoptosis. However, in muscles, apoptosis is diminished, which indicates the existence of a protective mechanism allowing that tissue to maintain homeostasis (Marycz et al. 2018a). Recent data also proved that EMS affects intestinal epithelial cells and pancreatic islets, contributing to their aging and senescence caused by excessive endoplasmic reticulum stress and ROS accumulation. As with ASC, autophagy/mitophagy may be a protective mechanism that allows those cells to maintain their physiological function.

Due to deterioration of EMS-derived ASC, attempts have been made to increase their stemness and therapeutic potential. Identification of mechanisms or agents that could reverse ASC deterioration and increase their therapeutic potential is especially important as it was shown that allogeneic cells are not fully immune privileged, as previously thought (Owens et al. 2016). Early attempts of EMS-derived ASC rejuvenation showed that pre-incubation of those cells with *Cladophora glomerata* (Chlorophyta) water extract enriched during a biosorption process in Cr(III) trivalent chromium and chromium picolinate improved their viability and decreased ROS accumulation (Marycz et al. 2017). It was also shown that combination of two substances—5-azacitidine and resveratrol—is able to rejuvenate these cells by modulation of mitochondrial dynamics, promoting mitochondrial fusion over fission (Kornicka et al. 2018b). What is more, combination of those chemicals improved osteogenic properties of EMS-derived ASC by modulation of autophagy and mitochondrial dynamics through PARKIN and RUNX-2. After 5-azacitidine and resveratrol preconditioning, those cells displayed increased matrix mineralization, enhanced expression of RUNX-2, collagen type I and osteopontin (Marycz et al. 2018b).

4　Algae as a Feed Adjective in Horses

Algae have been widely applied in the nutrition of many types of animals, including dogs, cats, cows and horses. Algae biomass was shown to provide natural vitamins, proteins, fatty acids and minerals. What is more, it positively affects fertility and immune response and helps to maintain proper body weight. Algae are also a rich source of bioactive molecules such as polyphenols, diphlorethohydroxycarmalol and phlorotannins, which makes them a favoured feed additive. There are several products on the market available for horses. Phycox®-EQ (Pharma Chemie, Inc. Syracuse, Nebraska, United States) is an extract of wild blue-green algae (Cyanobacteria) grown in the United States. The extract is especially rich in phycocyanin, which is recognized as an antioxidative, anti-inflammatory, hepatoprotective and neuroprotective factor. Phycox®-EQ is designed to improve joint health and mobility of animals. Another formula is AquaCid (Marigot Ltd. Strand Farm, Currabinny, Carrigaline, Co Cork, Ireland), a pelleted supplement for horses based on the red marine algae *Lithothamnion corallioides* (Rhodophyta). It is widely sold in Ireland and France, where it is abundant in the form of maerl beds. It was shown that supplementation with *L. corallioides* can reduce inflammation, increase bone turnover, reduce reactivity to stimuli, and have gastric buffering properties. It was also shown that green powder derived from freshwater blue-green algae *Arthrospira platensis* (formerly *Spirulina platensis*) exerts potent anti-histaminic, anti-inflammatory and immune system–moderating effects. Dried spirulina contains about 60% (51-71%) protein and less than 20% carbohydrates. Although it contains all the essential amino acids it is low on methionine and lysine. Thus, the daily diet of horse should be supplemented with these molecules. It was also shown that spirulina benefits horses with EMS. Its extracts were tested *in vitro* with ASC and intestinal epithelial cells (the first-line cells exposed to dietary compounds) from EMS horses. In the *in vivo* stage, horses were supplemented with the pelleted algae to investigate whether it affects insulin resistance and body weight. In cells, algae extract improved proliferation rate, diminished accumulation of ROS and enhanced mitochondria condition. Compared to the control diet, spirulina supplementation led to significant weight loss, reduced body condition scoring, and reduced fasting insulin levels. What is more, glucose and insulin tests administered before and after the experimental period revealed that five of the six EMS horses showed significantly reduced insulin levels and lost weight (Nawrocka et al. 2017).

A novel idea to improve bioactivity of algae is the process of biosorption, in which metal ions from aqueous solution are bound with functional groups on the surface on dead biomass. Algal cell wall is composed mainly of polysaccharides, lipids and proteins that play a pivotal role in the biosorption process. Functional groups such as hydroxyl, carboxyl, amino, ester and sulfhydryl allow for the exchange of metal ions and their binding. The advantages of biosorption process include high efficiency, low cost, possibility of metal recovery and regeneration of biosorbent. It is stated that biomass enriched with microelements may be applied as feed additives to supplement livestock diet with the recommended daily intake of given microelements (Michalak and Chojnacka 2008). These novel supplements will provide highly bioavailable and non-toxic form of microelements to animal diet. It was shown that marine macroalgae *Ulva prolifera* (formerly *Enteromorpha prolifera*) and *Cladophora* sp. (Chlorophyta) enriched with Mn(II), Zn(II), Cr(III), Cu(II) and Co(II) ions when added to the diet of laying hens increased microelement transfer to eggs, improved egg weight and eggshell, and increased body weight of hens (Michalak et al. 2011). Biosorption of algae can also be applied in horse nutrition as a method to deliver Cr(III), especially for animals with diagnosed EMS. Chameroy et al. (2011) have shown that Cr(III)-enriched biomass (in this case yeasts) enhanced glucose metabolism in ponies and horses, suggesting that Cr regulates insulin signaling. Thus, it is assumed that algal biomass, which is naturally rich in biologically active compounds when additionally enriched with Cr(III) ions, will also positively influence obese EMS horses. However, enriched algae have not been tested in horse nutrition, although it is a promising perspective. Another crucial element in the nutrition of horses with EMS is magnesium (Mg), as it was shown to regulate insulin sensitivity as well. Dietary intervention in case of EMS animals includes soaking hay in cold water as a way to decrease non-structural carbohydrate content. However, this process results in a loss of water-soluble macrominerals including magnesium, calcium, potassium, sodium, and phosphorus, which in consequence need to be delivered via supplements (Mack et al. 2014, Moore-Colyer et al. 2014). Thus, biosorption of Mg(II) ions by algae may be an interesting perspective and alternative mode of magnesium delivery in EMS horses. However, there are no data as yet on attempts at enrichment of algal biomass with that element. Incorporation of magnesium with biosorption was only proved for yeasts *Saccharomyces bayanus* (formerly *Saccharomyces uvarum*) (Fungi, Ascomycota) (Gniewosz et al. 2007). Yeasts contained 15.2 mg/g of Mg(II) in dry mass when kept under anaerobic conditions and 17.2 mg/g when kept under aerobic conditions.

5　MSC in the Treatment of Diabetes

Multiple studies have been performed to evaluate the potential utility of MSC in the treatment of metabolic disorders including diabetes. MSC are thought to exert multiple therapeutic effects including: (i) differentiation into insulin-producing cells (Xie et al. 2009); (ii) enhancement of angiogenesis and myogenesis through secretion of growth factors, e.g., insulin-like growth factor-1, vascular endothelial growth factor adrenomedullin, and hepatocyte growth factor (Zhang et al. 2008); (iii) synthesis of cardioprotective paracrine factor secretion, e.g., heat shock protein 20, hypoxia-regulated heme oxygenase-1, secreted frizzled-related protein 2, Bcl-2, hypoxic Akt-regulated stem cell factor, and stromal-derived factor (Wang et al. 2009); and (iv) differentiation into renal cells and regulation of immune response *in vivo* (Ezquer et al. 2008, Lee et al. 2006). It was also shown that MSC can mitigate diabetes-related delay in healing of tissues through stimulation of collagen expression and wound healing (Falanga et al. 2007). MSC therapeutic potential has been investigated in multiple studies in rodents and humans. There are no data about therapeutic application of MSC in horses affected by endocrinological disorder. This research area still needs to be evaluated. However, promising data coming from different animal models strongly indicate that such a therapy may exert beneficial effects. First it needs to be established whether autologous or allogenic MSCs should be applied, especially while taking into consideration deterioration of cells isolated from EMS individuals. Impairment of their cytophysiological properties may strongly limit therapeutic outcome.

6 Combined Diet—Stem Cell Therapy

Recently, algae have gained special attention in nutrition as they are a robust source of vitamins, minerals and fatty acids. Moreover, they may be an especially important dietary element for individuals suffering from metabolic disorders as they regulate immune response, increase insulin sensitivity, and help control body weight (Nawrocka et al. 2017). Because of their unique properties algae biomass may be applied as a promising food additive for EMS horses. It is expected that such a supplement will improve carbohydrate metabolism, intensify insulin sensitivity, reduce ROS and increase antioxidative protection of individuals. Another factor that may be used in the treatment of metabolic disorder not only in humans but also in animals is MSC. The knowledge gained in human and rodent studies should now be translated to horses suffering from EMS. It is tempting to speculate, based on existing data, that MSC therapy combined with appropriate nutrition with algal food adjectives can significantly improve therapeutic outcome and ameliorate EMS symptoms. MSC are known to exert immunomodulatory and insulin-sensitizing effects, so their application in the course of EMS is fully justified. On the other hand, algal supplements can effectively deliver microelements, decrease inflammation, increase antioxidative defence and help control body weight. Thus, combination of MSC and algal biomass may become a novel therapeutic strategy for EMS. However, it needs to be taken into account that MSC from EMS horses are characterized by decreased proliferation rate, increased apoptosis, mitochondria deterioration and accumulation of excessive ROS, which strongly limits their therapeutic potential. For that reason, approaches aimed at rejuvenation of those cells *in vitro* before their clinical applications are under investigation. It is hypothesized that, after pre-treatment of those cells with certain chemicals, they can be applied to EMS horses with probably higher therapeutic potential. A combination of stem cells and nutrition therapy may become a cutting edge approach in the treatment of metabolic disorders and a new gold standard in personalized veterinary medicine.

7 Future Perspectives for EMS Treatment

Equine metabolic syndrome becomes an endocrine disorder, which comprises elements of insulin dysregulation, excessive oxidative stress and elements of progressive aging especially in stem progenitor cells. More and more data underline the role of progenitor cells in insulin sensitivity, nutrient absorbance in small and large intestine, or hepatocyte failure under EMS conditions. In turn, it is well accepted that low glycemic index as a caloric restriction combined with particular bioactive substances such as polyphenols might serve as a long-term strategy for improvement of insulin sensitivity. In the course of EMS, it was shown, insulin resistance is linked with endoplasmic reticulum stress, which in consequence leads to mutation in a particular protein that negatively might affect body cell metabolism. Moreover, bearing in mind the facts of autophagy deterioration under insulin resistance state, the search for solutions that would reduce stress of endoplasmic reticulum together with improving autophagic flux might be reasonable pharmacological strategy. It is well established that algae under specific condition might supply an abundance of active agents, such as astaxanthin produced by *Haematococcus lacustris* (formerly *Hematoccocus pulvialis*) (Chlorophyta), which has been shown to possess a beneficial antioxidative effect, and that feature might be fundamental when combined with cellular therapies as therapeutic strategy for EMS treatment.

8 Conclusion

The impact of the large and small intestine biome in the course of insulin dysregulation must not be omitted. The impact of algae on the biome of an EMS horse might deliver in future valuable information regarding projection of novel therapies for EMS treatment.

Acknowledgments

This project is financed within the framework of a grant "Modulation mitochondrial metabolism and dynamics and targeting DNA methylation of adipose-derived mesenchymal stromal stem cell (ASC) using resveratrol and 5-azacitidin as a therapeutic strategy in the course of equine metabolic syndrome (EMS)" (2016/21/B/NZ7/01111) and a grant "The effect of bioactive algae enriched by biosorption on certain minerals such as Cr(III), Mg(II) and Mn(II) on the status of glucose in the course of metabolic syndrome horses. Evaluation *in vitro* and *in vivo*" (2015/18/E/NZ9/00607) attributed by The National Science Centre in Poland.

References Cited

Basinska, K., K. Marycz, A. Śmieszek and J. Nicpoń. 2015. The production and distribution of IL-6 and TNF-α in subcutaneous adipose tissue and their correlation with serum concentrations in Welsh ponies with equine metabolic syndrome. J. Vet. Sci. 16: 113–120.

Chameroy, K.A., N. Frank, S.B. Elliott and R.C. Boston. 2011. Effects of a supplement containing chromium and magnesium on morphometric measurements, resting glucose, insulin concentrations and insulin sensitivity in laminitic obese horses. Equine Vet. J. 43: 494–499.

Ezquer, F.E., M.E. Ezquer, D.B. Parrau, D. Carpio, A.J. Yañez and P.A. Conget. 2008. Systemic administration of multipotent mesenchymal stromal cells reverts hyperglycemia and prevents nephropathy in type 1 diabetic mice. Biol. Blood Marrow Transplant. 14: 631–640.

Falanga, V., S. Iwamoto, M. Chartier, T. Yufit, J. Butmarc, N. Kouttab, D. Shrayer and P. Carson. 2007. Autologous bone marrow-derived cultured mesenchymal stem cells delivered in a fibrin spray accelerate healing in murine and human cutaneous wounds. Tissue Eng. 13: 1299–1312.

Gniewosz, M., W. Duszkiewicz-Reinhard, S. Błażejak, J. Sobiecka and M. Zarzecka. 2007. Investigations into magnesium biosorption by waste brewery yeast *Saccharomyces uvarum*. Acta Sci. Pol. Technol. Aliment. 6: 57–67.

Johnson, P.J., C.E. Wiedmeyer, N.T. Messer and V.K. Ganjam. 2009. Medical implications of obesity in horses – lessons for human obesity. J. Diabetes Sci. Technol. 3: 163–174.

Kornicka, K., A. Smieszek, A.S. Wegrzyn, M. Rocken and K. Marycz. 2018a. Immunomodulatory properties of adipose-derived stem cells treated with 5-azacytydine and resveratrol on peripheral blood mononuclear cells and macrophages in metabolic syndrome animals. J. Clin. Med. 7: 1–18.

Kornicka, K., J. Szłapka-Kosarzewska, A. Śmieszek and K. Marycz. 2018b. 5-azacytydine and resveratrol reverse senescence and ageing of adipose stem cells *via* modulation of mitochondrial dynamics and autophagy. J. Cell. Mol. Med. 23(1): 237–259.

Lee, R.H., M.J. Seo, R.L. Reger, J.L. Spees, A.A. Pulin, S.D. Olson and D.J. Prockop. 2006. Multipotent stromal cells from human marrow home to and promote repair of pancreatic islets and renal glomeruli in diabetic NOD/scid mice. Proc. Natl. Acad. Sci. U.S.A. 103: 17438–17443.

Lu, D., B. Chen, Z. Liang, W. Deng, Y. Jiang, S. Li, J. Xu, Q. Wu, Z. Zhang, B. Xie and S. Chen. 2011. Comparison of bone marrow mesenchymal stem cells with bone marrow-derived mononuclear cells for treatment of diabetic critical limb ischemia and foot ulcer: a double-blind, randomized, controlled trial. Diabetes Res. Clin. Pract. 92: 26–36.

Mack, S.J., A.H. Dugdale, C.M. Argo, R.A. Morgan and C.M. McGowan. 2014. Impact of water-soaking on the nutrient composition of UK hays. Vet. Rec. 174: 452.

Marycz, K., J. Grzesiak, K. Wrzeszcz and P. Golonka. 2012. Adipose stem cell combined with plasma-based implant bone tissue differentiation *in vitro* and in a horse with a phalanx digitalis distalis fracture: a case report. Veterinarni Medicina 57(11): 610–617.

Marycz, K., K. Kornicka, K. Basinska and A. Czyrek. 2016a. Equine metabolic syndrome affects viability, senescence, and stress factors of equine adipose-derived mesenchymal stromal stem cells: new insight into EqASCs isolated from EMS horses in the context of their aging. Oxid. Med. Cell. Longev. Vol. 2016: Article ID 4710326, 17 pp.

Marycz, K., K. Kornicka, M. Marędziak, P. Golonka and J. Nicpoń. 2016b. Equine metabolic syndrome impairs adipose stem cells osteogenic differentiation by predominance of autophagy over selective mitophagy. J. Cell. Mol. Med. 20: 2384–2404.

Marycz, K., K. Kornicka, J. Grzesiak, A. Mieszek and A.J. Apka. 2016c. Macroautophagy and selective mitophagy ameliorate chondrogenic differentiation potential in adipose stem cells of equine metabolic syndrome: new findings in the field of progenitor cells differentiation. Oxid. Med. Cell. Longev. Vol. 2016: Article ID 3718468, 18 pp.

Marycz, K., I. Michalak, I. Kocherova, M. Marędziak and C. Weiss. 2017. The *Cladophora glomerata* enriched by biosorption process in Cr(III) improves viability, and reduces oxidative stress and apoptosis in equine metabolic syndrome derived Adipose Mesenchymal Stromal Stem Cells (ASCs) and their Extracellular Vesicles (MV's). Mar. Drugs 15(385): 1–18.

Marycz, K., K. Kornicka, J. Szlapka-Kosarzewska and C. Weiss. 2018a. Excessive endoplasmic reticulum stress correlates with impaired mitochondrial dynamics, mitophagy and apoptosis, in liver and adipose tissue, but not in muscles in EMS horses. Int. J. Mol. Sci. 19: 165.

Marycz, K., K. Kornicka, J.M. Irwin-Houston and C. Weiss. 2018b. Combination of resveratrol and 5-azacytydine improves osteogenesis of metabolic syndrome mesenchymal stem cells. J. Cell. Mol. Med. 22: 4771–4793.

Michalak, I. and K. Chojnacka. 2008. The application of macroalga *Pithophora varia* Wille enriched with microelements by biosorption as biological feed supplement for livestock. J. Sci. Food Agric. 88: 1178–1186.

Michalak, I., K. Chojnacka, Z. Dobrzański, H. Górecki, A. Zielińska, M. Korczyński and S. Opaliński. 2011. Effect of macroalgae enriched with microelements on egg quality parameters and mineral content of eggs, eggshell, blood, feathers and droppings. J. Anim. Physiol. Anim. Nutr. (Berl.) 95: 374–387.

Moore-Colyer, M.J.S., K. Lumbis, A. Longland and P. Harris. 2014. The effect of five different wetting treatments on the nutrient content and microbial concentration in hay for horses. PLoS One 9: 1–14.

Nawrocka, D., K. Kornicka, A. Śmieszek and K. Marycz. 2017. *Spirulina platensis* improves mitochondrial function impaired by elevated oxidative stress in Adipose-Derived Mesenchymal Stromal Cells (ASCs) and Intestinal Epithelial Cells (IECs), and enhances insulin sensitivity in Equine Metabolic Syndrome (EMS) horses. Mar. Drugs 15(237): 1–28.

Nicpoń, J., K. Marycz and J. Grzesiak. 2013. Therapeutic effect of adipose-derived mesenchymal stem cell injection in horses suffering from bone spavin. Pol. J. Vet. Sci. 16: 753–754.

Owens, S.D., A. Kol, N.J. Walker and D.L. Borjesson. 2016. Allogeneic mesenchymal stem cell treatment induces specific alloantibodies in horses. Stem. Cells Int. Vol. 2016: Article ID 5830103, 8 pp.

Peng, L., D. Xie, B.-L. Lin, J. Liu, H. Zhu, C. Xie, Y. Zheng and Z. Gao. 2011. Autologous bone marrow mesenchymal stem cell transplantation in liver failure patients caused by hepatitis B: short-term and long-term outcomes. Hepatology 54: 820–828.

Pileggi, A. 2012. Mesenchymal stem cells for the treatment of diabetes. Diabetes 61: 1355–1356.

Wang, X., T. Zhao, W. Huang, T. Wang, J. Qian, M. Xu, E.G. Kranias, Y. Wang and G.-C. Fan. 2009. Hsp20-engineered mesenchymal stem cells are resistant to oxidative stress *via* enhanced activation of Akt and increased secretion of growth factors. Stem Cells 27: 3021–3031.

Wang, X., Y. Wang, W. Gou, Q. Lu, J. Peng and S. Lu. 2013. Role of mesenchymal stem cells in bone regeneration and fracture repair: a review. Int. Orthop. 37: 2491–2498.

Xie, Q.-P., H. Huang, B. Xu, X. Dong, S.-L. Gao, B. Zhang and Y.-L. Wu. 2009. Human bone marrow mesenchymal stem cells differentiate into insulin-producing cells upon microenvironmental manipulation *in vitro*. Differentiation 77: 483–491.

Zhang, N., J. Li, R. Luo, J. Jiang and J.-A. Wang. 2008. Bone marrow mesenchymal stem cells induce angiogenesis and attenuate the remodeling of diabetic cardiomyopathy. Exp. Clin. Endocrinol. Diabetes 116: 104–111.

9

Seaweeds as Fish Feed Additives

Clélia Paulete Correia Neves Afonso* and Teresa Margarida Lopes da Silva Mouga

MARE – Marine and Environmental Sciences Centre, ESTM, Instituto Politécnico de Leiria, Santuário Nossa Senhora dos Remédios, 2520-641 Peniche, Portugal

1 Introduction

One of the biggest challenges facing humankind is the production of enough food to feed the world's growing population while ensuring its high nutritional quality, high security standards and an environmentally sustainable production system (Miranda et al. 2017). According to the United Nations (2017), the current world population of 7.6 billion is expected to reach 8.6 billion in 2030, 9.8 billion in 2050, and 11.2 billion in 2100. At the same time, there is a change in food consumption patterns globally, leading to an increase in the need for proteins of animal origin, and forecasts point to a scenario where it will be necessary to increase animal production further by 70% (Herman and Schmidt 2016). In 2015, fish accounted for about 17% of animal protein consumed worldwide, and global fish production reached a peak in 2016, with 171 million tonnes; aquaculture represented 47% of the total fish consumption (FAO 2018). In the past few decades, annual fish ingestion is consistently growing, from 9 kg in 1961 to over 20 kg in 2015, at an average annual rate of about 1.5%. The world aquaculture sector is growing more slowly than in the 1980s and 1990s, but still it grows faster than other major food production systems (FAO 2018) and faces important challenges when compared to terrestrial animal production. Because fish and shellfish species that are currently cultivated have been domesticated recently, they are not as well adapted as other domestic animals to farm conditions (Kibenge et al. 2012). High-density populations, regular handling, poor water quality, the presence of microorganisms and improper feed quality trigger stress that favours the emergence of pathogenic diseases (Vatsos and Rebours 2015, Reyes-Cerpa et al. 2018). Moreover, some pathogens tend to accumulate in the fish body, causing a threat to human health (Cavallo et al. 2013). Chronic disease also causes low food conversion rates, with resulting reduced growth and poor survival rates. The outbreak of an epidemic disease usually causes mass mortality, with significant economic losses (Vatsos and Rebours 2015). Thus, there is a need to improve aquaculture techniques such as fish feeding and try to assess the benefits of the presence of bioactive compounds.

Aquaculture includes more than 500 different animal species currently being produced worldwide, with around 80,000 tonnes produced in 2016, including finfish, crustaceans, bivalves, holothurians and other organisms (FAO 2018). Each of these has different culture needs in terms of facilities, nutrition, health and well-being. Estimates that point to the need to increase the production of feed for aquaculture are unrealistic, as resources are scarce (Watson et al. 2015). Because feed and feeding in the aquaculture sector are of utmost importance, nutritionists are continuously developing balanced diets that fulfil the nutritional and energy requirements of targeted species to optimize

---

*Corresponding author: clelia@ipleiria.pt

growth rates and survival of each species (Shields and Lupatsch 2012). Many commercially raised fish are carnivores (e.g., trout and salmon), while others are omnivores and herbivores (e.g., catfish, carp, and tilapia); therefore, different fish species vary in their capacity to effectively use different kinds of feed. Aquaculture currently over-relies on fishmeal to provide high-quality feed protein. An alternative to fishmeal would be for farmed fish to be fed with renewable plant-sourced protein and oil products harvested from terrestrial farms, such as soybean. While substituting plant protein for fishmeal is now routine at low levels, increasing the proportion of plant protein in fish feed, particularly in carnivorous fish species, limits fish growth rates and feed efficiency.

On the other hand, seaweeds are often included in human or animal feed because of the functional properties of polysaccharides or their high mineral content (Table 1). The nutritional requirements of fish are such that they probably offer less flexibility in diet formulation than do the requirements of most land animals. However, protein content in seaweeds differs according to species, and it is well known that some, like *Pyropia yezoensis* (formerly *Porphyra yezoensis*) can contain up to 47% of proteins on a dry mass basis (Fujiwara-Arasaki et al. 1984). Initial studies aiming to assess the nutritional potential of some red, green and brown seaweeds were carried out in the early 1980s (Montgomery and Gerking 1980), indicating that green seaweeds seemed to have a high nutritional potential, when compared to red or brown seaweeds. Later, the effects of algal biomass in fish diets were studied for several fish species, either herbivorous or carnivorous, indicating a potential increase in the growth rates and feed efficiency, associated with a higher resistance to diseases and stress, when fish were submitted to a diet containing small proportions of *Ulva* or *Porphyra/Pyropia* spp. (Mustafa and Nagawaka 1995, Mustafa et al. 1995). If the fish species selected for culture are carnivores, they will require a diet of high protein content, and they show a poor use of carbohydrate as an energy source. Apparently, the presence of certain types of carbohydrates in the diet is, in fact, detrimental.

Table 1. Composition of commercially available feed ingredients and fish requirements (% dry weight)

Composition of feed	% Crude protein	% Crude lipid	% Crude carbohydrates	Gross energy MJ/kg
Fishmeal	63.0	11.0	-	20.1
Soybean	44.0	2.2	39.0	18.2
Spirulina	58.0	11.6	10.8	20.1
Chlorella	52.0	7.5	24.3	19.3
Gracilaria sp.[1]	34.0	1.5	37.1	13.4
Gracilaria sp.[2]	10.0	0.9	50.1	11.2
Ulva lactuca[1]	37.4	2.8	42.2	15.7
Ulva lactuca[2]	12.5	1.0	57.0	11.2
Animal feed requirements	% Crude protein	% Crude lipid	% Crude carbohydrates	Gross energy MJ/kg
Salmon	37.0	32.0	15.0	13.5
Sea bream	45.0	20.0	20.0	19.1
Tilapia	35.0	6.0	40.0	13.5
Shrimp	35.0	6.0	40.0	13.5
Poultry	21.0	5.0	60.0	13.0
Cattle	12.0	4.0	65.0	10.1

Adapted from Shields and Lupatsch (2012).
[1] Cultured in integrated multi-trophic aquaculture system.
[2] Wild specimens.

Table 1 compares different algae with traditional fishmeal and with the nutritional requirements of different farmed animals. Both integrated multi-trophic aquaculture (IMTA) *Gracilaria* sp. and *Ulva* sp. seem to be adequate to nourish most aquaculture fish, carnivorous and herbivorous, although total crude lipids are rather low, being good alternatives to fishmeal without decreasing growth rates of aquaculture animals (Pereira et al. 2012, Shields and Lupatsch 2012). Feeds for aquatic animals are also more energy- and nutrient-dense than those for terrestrial animals. Fish therefore need to be fed less to support each unit of growth and present a lower feed conversion ratio (FCR) than terrestrial animals (1.0-2.0 to 2.2-5.8, respectively) (Shields and Lupatsch 2012). Common feed-stuffs contain protein ranging from 15% to 63%, guaranteeing the protein requirements for optimum growth of several fish species; however, proteins from vegetable sources such as soybean can lack several key amino acids (e.g., methionine, lysine and tryptophan). Diet must provide a suitable energy source and possess the correct amounts of proteins, lipids, carbohydrates, minerals, vitamins and growth factors, but also consider the water content, toxicity, palatability, physical form, stability during storage, and the bioavailability and stability of the nutrients.

Consequently, it has been stated that the inclusion of different percentages of seaweeds in fish feed, typically between 1% and 10%, results in a balanced source of nutrients that increases fish and growth rate, nutrient uptake, and total carotenoid content in the skin, enhances fish health including its metabolic rates and iodine concentration, improves response to stress, enhances the innate immune response of the fish, and minimizes the spread of pathogenic bacteria that affect fish (Marinho et al. 2013, Valente et al. 2015, Al-Asgah et al. 2016, Thanigaivel et al. 2016, Miranda et al. 2017, Moutinho et al. 2018).

2 Seaweeds as Nutritional Aquaculture Feed

Feed fish supply represents about 50% of production costs in intensive aquaculture, the most expensive component being the protein source (Halver and Hardy 2002). Fish feed for aquaculture is produced with fish parts, fish by-products or whole fish, very rich in protein (about 60%), and fish oil extracted from cooked fish, very rich in polyunsaturated fatty acids (PUFA) such as omega-3, with clear benefits to the nutritional value of the diet. Fishmeal and fish oil are still considered the most nutritious and digestible ingredients for farmed fish, but their inclusion rates in feeds for aquaculture have shown a clear downward tendency, largely as a result of supply and price volatility (FAO 2018). Alternative raw materials of vegetable origin such as soybean meal, have been used as major ingredients in aquaculture feed, replacing fish feed as a protein source. Yet, most terrestrial vegetables have an unbalanced amino acid composition and low digestibility, and no long chain PUFA. It is important to note that in fish, dietary lipids are an important source of essential fatty acids, vital for regular health, growth and reproduction; consequently, fish have an absolute requirement for both omega-6 and omega-3 PUFA in the diet (Turchini et al. 2009). Besides, soybean is also subject to price fluctuations (Cabanero et al. 2016) and requires large phosphorus inputs provided by mineral fertilizers; it is thus inefficient as a sustainable source of protein (Philis et al. 2018). Furthermore, soybean is needed to meet the demand for agricultural crops required for human nutrition.

Hence, other high-quality nutritive feeds are being evaluated to efficiently substitute for fishmeal and fish oil. Raw materials are being evaluated for their nutritive balance, digestibility, and absence of anti-nutritional factors, price and availability (Shields and Lupatsch 2012, Norambuena et al. 2015). Because seaweeds exhibit high protein percentage and high-quality protein, low total lipid percentage, but with a high PUFA content, besides many other bioactive compounds, they can be included in aquaculture feed as an alternative ingredient, determining good growth performance and improvement in nutrient utilization (Mustafa et al. 1995), better immune and stress response (Schleder et al. 2018). At the same time, they allow economic benefits beyond traditional feeds (Miranda et al. 2017).

2.1 Nutritional profile of seaweeds

Seaweeds are classified into three different taxa depending on the nature of the main accessory pigment present: Chlorophyta are green seaweeds and exhibit chlorophyll b, Rhodophyta are red seaweeds presenting the phycobilin phycoerythrin, and Ochrophyta (Phaeophyceae) are brown seaweeds and show the xanthophyll fucoxanthin.

They are phylogenetically distant from each other, with quite different morphologies, structures and chemical composition (Table 2). Nevertheless, in general, seaweeds exhibit high nutritional value due to the high concentrations of proteins, vitamins and minerals (Silva et al. 2015). Seaweeds also provide very low energy content due to the low levels of lipids, of which many are omega-3 and omega-6 PUFA, together with high concentration of polysaccharides of low digestibility (Wong et al. 2000).

The chemical composition, however, varies within the species and with geographical origin, environmental factors, life-cycle, season and other ecological conditions, such as temperature, nutrient concentration, and light intensity, which affect storage responses (Bocanegra et al. 2009, Francavilla et al. 2013, Vieira et al. 2018). Thus, there are considerable regional variations within the same taxon, according to the physicochemical conditions the seaweed has been exposed to during growth. These nutritional fluctuations may affect the quality of the fish feed due to fluctuations in bioactive properties of the seaweed biomass.

2.1.1 Polysaccharides

Common to all seaweeds is the large concentration of polysaccharides produced, mainly cell wall structural polysaccharides, ranging from 4% to over 75% (Holdt and Kraan 2011). Mucopolysaccharides and storage polysaccharides are also present (Cardoso et al. 2014). The algae carbohydrate synthesis is related to seasonal changes leading to maximum growth and increased photosynthetic activity (Paiva et al. 2014), thus environmental conditions greatly influence polysaccharide concentration.

Most carbohydrates produced by seaweeds are not digested by the gastrointestinal tract of humans or monogastric animals; thus, they can be regarded as dietary fibres and, therefore, seaweeds are usually low in calories. The dietary fibres are classified into two types: insoluble fibres such as cellulose, mannans and xylan, and water-soluble fibres such as agars, carrageenan, alginic acid, furonan, laminarin, porphyrin, ulvans and fucoidans. Insoluble fibre increases faecal bulk and decreases intestinal transit, contributing to weight control, improving cardiovascular and gastrointestinal health, and preventing cancer. Soluble fibre helps to increase viscosity and reduce glycaemic response and plasma cholesterol (Wong et al. 2000, Burtin 2003, Lahaye and Robic 2007, Holdt and Kraan 2011, Debbarma et al. 2016). These compounds also serve important anticoagulant, antimicrobial and anti-inflammatory functions (Freitas et al. 2012), thus presenting important properties for humans and for animals. They are also interesting prebiotic compounds that are being tested as food additives and fish feed supplements. Because the notorious presence of polysaccharides in seaweeds decreases the digestibility, namely of the protein fraction, there may be a need to treat them by fermentation processes or enzymatic digestion (Marrion et al. 2003, Mišurcová et al. 2010, Fleurence et al. 2012).

2.1.1.1 Phycocolloids

The three major groups of phycocolloids are alginates, carrageenans and agar, widely used in food industry as gelling, thickener, and emulsifier or stabilizer agents. The major polysaccharide of the brown seaweeds is alginic acid. It consists of unbranched chains of beta-1,4-linked D-mannuronic acid and blocks of beta-1,4-linked L-guluronic acid. Alginate is present as salts of several metals, namely sodium and calcium. It occurs in the cell wall and in the intercellular region (Percival 1979). Alginates are the main carbohydrates present in *Ascophyllum, Fucus* and *Sargassum* (de Jesus Raposo et al. 2016), among other species.

Table 2. Main distinctive characteristics of seaweed groups[1]

Taxa	Pigments	Cell wall polysaccharides	Cell wall mucilage	Reserve polysaccharides	Commercial species
Rhodophyta Red seaweeds	Chlorophylls a, d, phycocyanins, phycoerythrin, beta-carotenes, xanthophylls	Cellulose, xylans, mannans	Sulphated polysaccharides, e.g., agar, carrageenans, porphyrans sulphated galactans	Floridean starch	*Porphyra/Piropia* sp. *Gracilaria* sp. *Chondrus crispus Palmaria palmata Eucheuma* spp. *Kappaphyctus alvarezii*
Chlorophyta Green seaweeds	Chlorophylls a, b, beta-carotenes, lutein, other xanthophylls	Cellulose, xylans, mannans	Sulphated polysaccharides, e.g., sulphated galactans (ulvans)	Starch	*Ulva* sp. *Codium* sp. *Caulerpa* sp.
Phaeophyceae Brown seaweeds	Chlorophylls a, c, beta-carotenes, fucoxanthins, other xanthophylls	Cellulose, chitin	Alginic acid/alginates, sulphated polysaccharides, e.g.., fucans (fucoidans)	Laminarin Sargassan mannitol	*Laminaria* sp. *Saccharina japonica Undaria pinnatifida Sargassum fusiforme* (formerly *Hizikia fusiformis*)

[1] Adapted from Bocanegra et al. 2009, Holdt and Kraan 2011, Angell et al. 2016, FAO 2018.

Galactans (agaroids and carrageenans) are the major group of polysaccharides in Rhodophyta, and they consist of galactose or modified galactose units (Percival 1979). These molecules contain alternating repeating units of 1,3-alpha-gal and 1,4-beta-D-gal, and/or 3,6-anhydrogal (3,6-Agal). Agarans are common in *Porphyra/Pyropia, Gracilaria, Polysiphonia, Bostrychia* and *Cryptopleura* and carrageenans are present in *Chondrus, Kappaphycus, Eucheuma*, and *Phyllophora* (de Jesus Raposo et al. 2015). Other minor structural polysaccharides present in seaweeds are cell wall polysaccharides known as fucoidans (also characteristic of brown seaweeds), xylans (characteristic of red and green seaweeds), and ulvans (present in green seaweeds) (Percival 1979, Bocanegra et al. 2009, Holdt and Kraan 2011, Angell et al. 2016, de Jesus Raposo et al. 2016). Fucoidans are soluble homo- or heteropolymers, with L-fucose as the main sugar residue with varying proportions of galactose, mannose, xylose and uronic acids linked by different types of glycosidic bonds. These fucoidans are common in brown seaweeds such as *Laminaria* and *Saccharina* (Percival 1979, de Jesus Raposo et al. 2015, de Jesus Raposo et al. 2016). Galactofucans may appear in some brown macroalgae as well (e.g., *Laminaria, Undaria*) (de Jesus Raposo et al. 2016).

In green seaweeds, the polysaccharide fraction is present in *Codium fragile* as glucomannans and in *Ulva* sp. as ulvan. Water-soluble ulvans represent 8% to 29% of green algae dry weight and the main constituents are sulphate galactans, rhamnose, xylose, and glucuronic acid. Ulvan has high proportions of two rare monomers, rhamnose and iduronic acid, which can be used in pharmaceutical applications (Lahaye and Robic 2007, Angell et al. 2016). Also present are rhamnans in *Caulerpa* spp. and *Ulva* spp., galactans in *Caulerpa* spp., and pyruvylated galactans in *Codium* spp. (de Jesus Raposo et al. 2015, de Jesus Raposo et al. 2016).

2.1.1.2 *Storage polysaccharides*

Storage polysaccharides are species-specific. Notably, laminarin is characteristic of brown seaweeds such as *Laminaria* and *Saccharina* and it accumulates in the cell cytosol in specialized vacuoles. It is a heteropolymer of sugar alcohol, mannitol and a beta-1,3-linked glucan (Percival 1979). Seasonal variations of laminarin are common; it is most abundant during autumn and shows very low concentration or none in February and June (Holdt and Kraan 2011). Laminarin, along with fucoidans, exhibit promising potential in food and feed (Miranda et al. 2017) mainly due to their immunostimulant properties. Floridean starch (alpha-1,4-glucosidic linked glucose homopolymer with alpha-1,6-branches) is characteristic of red seaweeds and it also accumulates outside the plastid (Yu et al. 2002). The carbohydrate reserve of Chlorophyta is starch, which accumulates inside the chloroplast, although the starch granules are smaller and less structurally defined and present smaller molecules of amylose and amylopectin than land plants (Percival 1979).

2.1.1.3 *Cell wall polysaccharides*

Cell wall polysaccharides mainly consist of cellulose and hemicelluloses, neutral polysaccharides that produce the mechanical support of the cell. The cellulose and hemicellulose content of the seaweed species varies between 2% and 10% dry weight. Lignin is only found in *Ulva* sp. at concentrations of 3% dry weight (Holdt and Kraan 2011).

2.2 Lipids

Most lipids present in seaweeds are fatty acids. These are long carbon chains, with variable lengths and a carboxylic acid group. They can be classified according to the degree of saturation or the carbons. Thus, saturated fatty acids exhibit only single bonds, monounsaturated fatty acids exhibit one double covalent bond, and PUFA exhibit more than one double bond. Polyunsaturated fatty acids can still be divided according to location of the double bonds, related to the last carbon of the chain or the omega carbon: omega-3 or n-3 has the first double bond on carbon three counting from the methyl terminal end, omega-6 or n-6 has the first double bond on carbon six and omega-9 or n-9 has the first double bond on carbon nine counting from the methyl terminal (Sarojini and Uma Devi 2014).

Seaweeds are characterized by low levels of lipids, ranging from 1% to 5%, with the highest concentration registered of 7.8 in percentage of dry weight (Schmid et al. 2018), but their PUFA content is higher than those of terrestrial plants (Burtin 2003, Kumari et al. 2010). The lipid content is higher in the winter and spring and is lower in the summer, and it also varies according to the geographical location, climate and environmental conditions such as temperature, and nutrient content (Paiva et al. 2014). Vertebrates, including humans and finfish, cannot synthesize most PUFA, including n-6 family and most n-3; thus, these lipids are considered essential nutritional components for both humans and animals, enhancing growth, food utilization, health and reproductive viability (Balfry and Higgs 2001, Miranda et al. 2017).

Phaeophyceae tend to have the highest level of lipids, followed by Chlorophyta and finally Rhodophyta. Among the saturated fatty acids, palmitic acid (C16:0) is often the most dominant, ranging from 17.5% in *Undaria pinnatifida*, 44.5% in *Porphyra* sp., up to 57% in the green seaweed *Caulerpa racemosa,* and 79% in red seaweeds such a *Gracilaria corticata* (Dawczynski et al. 2007, Viera et al. 2011).

As to monounsaturated fatty acids, oleic acid (C18:1) is the most dominant and reaches 21% in brown seaweeds (Dawczynski et al. 2007). Marine algae are also very rich in PUFA of the n-3 and n-6 series. The content of PUFA in seaweeds varies from 34% in *Porphyra/Pyropia* sp. (Rhodophyta) to 74% in *Undaria pinnatifida* (Phaeophyceae) (Kendel et al. 2015). Phaeophyceae exhibited a fatty acid profile rich in the C18 and C20 length PUFA. The fatty acid profile of the Rhodophyta consists mainly of arachidonic acid 20:4n-6 and eicosapentaenoic acid 20:5n-3 (Kendel et al. 2015), but 16:0, 18:1n-9 are also common. The profile of Chlorophyta is quite different, showing a high degree of unsaturation within C16 and C18 fatty acids including small chain n-3 PUFA 16:4n-3 and 16:3n-3 along with 18:2 and 18:3, much like land plants (Fleurence et al. 1994, Kendel et al. 2015, Schmid et al. 2018). Although it varies considerably, the important n-3 PUFA are abundant in most red algae, and this is similar for the n-6 long-chain PUFA. For example, arachidonic acid varies from 0.1% in *Ulva lactuca* (Chlorophyta) and 6.14% in *Hypnea anastomosans* (formerly *Hypnea esperi*) to 58% in *Gracilaria dura* (Rhodophyta) (Kumari et al. 2010, Viera et al. 2011, Paiva et al. 2014, Sarojini and Uma Devi 2014, Debbarma et al. 2016), and the percentage of eicosapentaenoic acid in red algae *Hypnea spinella* (6.9%) is ten times that in *Ulva rigida* (0.7%) (Chlorophyta) (Viera et al. 2011).

Seaweeds also have essential fatty acids such as linoleic acid (C18:2n-6) and alpha-linolenic acid (C18:3n-3) in smaller quantities, respectively up to 14% in *Caulerpa veravalensis* and 13% in *Ulva fasciata* (Chlorophyta) (Sarojini and Uma Devi 2014, Debbarma et al. 2016). Essential PUFA are important for the development of vital structural lipids and elements of cell membranes. Thus, the cells of the immune system of the animal are affected by the ratios of PUFA and so the immune function and disease resistance are also leveraged by a good PUFA balance (Balfry and Higgs 2001).

2.3 Proteins

Seaweeds usually contain high levels of proteins, with mean values ranging from 10% to 30% of dry matter. Brown seaweeds such as *Ascophyllum nodosum, Fucus spiralis, Himanthalia elongata,* and *Laminaria digitata* show the lowest levels (5-15%), except for *Undaria pinnatifida*, whose levels may range from 11% to 24%. Green seaweeds possess an intermediate percentage (e.g., *Ulva* sp. ranges from 10% to 26%). Red seaweeds show the highest, for example, *Palmaria palmata, Gracilaria* sp., *Porphyra/Pyropia* sp. and *Grateloupia turuturu* (10-47%) (Fleurence 1999, Bocanegra et al. 2009, Fleurence et al. 2012, Vieira et al. 2018). These figures are higher than those found in high-protein vegetables, such as soybean, which exhibits protein content up to 42% (Dornbos and Mullen 1992).

Among all three main taxa, the content of protein depends on the seasonal period, being higher during winter and lower during summer (Fleurence 1999, Bocanegra et al. 2009, Fleurence 2017), since seaweeds exhibit higher ability to store nitrogen due to lower nitrogen metabolism (Vieira et al. 2018).

As stated, seaweed protein digestibility is also a matter to take into consideration. Digestion of seaweed is carried out by means of three enzymes: pepsin, pancreatin and pronase. Thus, the digestibility of algal proteins can differ according to the species and seasonal variations in antinutritional factor content. The compounds that limit the digestibility of algal proteins are either phenolic molecules or polysaccharides. In fact, these compounds inhibit enzyme activity and, therefore, have strong negative effects on protein digestibility (Fleurence 1999).

The green alga *U. lactuca* shows an interesting protein content (ca. 19%) and its digestible energy, although not very high, is similar to a good quality fodder. *Laminaria digitata* (Phaeophyceae) also has good level of protein (8-15%) and high *in vitro* organic matter digestibility (94%), suggesting it would be a good feed resource (Makkar et al. 2015).

2.3.1 Amino acids

Most vertebrates, including human and fish, require 10 essential amino acids (EAA), which are those that the animal body requires but does not produce (Hardy 2001). Red seaweeds typically have a higher quality of protein than green seaweeds and these are better than brown seaweeds. For most seaweeds, aspartic and glutamic acids constitute together a large part of the amino acid fraction, representing up to 40% of total amino acid content (Fleurence 1999, Lourenço et al. 2002). Many seaweeds exhibit high levels of acidic amino acids—aspartic acid and glutamic acid (Marrion et al. 2003)—as well as of asparagine, serine, histidine, arginine, lysine, valine, leucine and isoleucine (Vieira et al. 2018).

The amino acid composition of seaweeds has been widely studied in comparison with terrestrial plants such as soybean exhibiting all EAA, but with tryptophan and methionine frequently limiting, showing concentration below human and animal requirements (Fleurence 1999, Vieira et al. 2018). In seaweeds the concentrations of EAA such as threonine, lysine, tryptophan, methionine and cysteine are usually higher than those of soybean, up to 45% (Fleurence 1999, Lourenço et al. 2002). Other EAA such as leucine, isoleucine and valine show a profile close to that of terrestrial plants and other protein sources (Fleurence 1999, Vieira et al. 2018). In fact, many seaweeds exhibit a balanced amino acid profile, being an interesting source of food and feed protein (Wong et al. 2000, Vieira et al. 2018).

Most commercial species such as the red seaweeds *Gracilaria* and *Porphyra/Pyropia*, the brown seaweeds *Sargassum* (formerly *Hizikia*), *Laminaria*, *Saccharina*, and the green seaweed *Ulva* seem to have high-quality protein and relatively high concentrations of EAA. Red seaweeds have the highest average proportion of total EAA and lysine and the second highest average proportion of methionine compared to brown and green seaweeds (Angell et al. 2016). Despite these general data, EAA are species-dependent, vary throughout the year and show differences between cultured and wild specimens (Vieira et al. 2018).

Although the EAA content of seaweeds is compatible with fish requirements, namely carnivore species, using these protein sources as fish feed may interfere with protein quality, amino acid content and bioavailability. Hence, low protein content and an unbalanced amino acid profile can determine lower growth rates and diet conversion ratios (Hardy 2001). Thus, it is recommended that seaweeds be used not as a full meal, but as a feed supplement.

2.4 Minerals

Seaweeds are usually high in minerals because of their ability to capture macronutrients and metal contaminants from the environment, notably from marine habitat (Makkar et al. 2015, Neveux et al. 2018). The ash content includes macro-minerals and trace elements. Macro-minerals are those inorganic elements necessary in large quantities for key metabolic cell processes, namely calcium

(Ca), phosphorus (P), magnesium (Mg), sodium (Na), potassium (K) and chloride (Cl). Most seaweeds are rich in these mineral elements, red and brown seaweeds richer than green seaweeds (Table 3).

Table 3. Mineral requirements of selected fish aquaculture species compared to the mineral composition of selected seaweeds

| | *Recommended supplement* | | | | | | | |
| | *g/100 g diet* | | | | *mg/100 g diet* | | | |
	Ca	*K*	*Mg*	*Cu*	*Fe*	*Zn*	*Mn*	*Se*
Channel catfish *Ictalurus punctatus*[1]	1.5	0.26	0.04	1.5-5	30	20	2.4	0.25
Common carp *Cyprinus carpio*[1]	-	-	0.06	3	199	15-30	12-13	-
Rainbow trout *Oncorhynchus mykiss*[1]	-	-	0.06-0.07	3-3.5	-	15-30	12-13	0.15-0.38
Composition of macroelements and microelements in different seaweeds								
	g/100 g				*mg/100 g*			
Gracilaria corticata[2]	1.30	10.86	0.63	0.41	27.04	5.80	1.72	0.13
Sargassum swartzii[2]	1.86	7.71	0.70	0.58	28.26	5.28	3.84	0.16
Ulva rigida[2]	2.64	4.71	1.77	1.08	30.44	4.03	6.30	0.36

[1] NRC 2011,
[2] Kumar et al. 2011.

Again, there are large seasonal shifts in the concentration of minerals, the highest levels occurring during spring and the lowest during autumn. The ash content is larger in brown seaweeds, which have higher absorbent capacity from the environment, ranging from 15% to 45% dry weight, followed by green (11-55%) and red (7-27%) (Holdt and Kraan 2011, Paiva et al. 2014).

Regarding macrominerals, calcium accumulates in seaweeds at much higher levels than in terrestrial food-stuffs, e.g., an 8 g portion of *Ulva lactuca* provides 260 mg of calcium corresponding to 37% of Reference Nutrient Intake value recommended for humans (Macartain et al. 2007). Calcium is essential for skeletal tissue formation in fish and shellfish and for membrane permeability and it is required at 1.5 g/100 g diet (Table 3), but this element is usually present in the water. Phosphorus is also required in structural components such as bone, scales and teeth and its presence in diet is critical to fish and shellfish metabolism and growth (NRC 2011).

It has also been stated that *Porphyra/Pyropia* sp., *Osmundea pinnatifida* and *Fucus spiralis* contain high levels of potassium, magnesium and calcium (Paiva et al. 2014). Sodium and potassium are also present at relatively high levels, although Na:K ratios are usually below 1:5 (Macartain et al. 2007).

Seaweeds are also rich in trace elements, that is, those micronutrients required in much lower concentrations, but nevertheless indispensable to the organism's physiological functions. Important trace minerals such as zinc (Zn), iron (Fe), iodine (I), copper (Cu), manganese (Mn), and selenium (Se) are also found in high proportions in seaweeds (Table 3). In fact, essential mineral contents in seaweeds are at much higher levels than in many terrestrial mineral sources (Debbarma et al. 2016, Miranda et al. 2017) and all of these trace elements are also required for a balanced feed diet for fish and shellfish (NRC 2011). Selenium is an essential trace element with important biological functions either in humans or in animals. It is incorporated into selenoproteins with a wide range of health effects (Rayman 2017). Iodine is also an essential trace element to produce thyroid hormone, and iodine deficiency remains a significant public health problem worldwide (Macartain et al. 2007, Pearce et al. 2013). In animals it also causes thyroid hyperplasia (NRC 2011). Iron needs, along with copper needs, are significantly higher than other micromineral requirements and iron is present in seaweeds at higher levels than in many well-known terrestrial sources, such as meats and spinach (Macartain et al. 2007, Debbarma et al. 2016). Thus, seaweeds can be used as a source of iron to produce haemoglobin. Copper is included in metalloenzymes and its deficiency causes impaired growth and low enzyme activity (NRC 2011).

Seaweeds are regarded as important sources of such microminerals and can contribute to a balanced diet. Yet, trace elements, such as arsenic, may have negative health effects (Macartain et al. 2007) and, therefore, seaweeds must be cautiously used in feed supplementation. Also, there is not a single seaweed that meets entirely all the mineral requirements of humans or animals. However, seaweeds frequently exhibit minerals in organic form rather than in inorganic, which increases the availability of minerals (Makkar et al. 2015).

2.5 Vitamins

Vitamins are organic compounds that are required in trace amounts from exogenous sources for normal growth, health and reproduction (NRC 2011). These vitamins are usually obtained from diet and seaweeds are a good source of either water-soluble vitamins (B complex vitamins, as well as vitamin C) or fat-soluble vitamins (beta-carotene with vitamin A activity, and E) because of their ability to produce them as a response to stress induced by direct exposure to sunlight (Macartain et al. 2007).

Many water-soluble vitamins from the B complex play major physiological and metabolic roles in fish, although they are required in small amounts. Nutritional factors such as choline, inositol and ascorbic acid are required in larger amounts (Halver and Hardy 2002). Vitamin E (alpha-tocopherol), an important antioxidant, is abundant in *Undaria pinnatifida* (14.5 mg/100 g). High vitamin E content helps to protect PUFA in seaweed and to maintain their nutritional benefits (Macartain et al. 2007, Rajapakse and Kim 2011). The vitamin E content presents 27.34 and 4.86 mg/100 g of dry weight for *Porphyra* sp. and *Osmundea pinnatifida* (Rhodophyta), respectively. *Fucus spiralis* (Phaeophyceae) shows the highest vitamin E content, around 51.14 mg/100 g of dry weight, that is 34 times higher than in oats (1.49 mg/100 g) and four times higher than in olive oil (12 mg/100 g) (Paiva et al. 2014). The content of fat-soluble vitamins reveals a moderate content of vitamin A, 1.41, 1.27 and 1.20 mg/100 g for *F. spiralis, Porphyra* sp. and *O. pinnatifida*, respectively, when compared with vitamin A–rich vegetable such as carrot (16.9 mg/100 g) (Paiva et al. 2014). As to water-soluble vitamin C, Macartain et al. (2007) state that it is present in large amounts in *Ulva lactuca* (Chlorophyta)*, Undaria pinnatifida* (Phaeophyceae) and *Porphyra* sp. (Rhodophyta) presenting more than 125 mg/100 g, corresponding to more than 15% of the Reference Nutrient Intake. Also, red and brown seaweeds are rich in carotenes (pro-vitamin A) and vitamin C, and their amounts may range from 20 to 170 ppm and 500 to 3000 ppm, respectively (Rajapakse and Kim 2011).

Seaweeds are also among the few plant sources of vitamin B_{12}, which is not found in most land plants but is present in a few vegetables in considerable amounts, *Ulva lactuca* (Chlorophyta) being an important source (Macartain et al. 2007, Rajapakse and Kim 2011).

The deficiency of vitamins in fish diet causes important health problems, such as abnormal swimming, dark skin coloration, deformities, dermatitis, oedema, eye pathologies, loss of scales, muscle dystrophy, and organ pathological changes (NRC 2011).

2.6 Pigments

Chlorophyll a is the primary photosynthetic pigment responsible for triggering photosynthesis. Seaweeds also have other accessory photosynthetic pigments, which confer different absorption properties, therefore increasing each group's absorption spectrum. The accessory photosynthetic pigments, namely xanthophylls, also protect the photosystem from the deleterious effect of high-intensity light (Valentão et al. 2010). The amount of pigment present in the cells also varies according to environmental conditions, such as light intensity (increasing with depth) and nutrient availability (increasing with higher amounts of nutrients), as an acclimation mechanism. Green seaweeds have both chlorophyll a and b (Table 2), beta-carotene, lutein and other derivative xanthophylls, such as siphonoxanthin, occurring in siphonaceous genera, such as *Codium* spp. (Chlorophyta) (Lee 2008).

Red seaweeds contain pigments called "phycobilins", namely phycoerythrin, a pink pigment. Other phycobilins are present in smaller concentrations, such as phycocyanin, allophycocyanin and phycoerythrocyanin. The concentration of phycobilins differs according to environmental conditions. In low nitrogen concentration, the thalli of red algae turn palein, whereas well-fertilized thalli are dark, because they are able to store nitrogen in the phycoerythrin and can remobilize the nitrogen in the pigment to support growth in case of nitrogen deficiency (Rabiei et al. 2016). In brown algae chlorophyll c is present in two forms, c1 and c2. The major yellow accessory pigment in Phaeophyceae is beta-carotene; the major brown pigment is fucoxanthin, but they also exhibit violaxanthin (Angell et al. 2016).

Along with other additives, such as protein, phycocolloids, PUFA and fucoidan, pigments play an important role in animal feeding, namely phycobilins and carotenoids (Chojnacka et al. 2012). The beneficial effects of pigments, namely carotenoids, are thought to be due to their role as antioxidants (Miranda et al. 2017). Other activities include the pro-vitamin A ability of beta-carotene (Rajapakse and Kim 2011) and anti-tumoral potential of fucoxanthin (Holdt and Kraan 2011).

2.7 Phenols

Phenolic compounds are secondary metabolites composed by one or more aromatic rings and at least two hydroxyl groups. Simple phenols include a single aromatic group and polyphenols include many aromatic rings (Fabrowska et al. 2015). Phenols are a highly heterogenous group of molecules, including phlorotannins, bromophenols, terpenoids and other phenolic compounds that occur in all terrestrial plants and seaweeds (Targett and Arnold 1998, Burtin 2003). Algal polyphenols derive from phloroglucinol units (1,3,5-trihydroxybenzene). They are most abundant in brown seaweeds mainly in the form of phlorotannins ranging from 5% to 15%. The phenolic fraction present in red seaweeds and green seaweeds is mainly composed of bromophenols (Liu et al. 2011). These are polyhydroxylated and polybromated structures derived from polyphenols by the action of bromoperoxidases (Li and Kim 2011). Phlorotannins and bromophenols are specific to seaweeds and not found in any other natural sources. Although a high phenolic content may limit protein availability, phenols exhibit high bioactive properties. For example, phlorotannins inhibit enzymes such as phospholipase A, lipoxygenase and cyclooxygenase-1,128 hyaluronidase,129 and tyrosinase (Bocanegra et al. 2009), and thus exhibit great antioxidant activity (Lopes et al. 2012, Peixoto et al. 2018).

3 Seaweeds as Fish Feed

The inclusion of up to 10% of green, red or brown seaweeds in fish feed meal is considered a good source of crude protein, crude lipids, fatty acids, essential amino acids, essential lipids, minerals, vitamins and other micronutrients. Hence, alternative algae-derived products as a partial substitution for traditional fishmeal may be an economical and sustainable alternative source of protein and lipid, besides being more available than traditional crop plants (Norambuena et al. 2015). Many authors have studied the effect of different seaweeds on growth performance and health of aquaculture animals (Table 4) and proved that supplementation works as a growth promoter, besides their nutraceutical effect.

Table 4. Effects of seaweeds, seaweed extract or seaweed molecules on various aquaculture organisms

Seaweed species	Adequate % supplementation	Purpose	Fish and shellfish species targeted	References
Phaeophyceae (brown algae)				
Alginate from *Sargassum* sp.	4 g kg⁻¹	IS	Walking catfish, *Clarias* sp.	Isnansetyo et al. 2015
Alginic acid from *Macrocystis pyrifera*	100 µg mL⁻¹	AB, IS	Atlantic cod, *Gadus morhua*	Caipang et al. 2011
Fucoidan from *Fucus vesiculosus*				
Alginic acid	Alginic acid at 1%	IS	Red drum, *Sciaenops ocellatus*	Mendoza-Rodriguez et al. 2017
Sodium alginate	(3 g L⁻¹) diet	IS	Red swamp crayfish, juveniles, *Procambarus clarkii*	Mona et al. 2015
	10 g kg⁻¹	IS	Nile tilapia, *Oreochromis niloticus*	Van Doan et al. 2016
Ergosan, extract with 1% alginic acid from *Laminaria digitata*	10 mL kg⁻¹ 0.5% 4 g kg⁻¹ alginic acid	IS	Snakehead murrel, *Channa striata* European seabass, *Dicentrarchus labrax* Beluga, juvenile, *Huso huso*	Miles et al. 2001 Bagni et al. 2005 Ahmadifar et al. 2009
Fucoidan from *Padina* sp.	2000 mg kg⁻¹	IS	Striped catfish, *Pangasius hypophthalmus*	Purbomartono et al. 2018
Fucoidan from *Sargassum swartzii*	1-6%	N, AB, IS	Roho, *Labeo rohita*	Gora et al. 2018
Fucoidan from *Sargassum ilicifolium*	0.4-0.6 mg kg⁻¹	IS	Nile tilapia, *Oreochromis niloticus*	Isnansetyo et al. 2016
Fucoidan	1.0 g kg⁻¹	AB, IS	Pacific white shrimp, juvenile, *Penaeus vannamei* Yellow catfish, juvenile, *Pelteobagrus fulvidraco*	Kitikiew et al. 2013 Yang et al. 2014
Laminarin from *Laminaria hyperborea*	1-6 mg laminarin kg⁻¹	IS	Atlantic salmon, *Salmo salar*	Dalmo and Seljelid 1995
Polysaccharides from *Padina gymnospora*	0.01%, 0.1% or 1% (w/w of the feed)	IS	Common carp, *Cyprinus carpio*	Rajendran et al. 2016
Ascophyllum nodosum	2.5-5%	N	Red seabream, *Pagrus major*	Nakagawa et al. 1997 Yone et al. 1986

(Contd.)

Table 4. (*Contd.*)

Seaweed species	Adequate % supplementation	Purpose	Fish and shellfish species targeted	References
Fucus spp.	2.5-7.5%	IS	European seabass, *Dicentrarchus labrax*	Peixoto et al. 2016
Saccharina japonica (formerly *Laminaria japonica*)	50%	N	Sea cucumber, *Apostichopus japonicus*	Xia et al. 2012
Macrocystis pyrifera	Injection (10 µg) Immersion (350 mg L⁻¹)	AB, IS	Pacific white shrimp, *Penaeus vannamei*	Campos et al. 2014
Petalonia binghamiae	Extract at 6 and 10 µg g⁻¹	AV, IS	Pacific white shrimp, *Penaeus vannamei*	Chen et al. 2014
Sargassum angustifolium	Hot water extract 400 mg kg⁻¹	AB, IS	Rainbow trout, *Oncorhynchus mykiss*	Zeraatpisheh et al. 2018
Sargassum ilicifolium (formerly *Sargassum duplicatum*)	Immersed in 300 mg L⁻¹ Extract and injection with 10 µg g⁻¹ extract	IS	Pacific white shrimp, *Penaeus vannamei*	Yeh et al. 2006
Sargassum fusiforme	0.5% and 1% supplemented diet	AB, IS	Fleshy prawn, *Fenneropenaeus chinensis*	Huang et al. 2006
Sargassum hemiphyllum var. *chinense*	Powder at 500 mg L⁻¹, Extract at 300 mg L⁻¹	IS	Pacific white shrimp, *Penaeus vannamei*	Huynh et al. 2011
Sargassum ilicifolium (formerly *Sargassum duplicatum*)	10%	N, AB, IS	Beluga, *Huso huso*	Yeganeh and Adel 2018
	2.5 g kg⁻¹ of a heat-treated seaweed meal	AB, IS	Pacific white shrimp, *Penaeus vannamei*	Lin et al. 2017
Anthophycus longifolius and *Padina gymnospora*	100 mg L⁻¹ to 500 mg L⁻¹ of dried solvent extract in 10 L of water	IS	Sea cucumber, *Apostichopus japonicus*	Thanigaivel et al. 2015
Sargassum pallidum Saccharina japonica (formerly *Laminaria japonica*)	100% pure Mixed	N	Japanese abalone, *Haliotis discus hannai*	Qi et al. 2010
Sargassum sp. *Padina* sp.	Mannan oligosaccharide supplemented at 2000 mg kg⁻¹ diet	IS	Mud crab, *Scylla serrata*	Traifalgar 2017

(*Contd.*)

Sargassum swartzii	10 g kg⁻¹ feed diet	IS	Tiger prawn, *Penaeus monodon*	Felix et al. 2004
	1% kg diet⁻¹ water extract	AB	Flathead grey mullet, *Mugil cephalus*	Kanimozhi et al. 2013
Undaria pinnatifida	2–3%	IS	Tiger prawn, juvenile, *Penaeus monodon*	Niu et al. 2015
	5%	N	Red seabream, *Pagrus major*	Yone et al. 1986
	100% Pure and mixed	N	White sea urchin, *Tripneustes gratilla*	Floreto et al. 1996
Stoechospermum polypodioides	30%	AB, IS	Common carp, *Cyprinus carpio*	Radhika and Mohaideen 2016
Undaria pinnatifida *Sargassum filipendula*	4%	N, AV, IS	Pacific white shrimp, *Penaeus vannamei*	Schleder et al. 2018
Chlorophyta (green algae)				
Polysaccharides from *Acrosiphonia orientalis*	Fed medicated diet (4 g kg⁻¹)	AV, IS	Tiger prawn, juvenile, *Penaeus monodon*	Manilal et al. 2009
Polysaccharides from *Ulva rigida*	Injected intraperitoneally with 1 mL of thioglycolate (3%)	IS	Turbot, *Psetta maxima*	Castro et al. 2006
	10 mg kg⁻¹		Grey mullet, *Mugil cephalus*	Akbary and Aminikhoei 2018
Ulvan from *Ulva clathrata*	0.5 and 1% ulvan	N, IS	Nile tilapia, *Oreochromis niloticus*	Quezada-Rodríguez and Fajer-Ávila 2017
Ulvan from *Ulva intestinalis*	0.15 to 0.21% kg⁻¹ diet	IS	Pacific white shrimp, *Penaeus vannamei* and Tiger prawn, juveniles, *Penaeus monodon*	Lauzon and Serrano 2015
Ulvan from *Ulva* sp.	1000 to 1500 mg kg⁻¹ diet	AV, IS	Tiger prawn, *Penaeus monodon*	Declarador et al. 2014
Verdemin, derived from macroalgae *Ulva ohnoi*	2.5% and 5%	N	Atlantic salmon, *Salmo salar*	Norambuena et al. 2015
Rhizoclonium riparium (formerly *Rhizoclonium implexum*)	12.6%	N	Nile tilapia, *Oreochromis niloticus*	Cabanero et al. 2016
	10.5%		Pacific white shrimp, *Penaeus vannamei*	Serrano et al. 2017
Ulva lactuca	5%	N	White spotted snapper, juvenile, *Lutjanus stellatus*	Zhu et al. 2016
	50%		Gilthead seabream, *Sparus aurata*	Magnoni et al. 2017
			Pacific white shrimp, *Penaeus vannamei*	Pallaoro et al. 2016
	30%	AB, IS	Common carp, *Cyprinus carpio*	Radhika and Mohaideen 2016

(Contd.)

Table 4. (*Contd.*)

Seaweed species	Adequate % supplementation	Purpose	Fish and shellfish species targeted	References
Ulva linza	1 g kg^{-1} extract	AB	Tiger prawn post-larvae, *Penaeus monodon*	Selvin et al. 2011
Ulva australis (formerly *Ulva pertusa*)	100% pure and mixed	N	White sea urchin, *Tripneustes gratilla*	Floreto et al. 1996
Ulva rigida	5%	N	Common carp, *Cyprinus carpio* Salmon, *Salmo salar* Nile tilapia, *Oreochromis niloticus* European seabass, *Dicentrarchus labrax* Senegalese sole, *Solea senegalensis*	Diler et al. 2007 Moroney et al. 2017 Ergün et al. 2009 Valente et al. 2006 Moutinho et al. 2018
	10% seaweed meal			
Ulva sp.	Mannan oligosaccharide supplemented at 2000 mg kg^{-1} diet	IS	Mud crab, *Scylla serrata*	Traifalgar 2017
	2.5-7.5%		European seabass, *Dicentrarchus labrax*	Peixoto et al. 2016
Rhodophyta (red algae)				
ι-Carrageenan	10 g kg^{-1} diet	IS	Rohu, *Labeo rohita*	Kumar et al. 2014
κ-Carrageenan	Intraperitoneal injection of κ-carrageenan	IS	Common carp, *Cyprinus carpio*	Fujiki et al. 1997
κ-Carrageenan from *Hypnea musciformis*	5 g kg^{-1}	IS	Nile tilapia, *Oreochromis niloticus*	Villamil et al. 2018
Carrageenan	0.5%	IS	Red drum, *Sciaenops ocellatus*	Mendoza-Rodriguez et al. 2017
Polysaccharides from *Acrosiphonia orientalis*	Fed medicated diet (4 g kg^{-1})	AV, IS	Tiger prawn, juvenile, *Penaeus monodon*	Manilal et al. 2009
Polysaccharide from *Kappaphycus alvarezii*	Polysaccharide extract treated with 0.5%	AB, IS	Asian seabass, *Lates calcarifer*	Sakthivel et al. 2015
Sulfated galactans from *Gracilaria fisheri*	500 µg mL^{-1}	AV, IS	Tiger prawn, *Penaeus monodon*	Rudtanatip et al. 2014
Sulphated polysaccharides from *Chondrus crispus*	10 µg mL^{-1}, 20 µg mL^{-1}, 50 µg mL^{-1}	IS	Mussel, *Mytilus* spp.	Rudtanatip et al. 2018

(*Contd.*)

Asparagopsis armata	0.5% of extract 0.85 and 1.1 g kg⁻¹	AB, IS	Gilthead seabream, larvae, *Sparus aurata* Tiger prawn, post-larvae, *Penaeus monodon*	Castanho et al. 2017 Manilal et al. 2012
Eucheuma denticulatum	3%	N	Japanese flounder, juvenile, *Paralichthys olivaceus*	Ragaza et al. 2013
Gelidium amansii	Hot-water extract at 400 and 600 mg L⁻¹, Injected with hot-water Extract at 6 µg g⁻¹ Fed with 2.0 g kg⁻¹	AB, IS	Pacific white shrimp, *Penaeus vannamei*	Fu et al. 2007
Gracilaria arcuata	10%	N	North African catfish, *Clarias gariepinus*	Al-Asgah et al. 2016
Gracilaria bursa-pastoris and *Crassiphycus corneus* (formerly *Gracilaria cornea*)	5%	N	European seabass, *Dicentrarchus labrax*	Valente et al. 2006
Gracilaria cervicornis	50%	N	Pacific white shrimp, *Penaeus vannamei*	Marinho-Soriano et al. 2007
Crassiphycus corneus (formerly *Gracilaria cornea*)	Ethanolic extract, at 10 mg mL⁻¹	IS	Senegalese sole, *Solea senegalensis*	Diaz-Rosales et al. 2007
Gracilaria corticata	1 g kg⁻¹ 30%	AB, IS	Common carp, *Cyprinus carpio*	Nasaran and Huxley 2013 Radhika and Mohaideen 2016
Gracilaria fisheri	Ethanol, methanol and chloroform extracts 90-190 µg ml⁻¹	AB, IS	Tiger prawn, *Penaeus monodon*	Kanjana et al. 2011
Gracilaria gracilis	4 g kg⁻¹ of feed 0.25 and 0.5%	AB, IS IS	Pacific white shrimp, *Penaeus vannamei* Zebrafish, *Danio rerio*	Zahra et al. 2017 Hoseinifar et al. 2018
Gracilaria vermiculophylla	5%	N IN	Rainbow trout, *Oncorhynchus mykiss* Gilthead seabream, *Sparus aurata*	Valente et al. 2015 Magnoni et al. 2017
Gracilaria sp.	2.5-7.5% 0.5% methanolic extract	N, IS	European seabass, *Dicentrarchus labrax*	Peixoto et al. 2016 Peixoto et al. 2018
Gracilariopsis lemaneiformis	33% 15%	N	White-spotted rabbitfish, *Siganus canaliculatus* Black seabream, juvenile, *Acanthopagrus schlegelii*	Xu et al. 2011 Xuan et al. 2013
	100% pure or mixed		Japanese abalone, *Haliotis discus hannai*	Qi et al. 2010

(Contd.)

Table 4. (*Contd.*)

Seaweed species	Adequate % supplementation	Purpose	Fish and shellfish species targeted	References
Kappaphycus alvarezii	0.5 g L^{-1}	AB, IS	Pacific white shrimp, *Penaeus vannamei*	Suantika et al. 2017
Gloiopeltis furcata	100% pure and mixed	N	White sea urchin, *Tripneustes gratilla*	Floreto et al. 1996
Laurencia snyderae	200, 400 and 600 mg mL^{-1}	AB, IS	Pacific white shrimp, juvenile, *Penaeus vannamei*	Dashtiannasab et al. 2016
Porphyra dioica	10%	N	Rainbow trout, *Oncorhynchus mykiss*	Soler-Vila et al. 2009
Porphyra purpurea	9%	N	Thicklip grey mullet, *Chelon labrosus*	Davies et al. 1997
Portieria hornemannii	Seaweed extract 25 mg/0.1 mL	IS, AB	Sebae anemonefish, *Amphiprion sebae*	Karuppiah 2014
Pterocladiella capillacea (formerly *Pterocladia capillacea*)	5%	N	European seabass, *Dicentrarchus labrax*	Wassef et al. 2013
Pyropia haitanensis	2.51-3.14%	N, IS, AB	Pacific white shrimp, *Penaeus vannamei*	Niu et al. 2018

N – Nutrition, growth and physiology; IS – Immuno-stimulant; AB – Antibacterial; AV – Antiviral.

Studies of the effect of seaweeds on aquaculture species started long ago. Mustafa et al. (1995) studied the effect of seaweeds on red seabream (*Pagrus major*), proving that they improve feed efficiency, protein efficiency ratio, and muscle protein deposition. In their study, small amounts of various kinds of macroalgal supplementation—the most efficient being *Pyropia yezoensis* (Rhodophyta), then *Ascophyllum nodosum* (Phaeophyceae) and *Ulva australis* (formerly *Ulva pertusa*) (Chlorophyta)—increased fish growth, mean body weight and body length because they seemed to promote protein and lipid metabolism and disease resistance. So, all the three groups of seaweeds showed important effects on fish. In later studies, brown seaweeds such as *Ascophyllum nodosum, Sargassum swartzii, Saccharina japonica* (formerly *Laminaria japonica*) and *Undaria pinnatifida* were fed to finfish, shrimps, abalone, and sea cucumber to evaluate their nutritional potential with very promising results. Similarly, red seaweeds such as *Eucheuma denticulatum*, *Gracilaria* spp., *Porphyra/Pyropia* spp., and *Pterocladiella capillacea* (formerly *Pterocladia capillacea*) have been tested to feed finfish and shrimp, with very interesting results. Likewise, green seaweed species such as *Ulva* spp. were extensively studied, the results indicating that their inclusion in feed can induce positive effects on growth, body composition, and resistance to stress and diseases (Pereira et al. 2012, Natify et al. 2015, Silva et al. 2015). However, as with other seaweeds, they can contain substances that may be toxic or have an anti-nutrient activity and reduce digestive enzyme activity, leading to a reduction in nutritional quality of the feed and feed utilization, and consequently to negative effect on fish growth (Omnes et al. 2017). But further studies are needed, since feeding gilthead seabream (*Sparus aurata*) with *Ulva lactuca* showed no negative effects on growth performance, probably because *U. lactuca* is part of its natural diet (Shpigel et al. 2017).

Regarding marine shrimps fed with seaweeds, different results have been achieved, probably due to different methods of analysis employed, distinct species, and seasonal and geographical variations. Some interesting results were obtained in marine white shrimp (*Litopenaeus vannamei*) fed with 4% brown seaweeds *Sargassum filipendula* and *Undaria pinnatifida.* The growth performance remained unchanged, but the digestive capacity was higher because of the increase in the gut epithelia surface (allowing better absorption) and increasing amylase and lipase activity (Schleder et al. 2018). The increase of amylase activity is beneficial, since it promotes the digestion of polysaccharides, thus making protein and lipids available for digestion and adsorption. The increase in lipase activity is also significant because it improves the absorption of lipids from feed ingredients. Regarding the immune response, the authors noticed an increased resistance to white spot syndrome virus and lower counts of *Vibrio* spp. in the shrimp digestive system. The authors suggest that these results may be due to supplementation with brown seaweeds, which have the highest content of bioactive compounds, improving immune and antimicrobial responses. Similar studies using *Ulva lactuca* (Chlorophyta), *Eisenia* sp. (Phaeophyceae) and *Porphyra/Pyropia* sp. (Rhodophyta) extracts in a feeding trial on white shrimp showed that an inclusion level of 5% of all types of seaweed resulted in an increase of the chymotrypsin, lipase and amylase enzyme activities. Therefore, the use of any of the three proposed seaweed extracts in balanced feed, especially *Ulva*, might promote shrimp growth and productivity (Omont et al. 2018).

Other valuable aquaculture organisms are abalone species (*Haliotis* spp.), marine gastropods, usually fed with terrestrial crops or wild seaweeds. The use of cultivated seaweeds in place of such feeds could provide for better environmental, nutritional and/or economic outcomes, decreasing environmental impacts and nutritional volatility of wild specimens, which affect abalone quality. To substitute for wild seaweed specimens, it has been suggested that these organisms can efficiently be grown in IMTA systems, fed with different cultivated species of seaweeds, namely *Ulva lactuca*, improving their dietary value (Viera et al. 2011). Other species used in all-vegetable-based formulated feeds for abalone such as *Laminaria digitata* (Phaeophyceae) showed improved growth, condition index and dietary protein utilization, but also reduced efficiency of dietary protein. Also, the n-6/n-3 ratio was much lower in the fresh algae, suggesting that abalone elongates fatty acids to larger carbon chains (Viera et al. 2015). *Grateloupia turuturu* (Rhodophyta), an invasive species in Europe, also provided for significantly higher specific growth rates, both length and weight, of abalone compared to formulated feeds, with higher carbohydrate/protein ratio, higher ash content and lower lipid content (Mulvaney et al. 2013).

Some negative effects on fish growth have been reported that may be attributable to the anti-nutrients present in the seaweed species (Yeganeh and Adel 2018), but it is worth noting that these studies have proved that the presence of seaweeds in fish diets has positive effects on health and well-being, although the presence of seaweeds in the fish feed is dose-dependent and species-specific (Valente et al. 2006, Norambuena et al. 2015). Small amounts of algae in fish diets resulted in some positive effect in growth performance, feed utilization efficiency, carcass quality, physiological activity, intestinal microbiota, disease resistance, improved stress response, immune system, and feed intake, among other parameters, demonstrating the potential of such organisms as feed supplement for aquaculture fish (Shields and Lupatsch 2012, Wassef et al. 2013, Cárdenas et al. 2015, Makkar et al. 2015, Niu et al. 2018, Rudtanatip et al. 2018, Schleder et al. 2018).

4 Seaweeds as Functional Feed Additives

The scientific community has demonstrated its commitment to the discovery and inclusion of new functional ingredients and the development of novel aquaculture fish feed formulations. These new components, once incorporated in the diet, can bring benefits to animal health and welfare status in addition to the basic nutritional effect (Martirosyan and Singh 2015). Recently, the concept of functional feed ingredient evolved to include also aspects related to environmental protection and economic gains (Olmos-Soto et al. 2015). This development of novel feed formulations and new ingredients is of utmost importance and represents a promising field from both the industrial and scientific point of view (Miranda et al. 2017). The inclusion of seaweed biomass in the fish diet, in different percentages, showed promising results in several parameters, and it must be emphasized that in this concept of functional food, the presence of bioactive substances is of fundamental importance. These are secondary metabolism products, which have beneficial effects on the health of the animal and can act synergistically, when added to food or feed, usually in small amounts. They represent the basis of the functional food and feed concepts and may include compounds such as flavonoids, lignin, isoprenoids, pigments such as carotenoids and chlorophylls, vitamins, phenolic acids, lipids such as fatty acids and sterols, polysaccharides, namely fibres, and some proteins and peptides.

Aquaculture feed has been evaluated for its productive parameters, regarding fish growth performance, survival rates and nutrient retention. Yet, it is quite clear that other parameters should be evaluated so as to assess the health condition of the cultivated fish, which will increase production yield, meat quality and, thus, market price. The use of functional foods, defined as "foods with dietary ingredients that provide healthy and economic benefits beyond basic nutrition", namely seaweed biomass, which enclose important bioactive compounds, represents a great opportunity in animal nutrition (Olmos-Soto et al. 2015). The wide variety and biological properties of the compounds discovered can give us a clear idea of the potential associated with the incorporation of seaweed extracts or isolated molecules in functional fish feed. In aquaculture, it is well known that seaweeds contain a number of bioactive compounds and consequently can be used as a fish feed additive to treat bacterial disease in a prophylactic and preventive way (Thanigaivel et al. 2016). There is, therefore, an increase in the use of balanced nutrition feed additives to improve cultured fish and shellfish health and disease resistance (Meena et al. 2013).

Feeding the red algae *Gracilaria pulvinata* as a protein source in fishmeal-based diets for Asian seabass (*Lates calcarifer*), a carnivorous species, showed that growth and feeding performance, including specific growth rate, feed conversion, protein efficiency ratios, feed intake and nutrient contents in the carcass, did not change significantly. However, serum lysozyme, alternative complement activities, serum immunoglobulin, and total protein content significantly decreased in fish fed with 9% supplementation. The levels of intestinal total protease and amylase activities remained unchanged; however, intestine lipase activity was significantly higher in fish fed with diets that contained 6% *G. pulvinata*. Globally, the inclusion of *Gracilaria* at 3%, the lower percentage studied, did not show negative effects on the growth performance, body composition, and health parameters and it improved the immune system of the fish (Morshedi et al. 2018). Effects of dietary seaweed supplementation with *Gracilaria* sp. (Rhodophyta) and *Alaria* sp. (Phaeophyceae) were also assessed in meagre (*Argyrosomus regius*) subjected to bacterial infection, using *Photobacterium damselae* subsp. *piscicida* (Bacteria, Proteobacteria) (Peixoto et al. 2017), and no effects on the fish growth performance were registered. Curiously, lipid peroxidation levels were significantly higher in fish fed control diet than in fish fed supplemented diets, and an interaction between infection and diet was found for glutathione peroxidase and reduced glutathione activities, indicating that the inclusion of seaweeds may confer advantages in coping with biotic stressors. In Atlantic salmon (*Salmo salar*), supplementation with *Palmaria palmata* (Rhodophyta) led to an improved hepatic function and a similar growth performance when compared to a control diet (Wan et al. 2016). In addition, the inclusion of this seaweed in the salmon diets enhances the surface colour of the fillets due to deposition of *P. palmata* pigments; it may, therefore, be considered as a replacement functional ingredient for farmed Atlantic salmon feed (Moroney et al. 2015).

Many other seaweed species were included in fish, shrimp, abalone, sea cucumber and other aquatic animal diets, proving that seaweed inclusion could be an important and cost-effective measure for intensive culture, showing important growth and health benefits to aquaculture animals.

5 Seaweeds as Fish Prebiotics and Antimicrobials

In recent years, we have become more aware of the fact that animal microbiota, in general, and gut microbiota, in particular, play a major role in health and disease prevention. Evidence regarding the beneficial effects, for either the host or the gut microbiota, of some feed ingredients or isolated biochemical compounds is now clear. Among these, a significant role is played by polysaccharides, such as alginates, fucoidans, carrageenans and exopolysaccharides, which are present in seaweeds. The possibility of using these prebiotic compounds to modulate the microbiome, and, consequently, prevent diseases is close to reality (Miranda et al. 2017). Prebiotics are defined as "selectively fermented ingredient that allows specific changes, both in the composition and/or activity in the gastrointestinal microflora that confers benefits upon host well-being and health" (Gibson et al. 2004). Prebiotics act by promoting the growth of specific commensal bacteria, which show the capacity to deter the adhesion and invasion of pathogenic microorganisms in the gastrointestinal epithelium. They also show additional effects, such as the stimulation of mucus production, production of short chain fatty acids, changes in the pH and cytokine induction. Altogether, the effect of prebiotics directly enhances the innate immunity response, showing effects on phagocytosis, macrophage activation and lysozyme production, and increasing antibody production. Aquatic animals growing in high-density systems are exposed to stress conditions that affect their immune system, thus becoming susceptible to a number of pathogens. This is a major problem in aquaculture, causing considerable economic losses necessitating the routine use of toxic chemicals and antibiotics. Excessive and incorrect use of chemicals such as tetracyclines and sulfomides, and anthelmintic agents such as pyrethroid insecticides and avermectins, has been reported in the past, with negative impacts including the leaching of residues to water and sediment, malfunctions of immune system, suppression of immune response, and the spread of antibiotic resistance (Romero et al. 2012, Hindu et al. 2018, Novais et al. 2018). The last, if transferred to human pathogens, has obvious impacts on public health (Vatsos and Rebours 2015).

Prebiotics, or non-digestible forage additives, stimulate the activity or abundance of beneficial gastrointestinal fish bacteria, improving the health and disease resistance of aquatic animals, as reviewed by Akhter et al. (2015). Additionally, the antimicrobial activity of seaweeds and seaweed extracts is well known, but its effects vary with environmental factors such as habitat, season, geographical area, growth stage, sexual maturity, cultivation method and extraction techniques used (Narayani et al. 2011, Vatsos and Rebours 2015).

Prebiotics are carbohydrates that can be classified according to molecular size or degree of polymerization and can be found in seaweeds as cell-wall constituents (such as cellulose, hemicellulose, lignin, phycocolloids, and some other polysaccharides), but also as storage products, including floridean starch and laminarin (Kraan 2012), as previously discussed. According to Zaporozhets et al. (2014), the prebiotic activity of extracts or polysaccharides from marine seaweeds, combined with the biological properties so far discovered, shows great potential for use as functional nutrition ingredient. The results show that these compounds play a role in the modulation of intestinal microbiota, limiting gastrointestinal tract inflammation and normalizing the immune system. The *in vitro* effects of digested seaweeds from the species *Sargassum muticum* (Phaeophyceae) and *Osmundea pinnatifida* (Rhodophyta), simulating the physiological conditions that occur during the digestive process, confirm the effect in the modulation of the gut microbial ecology and the potential application as functional feed and bioactive ingredient source (Rodrigues et al. 2016). However, this study, like many others, focuses mainly on the human gut microbiota, and studies focusing in fish microbiota are rather scarce. Lee et al. (2016a) used olive flounder (*Paralichthys olivaceus*) to evaluate the effect of the brown algae *Ecklonia cava* and determine how it affects the growth rate and immune response to pathogenic bacteria. The results show that supplementation improved the growth and enhanced the innate immune response, leading to lower mortality when pathogenic bacteria infected the fish. The same algae, *Ecklonia cava*, fed to zebrafish (*Danio rerio*), led to improved growth and survival of fish infected with *Edwardsiella tarda*. *Ecklonia cava* was also shown to stimulate the growth of lactic acid bacteria in a dose-dependent manner, have antimicrobial activity against fish pathogenic bacteria, and have uses as a prebiotic (Lee et al. 2016b).

It is widely accepted that healthy gut microbiota is crucial to support host health and well-being, and the gut microbiota must be further investigated since the gastrointestinal tract is one of the foremost routes of infection in fish. Strengthening the defence mechanisms by supplying prebiotics present in marine seaweeds, through diet, is a most promising method of controlling disease in aquatic animals. A recent study on characterization and *in vitro* testing of polysaccharides from *Grateloupia filicina* and *Eucheuma denticulatum* (formerly *Eucheuma spinosum*) (Rhodophyta) (Chen et al. 2018) promoted *Bifidobacterium* (Bacteria, Actinobacteria) proliferation. According to the authors, the different behaviours of the seaweed-origin polysaccharides might reflect differences in monosaccharide composition and structure, but more studies must be carried out *in vivo*. It is, indeed, necessary to look for a better understanding of the benefits of prebiotics in aquatic organisms, since these compounds seem to exhibit promising effects, but the research is still limited.

The potential use of seaweeds and their extracts as therapeutic or prophylactic agents in aquaculture is regarded as promising, since most of the groups of seaweeds analysed so far exhibit antimicrobial properties against pathogens of fish and shrimp. As a general rule, the bacterial infections are the most prevalent (around 50%), followed by viral, parasitic and fungal infections. Some of the most common fish infectious bacteria are furunculosis (*Aeromonas salmonicida*) and vibriosis (*Vibrio salmoninarum, V. angullarum, V. alginoliticus*). Fungus diseases include water moulds (*Saprolegniaceae*) which affect skin and gills, and other parasites that live on fish and other aquaculture animals, such as copepods and trematodes (Pillay and Kutty 2005). Shrimp farmers also find disease outbreaks caused by virus or vibriosis, including *V. alginoliticus*, that may result in high mortality (Chen et al. 2014). Bacterial species that cause diseases in fish and shrimp are quite common in the aquatic environment, hence the evaluation of the capacity of different seaweed species to inhibit the growth of fish pathogenic bacterial species makes it possible to assess their potential as alternatives to the use of antibiotics in aquaculture.

Most of the data collected so far in fish refers to *in vitro* studies, and no studies are published with reference to salmon or carp, but there are some studies in shrimp that focus on the antimicrobial effects of seaweed extracts mainly against *Vibrio* species (Kanjana et al. 2011, Thanigaivel et al. 2014). Crude extract of the red seaweed *Asparagopsis* sp. were evaluated for *in vivo* activity against

Vibrio in tiger prawn *Penaeus monodon* and proved to be highly effective in controlling the infection (Manilal et al. 2012). Regarding the viral pathogens, fungus and parasites affecting fish, there is little information and *in vivo* studies are limited (Manilal et al. 2009, Hutson et al. 2012).

The potential use of marine seaweeds in aquaculture has driven a number of studies, among them studies on *Asparagopsis taxiformis* (Rhodophyta)*,* considered one of the most promising species for the production of bioactive metabolites. The *in vitro* antibacterial activity, easy handling and absence of adverse effects on marine fish species were reported, although the bioactivities depend on the seasonal period of sampling (Marino et al. 2016). Ethanol extracts of *A. taxiformis* exhibited significantly inhibitory activity against *Aeromonas salmonicida* subsp. *salmonicida*, *Vibrio alginolyticus*, and *V. vulnificus*, and moderate activity against *Photobacterium damselae* subsp. *damselae*, *P. damselae* subsp. *piscicida*, *V. harveyi* and *V. parahaemolyticus*. European seabass (*Dicentrarchus labrax*) and gilthead seabream (*Sparus aurata*) were fed with pellets supplied with the alga and algal extracts and there were no negative effects on the fish. The bioactivy of seaweed extracts was also analysed in *Gracilariopsis longissima* (Rhodophyta) lipidic extract, showing a broad spectrum against *Vibrio* species (Cavallo et al. 2013). Agra Wijnana et al. (2018) and Kasanah et al. (2018) reported the bioactivity of *Gracilaria arcuata* and *Hydropuntia edulis* (Rhodophyta) against *Aeromonas hydrophila* and *Vibrio* sp. and identified the antibacterial compounds as hexadecanoic acid and sterol compounds. Additionally, ethanol extract of *Anthophycus longifolius* (formerly *Sargassum longifolium*) (Phaeophyceae) was found to be effective against *A. salmonicida* and the compounds with antibacterial activity were isolated and purified (Thanigaivel et al. 2018).

When antimicrobial properties are studied *in vivo*, the seaweed extracts are either incorporated in the feed or added to the water directly (Vatsos and Rebours 2015). When algal biomass is used directly in the feed it is dried beforehand to reduce the water content present in the final formulation. The drying method must be efficient, fast (preventing deterioration of biomass) and low in cost, but also adaptable to an industrial scale. Lyophilization normally exhibits excellent advantages in terms of the conservation of bioactive properties, but cost renders its use unacceptable. The alternative is to dry the seaweeds in chambers with forced ventilation, at temperatures of 20-25°C. Another option is the delivery of extracted bioactive compounds, either dissolved in the fish tank water (thus they must be soluble) or included in feed as additives. In this case, the doses can be calculated accurately, and extracts seem to act directly against the pathogens and/or stimulate the fish immune system, thus being a more efficient mechanism than direct biomass use (Vatsos and Rebours 2015). Organic solvents tend to extract active substances with higher antimicrobial activities, for instance, and water is rarely used, but different solvent and extraction methods were used in studies, making it difficult to compare results. Extracts incorporated in the feed seem to be an effective delivery method, allowing for the treatment and prevention of pathogenic diseases, but the costs are still too high for the industry to use them extensively. Based on the current state of the art, seaweed extracts show promising properties as antimicrobial agents, but further research is needed.

6 Seaweeds as Immunostimulants

Disease in aquaculture fish, shrimp and other shellfish can be caused by a biological agent—pathogen—or by the environment. Potential pathogens are always present in water, including virus, bacteria, fungus, protozoans, helminths and other worms, and parasitic crustaceans. The host-pathogen relationship usually undergoes several stages, beginning with an incubation period of a few days, after which symptoms become evident. Whether the host survives the infection depends on its immunity condition. Survivors may totally recover from the outbreak or may become asymptomatic carriers of the pathogen (Pillay and Kutty 2005).

Outbreaks tend to occur when the immunity system activity is low, and when fish are under stressful conditions. The control of such infectious diseases in farmed fish depends, as stated, on the use of different chemotherapeutic agents with antimicrobial activity; therefore, the uncontrolled growth of resistant pathogens in aquaculture resulting from the extensive use of antibiotics is a major health issue and the main concern of researchers and farmers (Meena et al. 2013, Akbary and Aminikhoei 2018).

To prevent disease outbreaks there is a need to increase fish health status by enhancing innate humoral and cellular defence mechanisms (Meena et al. 2013), which demand a balanced diet, appropriate environmental conditions, implementation of biosecurity measures and preventive methods (efficient, durable and innocuous), including the use of vaccines, probiotics, and immunostimulants shown to significantly reduce the need for antibiotics (World Health Organization 2011). In many countries, legislation has already been approved that regulates the use of antibiotics in fish farms, significantly decreasing their use. Therefore, the use of immunostimulants, that is, a natural or chemical protective compound, has increased to induce non-specific immune responses promoting disease resistance (Thanigaivel et al. 2015, Vallejos-Vidal et al. 2016, Yeganeh and Adel 2018).

Immunostimulants also improve health and growth of farmed animals, reduce mortality due to opportunistic pathogens, prevent virus diseases, enhance resistance to parasites, virus, bacteria and fungus, reduce mortality, and enhance the efficacy of vaccines and anti-microbial substances (Raa 2000). Evaluation of the immune response of farmed fish or shellfish should focus on different parameters. These include haematocrit and leucocyte count, phagocytic activity, bactericidal activity, oxidative radical production including the generation of superoxide and nitric oxide radicals, peroxidase production, immunoglobulin concentration and *in vitro/in vivo* measurements (Barman et al. 2013, Vallejos-Vidal et al. 2016). The evaluation of the immune response also depends on many other mechanisms such as enzymatic action. For example, lysozyme is a lytic enzyme present in saliva and it is important in the defence against pathogens. Other important molecules include myeloperoxidase, immunoglobulin IgM, and cytokines (Vallejos-Vidal et al. 2016).

Among the immunostimulants currently tested in aquaculture, seaweeds have proved to be important ingredients as they promote growth, stimulate the appetite, enhance tonicity, exhibit immunostimulant properties, and also exhibit anti-pathogen properties (Reverter et al 2014). Most immunostimulants currently used in aquaculture are structural bacterial lipopolysaccharides, bacterial glucans and other bacterial derivatives, algae carbohydrates including glucans, sulphated polysaccharides, peptides, vitamins, carotenoids, fatty acids, growth hormones, minerals, prebiotics and probiotics, amongst other molecules (Barman et al. 2013, Meena et al. 2013, Telles et al. 2018). The use of immunostimulants increases resistance and thus protects fish against many infectious bacterial diseases, such as *Vibrio* spp., *Aeromonas salmonicida* and *Streptococcus* sp., as well as viral infections such as infectious hematopoietic necrosis and yellowhead disease and parasitic infections such as white spot disease and sea lice (Barman et al. 2013).

One of the most interesting immunostimulants found in seaweeds is laminarin, present in brown seaweed and usually obtained from *Laminaria hyperborea*. This polymer directly activates macrophages by binding directly to them and to other white blood cells (Meena et al. 2013). This property, along with the increase in respiratory burst activity, has been proved in Atlantic salmon (*Salmo salar*) (Dalmo and Seljelid 1995). Additionally, feeding laminarin to mussels showed improved immune response and modulation of the haemocyte activity and restricted the growth of *Vibrio* spp. and *Escherichia coli* (Mar Costa et al. 2008). This resistance enhancement has also been reported for other bacterial infections in several species of fish such as carp (*Cyprinus carpio*), Atlantic salmon (*Salmo salar*), rainbow trout (*Onchorhynchus mykiss*), yellow tail (*Seriola quinquradiata*) and African catfish (*Clarias gariepinus*) (Barman et al. 2013). Finally, rainbow trout (*Oncorhynchus mykiss*) treated with beta-glucans also possessed higher gene expression with regard to pro-inflammatory cytokines and anti-inflammatory cytokines (Zhang et al. 2009). Carrageenan extracted from *Kappaphycus alvarezi* (Rhodophyta) also seems to exhibit immunostimulant potential against bacterial attacks, such as *Vibrio alginolyticus* on white shrimp. Besides, the fermentation process of the seaweed reduces carbohydrate content, maintains carrageenan content and increases some amino acids such as glycine, alanine, proline and serine that together contribute to the immunostimulant effect of the red seaweed. Thus, the use of 10% supplement is important either nutritionally or as a probiotic (Hardjani et al. 2017). As to alginates, Skjermo et al. (1995) states that mannuronic acid of alginate extracted from *Ascophyllum nodosum* (Phaeophyceae) is an active component in immune system, showing a strong stimulatory effect on cytokine production by human monocytes. More recent work also using alginate and cellulose on fish feed demonstrated an increased in lysozyme activity of beluga (*Huso huso*) and weight of fish (Heidarieh et al. 2011). Also, Atlantic cod (*Gadus morhua*) fed with alginate showed an increase in specific growth rate, but when the same supplements were used in spotted wolffish (*Anarhichas minor*), the growth rate increased only in low doses of algae-derived alginate. Hence, the immunostimulant effect seems to be species-specific (Vollstad et al. 2006). The immune stimulant effect of seaweed polysaccharides seems to be linked to their ability to modulate the intestinal metabolism, namely fermentation, and also to inhibit the pathogen evasion and adhesion (Hindu et al. 2018).

Another algae-derived compound evaluated as an immunostimulant in fish is fucoxanthin, which has anti-inflammatory activity and antioxidant benefits, amongst others. In gilthead seabream (*Sparus aurata*) fed with a fucoxanthin-supplemented diet formulation a significant increase in haemolytic complement activity was detected, as well as phagocytic capacity, and expression level of beta-defensin, thus proving the immunostimulant effect of astaxanthin (Cerezuela et al. 2012).

Immunostimulants seem, in fact, to be useful tools for prophylactic treatment of farmed fish and shellfish. They are safer than chemotherapeutants and their range of effectiveness is wider than vaccination. The strength of these compounds seems to rely on their ability to enhance juvenile and larval cultures before their specific immune system matures, and their ability to improve nonspecific immune function against a broad range of pathogens (Barman et al. 2013). However, vaccines, balanced diet nutrition and good management techniques are still mandatory.

7 Seaweeds as Antioxidants

In the metabolism of a seaweed, as in all living organisms, reactive oxygen species (ROS) are normally produced and eliminated by enzymatic and non-enzymatic defence systems (Cavas and Yurdakoc 2005). ROS are produced in different molecular forms, such as superoxide radical (O_2^-), hydroxyl radical (˙OH), hydrogen peroxide (H_2O_2) and nitric oxide (NO). These may be harmful to the cell because of their involvement in the initiation of molecular oxidations and enzyme activation, generating oxidative damage that leads to cell injury and ultimately to death. ROS are produced from two different sources: endogenous sources such as mitochondria, peroxisome, photosynthesis, phagocytosis, enzymatic reactions and auto-oxidation reactions, and exogenous sources such as pollution, heavy metals and pesticides. They have a relatively short but highly reactive life, since they readily combine with biomolecules such as DNA, proteins, lipids and carbohydrates, producing changes in their structure and functioning. Under stressful conditions, the accumulation of ROS and other free radicals can cause irreversible damage to cellular components. The tissues with greater metabolic activity are those that suffer the most oxidative damage; in fish these are the gills, liver and kidney, the main damage being the oxidation of lipid membranes (lipid peroxidation), proteins and nucleic acids (Faheem and Lone 2005, Javed et al. 2017, Kaur 2017). Lipid peroxidation in fish is due to the contamination of water or the redox cycles of cellular enzymes, and in response living organisms produce catalases, superoxide dismutase and glutathione peroxidase to defend against the damage associated with the presence of these compounds. In addition to these antioxidants of enzymatic nature, there are several antioxidant compounds responsible for the protection of biological systems against the adverse effects of excessive oxidation. These molecules are of great interest in both food and feed industries, particularly for their ability to improve the stability of oils and fats rich in PUFA. This group of antioxidants include compounds obtained from the secondary metabolism of plants and algae, which can be understood as natural antioxidant substances and are extensively studied. These compounds belong to the phenolic and polyphenolic class as well as carotenoids and antioxidant vitamins, among others (Al-Saif et al. 2014, Al-Asgah et al. 2016). Some of these metabolites, such as carotenoids, glycerol, alginates, and carrageenan, are extensively used in the pharmaceutical and food industries. The activity of antioxidants and their mechanism of action are dependent on the structure of the molecules involved, but also on the system in which they are present.

Seaweeds are potential sources of bioactive secondary metabolites with antioxidant properties, probably due to growth conditions (high light intensities, tide fluctuations, and oxygen concentration) that result in photodamage and free radical production, thus stimulating the presence of anti-oxidative mechanisms and compounds with protecting effects (Narasimhan et al. 2013). Accordingly, the variations on the total phenolic content in marine seaweeds are dependent on extrinsic factors, such as light irradiance, depth of growth, salinity and nutrients, but also on intrinsic factors, such as seaweed age, morphology and reproductive stage.

The antioxidant activity of different extracts from marine seaweeds has been extensively studied, and several species already demonstrate high potential for application as antioxidants in the food, feed and pharmaceutical industries. The antioxidant activity of extracts of *Ulva lactuca, Ulva fasciata, Ulva intestinalis, Acrosiphonia orientalis, Ulva compressa* and *Caulerpa racemosa* (Chlorophyta)*, Laurencia obtusa* (Rhodophyta)*,* and others assessed using various antioxidant bioassays correlates with the total phenolic content in the algal extracts, highlighting the ability of these compounds to scavenge free radicals such as singlet oxygen, superoxide and hydroxyl radicals (Demirel et al. 2011, Siva Kumar and Rajagopal 2011, Kosanić et al. 2015). The supplementation of European seabass juvenile (*Dicentrarchus labrax*) diet with the seaweeds *Gracilaria* spp. (Rhodophyta), *Ulva* spp. (Chlorophyta), or *Fucus* spp. (Phaeophyceae), at different levels, showed no effect on the voluntary feed intake, feed conversion ratio, protein efficiency ratio, and growth performance but improved the overall immune and antioxidant responses (Peixoto et al. 2016). Some studies indicate that this activity may also be due to compounds such as pigments (chlorophylls, carotenoids), essential oils, and low molecular weight polysaccharides (Heo et al. 2005).

Previous research (Abd El-Baky et al. 2009, Al-Amoudi et al. 2009) found antioxidant activity for the green algae *U. lactuca* and (Heo et al. 2005, Zubia et al. 2007) found strong antioxidant capacity for an extract of *U. intestinalis*. These authors determined antioxidant activity for the above-mentioned species but by using other extraction solvents such as aqueous, methanol, hexane, and dichloromethane/methanol. Different extraction solvents, according to their polarity, may have extracted various compounds including pigments (chlorophyll a and b and carotenoids), alkaloids, and phenolic compounds, as well as essential oil, which can participate in the great antioxidant activity (Shanab et al. 2011). This means that synergistic effects may occur between these constituents leading to the pronounced antioxidant activity of algal extract (containing the antioxidant active components). In contrast, extracts that have not only lower pigment contents but also a lower content of phenolic compounds have reduced antioxidant activity. Also, in experiments with other algae carried out by numerous researchers (Shanab et al. 2011, Siva Kumar and Rajagopal 2011), it was found that the tested species exhibit different activities depending on the extraction solvents used. The total antioxidant activity greatly depends on the content of antioxidant active compounds in extracts because the type of solvent used in the extraction of the phenolic compounds can utterly change the results obtained, since the polarity of the solvent determines the substances to be dissolved and measured in the extract.

Likewise, some studies indicate that seaweeds may be a naturally rich source of macromolecular antioxidants, which are normally overlooked. These macromolecular antioxidants or non-extractable polyphenols are associated with insoluble dietary fibre. They are commonly ignored in studies of bioactive compounds but exhibit significant biological activity and show specific features that are distinct from both dietary fibre and extractable polyphenols isolated (Sanz-Pintos et al. 2017). Results indicate that macromolecular antioxidants represent a considerable fraction of the total polyphenol content, with hydroxycinnamic acids, hydroxybenzoic acids and flavonols being the main constituents. For a long time, dietary fibre and polyphenols were regarded as separate compounds, but they actually have a close relation, and they can be considered as a single unit (Saura-Calixto 2011). This insoluble macromolecular fraction showed remarkable antioxidant capacity, as determined by complementary assays, indicating that seaweed could be an important source of commonly ignored antioxidants.

Additionally, ROS not only can cause damage to cellular components, but also promote the degradation of oils and fats present in food and feed, leading to the appearance of odours and rancid flavour, thus contributing to a lower quality and nutritional composition and allowing for the formation of potentially toxic metabolites. The use of seaweed extracts as preventers of lipid peroxidation in fish feeds, a process that causes the rancidity of fats and the destruction of liposoluble vitamins (A, D and E), amino acids and carotenoids, is an interesting option. If the oxidative process can proceed in the feed or in any of its ingredients, the nutritional value of the feed might be seriously reduced, causing nutrient deficiencies. Oxidation can be controlled in several different ways, primarily by avoiding the use of highly unstable fats and oils, but also using antioxidant additives, either synthetic (such as BHA or BHT) or natural (such as natural tocopherols or plant extracts). These must be non-toxic for the fish, effective at low concentration and of low cost. A study of the presence of synthetic antioxidants BHA, BHT and ethoxyquin in several commercially important species of farmed fish, namely Atlantic salmon (*Salmo salar*), halibut (*Hippoglossus hippoglossus*), Atlantic cod (*Gadus morhua*) and rainbow trout (*Oncorhynchus mykiss*) indicates that the consumption of 300 g of farmed Atlantic salmon contributes to 75% of the acceptable daily intake for BHT (Lundebye et al. 2010). Consumer awareness of the risks associated with the presence of these synthetic additives in the fish carcass, and recent changes in the European regulation, are pushing the feed industry towards the use of natural antioxidants, such as vitamin E. These have a positive image and general acceptance, show few limitations on their use and have potential health benefits. On the other hand, they are costlier and have potential undesired flavours or odours, and they have not yet been extensively tested. The antioxidant effect of plant extracts, such as rosemary extracts, added to extruded fish feed stored at different temperatures showed promising results in delaying the lipid peroxidation process and they are currently used as a natural alternative (Hernández et al. 2014). The addition of these compounds can also have beneficial effects on fish, as shown when alpha-tocopherol combined with ascorbic acid, iron and selenium were added to the diet of Chinook salmon (*Oncorhynchus tshawytscha*) (Welker and Congleton 2009). The oxidative stress decreased, and the results indicated that tocopherols and ascorbic acid reduced oxidative stress levels, with the minimum required of both being 50 mg.kg^{-1} of feed. In serpae tetra (*Hyphessobrycon eques*) the addition of astaxanthin, beta-carotene, and a combination of the two reduced the oxidative stress in high ammonia conditions (Pan et al. 2011). Gao et al. (2012) analysed the effects of the supplementation of vitamin E in fish feed elaborated with oxidized oils on growth and on the level of lipid peroxide in liver and muscle of red seabream (*Pagrus major*). They concluded that supplementation with vitamin E leads to better growth and reduced lipid peroxide levels. Lu et al. (2016) also studied the effects of dietary vitamin E on growth performance and antioxidant status, and additionally looked at the innate immune response and resistance to *Aeromonas hydrophila* of juvenile yellow catfish (*Pelteobagrus fulvidraco*). They found that the dietary vitamin E levels significantly influenced weight gain, specific growth rate and feed efficiency. Moreover, fish diets supplemented with vitamin E had higher activities of superoxide dismutase and catalase in serum, when compared to control, and had a higher survival rate.

8 Conclusion

Aquaculture is a sector that has a high economic and social weight. The costs associated with feeding marine organisms are one of the most important productive factors; however, animal health and welfare are increasingly at the centre of attention and legislative changes have recently been observed, in order to safeguard the well-being and health of aquatic organisms as well as consumers, avoiding, for example, the massive use of chemotherapeutics.

Macroalgae may be incorporated into animal feed as a source of macronutrients (e.g., protein), but also as supplements, with beneficial effects as prebiotics, antioxidants, antimicrobials, or immunostimulants. Extraction of components present in seaweeds, such as polysaccharides, polyphenols, pigments, vitamins, and other compounds, has been shown to be a promising pathway. Currently, *in vivo* studies have increased and there is already ample evidence of the beneficial effects of the incorporation of various species of macroalgae into marine fish feed, particularly as regards its effects on growth and immune response. Some species show high potential as sources of protein or PUFA and may be suitable as substitutes for fishmeal and fish oils in fish feed formulation. There is also clear evidence of the antimicrobial effect of seaweeds against aquaculture fish pathogens that lead to mass losses. The use of natural extracts, namely from seaweeds, to substitute for antibiotics and other chemotherapeutics and thereby prevent resistance and infection and promote growth and general health is a promising route, and evidence of the associated advantages is now accumulating. However, regarding the antioxidant effects, there is still a long way to go, since *in vivo* studies are very limited, although *in vitro* studies demonstrate an enormous potential of macroalgae in preventing the harmful effects of ROS in cells. Identification and purification of bioactive compounds have been reported, thus allowing the combination of molecules and the study of potential synergistic effects. It must be remembered that purification sometimes prevents the occurrence of unexpected interactions between the compounds, which may result in deleterious effects on the growth and health of the animal.

The advantages associated with the use of macroalgae are undoubtedly linked to their enormous biodiversity, with more than 10,000 species referenced, translated into different abilities to adapt to a wide range of environmental conditions. This diversity necessarily results in the presence of natural compounds and a unique composition, for example, at the level of proteins, distinct from terrestrial plants. Macroalgae are an abundant resource in many areas of the globe, presenting high rates of photosynthesis and growth, and some species are easy to grow. This precious resource, whether in the form of biomass or in the form of extracted compounds, has naturally aroused interest in the feed sector, particularly for aquaculture, and an extensive use of seaweeds is foreseeable in the future.

Acknowledgments

The authors thank the Fundação para a Ciência e Tecnologia (FCT) for its support, through the strategic project UID/MAR/04292/2019 granted to MARE.

References Cited

Abd El-Baky, H.H., F.K. El Baz and G.S. El-Baroty. 2009. Evaluation of marine alga *Ulva lactuca* L. as a source of natural preservative ingredient. Elec. J. Env. Agricult. Food Chem. 7: 3353–3367.

Agra Wijnana, A.P., N. Kasanah and Triyanto. 2018. Bioactivity of red seaweed *Gracilaria arcuata* against *Aeromonas hydrophila* and *Vibrio* sp. Nat. Prod. J. 8: 147–152.

Ahmadifar, E., G. Azari Takami and M. Sudagar. 2009. Growth performance, survival and immunostimulation, of Beluga (*Huso huso*) juvenile following dietary administration of alginic acid (Ergosan). Pak. J. Nutr. 8: 227–232.

Akbary, P. and Z. Aminikhoei. 2018. Effect of water-soluble polysaccharide extract from the green alga *Ulva rigida* on growth performance, antioxidant enzyme activity, and immune stimulation of grey mullet *Mugil cephalus*. J. Appl. Phycol. 30: 1345–1353.

Akhter, N., B. Wu, A. Mahmood and M. Mohsin. 2015. Fish & Shellfish Immunology Probiotics and prebiotics associated with aquaculture: a review. Fish Shellfish Immunol. 45: 733–741.

Al-Amoudi, O.A., H.H. Mutawie, A.V. Patel and G. Blunden. 2009. Chemical composition and antioxidant activities of Jeddah corniche algae, Saudi Arabia. Saudi J. Biol. Sci. 16: 23–29.

Al-Asgah, N.A., E.S.M. Younis, A.W.A. Abdel-Warith and F.S. Shamlol. 2016. Evaluation of red seaweed *Gracilaria arcuata* as dietary ingredient in African catfish, *Clarias gariepinus*. Saudi J. Biol. Sci. 23: 205–210.

Al-Saif, S.S.A., N. Abdel-Raouf, H.A. El-Wazanani and I.A. Aref. 2014. Antibacterial substances from marine algae isolated from Jeddah coast of Red sea, Saudi Arabia. Saudi J. Biol. Sci. 21: 57–64.

Angell, A.R., S.F. Angell, R. de Nys and N.A. Paul. 2016. Seaweed as a protein source for mono-gastric livestock. Trends Food Sci. Technol. 54: 74–84.

Bagni, M., N. Romano, M.G. Finoia, L. Abelli, G. Scapigliati, P.G. Tiscar, M. Sarti and G. Marino. 2005. Short- and long-term effects of a dietary yeast β-glucan (Macrogard) and alginic acid (Ergosan) preparation on immune response in sea bass (*Dicentrarchus labrax*). Fish Shellfish Immunol. 18: 311–325.

Balfry, S.K. and D.A. Higgs. 2001. Influence of dietary lipid composition on the immune system and disease resistance of finfish. pp. 213–234. *In*: Lim, C. and Webste, C.D. (eds.). Nutrition and Fish Health. CRC Press, an Imprint of Taylor and Francis Group, Boca Raton FL.

Barman, D., P. Nen, S.C. Mandal and V. Kumar. 2013. Immunostimulants for aquaculture health management. J. Mar. Sci. Res. Dev. 3: 134.

Bocanegra, A., A. Bastida, J. Benedí, S. Ródenas and F.J. Sánchez-Muniz. 2009. Characteristics and nutritional and cardiovascular-health properties of seaweeds. J. Med. Food. 12: 236–258.

Burtin, P. 2003. Nutricional value of seaweeds. Electron. J. Environ. Agric. Food Chem. 2: 498–503.

Cabanero, P.C., B.L.M. Tumbokon and A.E. Serrano Jr. 2016. Nutritional evaluation of *Rhizoclonium riparium* var. *implexum* meal to replace soybean in the diet of Nile Tilapia Fry. Isr. J. Aquac. 68.2016.1278.

Caipang, C.M.A., C.C. Lazado, I. Berg, M.F. Brinchmann and V. Kiron. 2011. Influence of alginic acid and fucoidan on the immune responses of head kidney leukocytes in cod. Fish Physiol. Biochem. 37: 603–612.

Campos, L.N.S., F.D. Herrera, A.D.R. Araujo, R.A.G. Sánchez, M.L.L. Partida, M.J.A. Ruiz and A.F.L. Navarro. 2014. *Litopenaeus vannamei* immune stimulated with *Macrocystis pyrifera* extract: improving the immune response against *Vibrio campbellii*. J. Coast. Life Med. 2: 617–624.

Cárdenas, J.V., A.O. Gálvez, L.O. Brito, E.V. Galarza, D.C. Pitta and V.V. Rubin. 2015. Assessment of different levels of green and brown seaweed meal in experimental diets for whiteleg shrimp (*Litopenaeus vannamei*, Boone) in recirculating aquaculture system. Aquac. Int. 23: 1491–1504.

Cardoso, S.M., L.G. Carvalho, P.J. Silva, M.S. Rodrigues, O. Pereira and L. Pereira. 2014. Bioproducts from seaweeds: a review with special focus on the Iberian Peninsula. Curr. Org. Chem. 18: 896–917.

Castanho, S., G. Califano, F. Soares, R. Costa, L. Mata, P. Pousão-Ferreira and L. Ribeiro. 2017. The effect of live feeds bathed with the red seaweed *Asparagopsis armata* on the survival, growth and physiology status of *Sparus aurata* larvae. Fish Physiol. Biochem. 43: 1043–1054.

Castro, R., M.C. Piazzon, I. Zarra, J. Leiro, M. Noya and J. Lamas. 2006. Stimulation of turbot phagocytes by *Ulva rigida* C. Agardh polysaccharides. Aquaculture 254: 9–20.

Cavallo, R.A., M.I. Acquaviva, L. Stabili, E. Cecere, A. Petrocelli and M. Narracci. 2013. Antibacterial activity of marine macroalgae against fish pathogenic *Vibrio* species. Cent. Eur. J. Biol. 8: 646–653.

Cavas, L. and K. Yurdakoc. 2005. A comparative study: assessment of the antioxidant system in the invasive green alga *Caulerpa racemosa* and some macrophytes from the Mediterranean. J. Exp. Mar. Biol. Ecol. 321: 35–41.

Cerezuela, R., F.A. Guardiola, J. Meseguer and M.Á. Esteban. 2012. Enrichment of gilthead seabream (*Sparus aurata* L.) diet with microalgae: effects on the immune system. Fish Physiol. Biochem. 38: 1729–1739.

Chen, X., Y. Sun, L. Hu, S. Liu, H. Yu, R. Xing, R. Li, X. Wang and P. Li. 2018. *In vitro* prebiotic effects of seaweed polysaccharides. J. Oceanol. Limnol. 36: 926–932.

Chen, Y.Y., J.-C. Chen, Y. Lin, S. Yeh, K. Chao and C. Lee. 2014. White shrimp *Litopenaeus vannamei* that have received *Petalonia binghamiae* extract activate immunity, increase immune response and resistance against *Vibrio alginolyticus*. J. Aquac. Res. Dev. 5: 268.

Chojnacka, K., A. Saeid, Z. Witkowskaand and L. Tuhy. 2012. Biologically active compounds in seaweed extracts – the prospects for the application. Open Conf. Proc. J. 3: 20–28.

Dalmo, R.A. and R. Seljelid. 1995. The immunomodulatory effect of LPS, laminaran and sulphated laminaran [β(l,3)-D-glucan] on Atlantic salmon, *Salmo salar* L., macrophages *in vitro*. J. Fish Dis. 18: 175–185.

Dashtiannasab, A., M. Mesbah, R. Pyghan and S. Kakoolaki. 2016. The efficacy of the red seaweed (*Laurencia snyderiae*) extract on growth performance, survival and disease resistance in white shrimp. Iran. J. Aquat. Anim. Health 2: 1–10.

Davies, S.J., M.T. Brown and M. Camilleri. 1997. Preliminary assessment of the seaweed *Porphyra purpurea* in artificial diets for thick-lipped grey mullet (*Chelon labrosus*). Aquaculture 152: 249–258.

Dawczynski, C., R. Schubert and G. Jahreis. 2007. Amino acids, fatty acids, and dietary fibre in edible seaweed products. Food Chem. 103: 891–899.

de Jesus Raposo, M.F., A.M.M.B. De Morais and R.M.S.C. De Morais. 2015. Marine polysaccharides from algae with potential biomedical applications. Mar. Drugs 13: 2967–3028.

de Jesus Raposo, M.F., A.M.M.B. De Morais and R.M.S.C. De Morais. 2016. Emergent sources of prebiotics: seaweeds and microalgae. Mar. Drugs 14: 1–27.

Debbarma, J., B. Madhusudana Rao, L. Narasimha Murthy, S. Mathew, G. Venkateshwarlu and C.N. Ravishankar. 2016. Nutritional profiling of the edible seaweeds *Gracilaria edulis*, Ulva lactuca and *Sargassum* sp. Indian J. Fish. 63: 81–87.

Declarador, R.S., A.E. Serrano and V.L. Corre Jr. 2014. Ulvan extract acts as immunostimulant against white spot syndrome virus (WSSV) in juvenile black tiger shrimp *Penaeus monodon*. AACL Bioflux 7(3): 153–161.

Demirel, Z., F.F. Yilmaz-Koz, N.U. Karabay-Yavasoglu, G. Ozdemir and A. Sukatar. 2011. Antimicrobial and antioxidant activities of solvent extracts and the essential oil composition of *Laurencia obtusa* and *Laurencia obtusa* var. *pyramidata*. Rom. Biotechnol. Lett. 16: 5927–5936.

Díaz-Rosales, P., C. Felices, R. Abdala, F.L. Figueroa, J.L. Gómez Pinchetti, M.A. Moriñigo and M.C. Balebona. 2007. *In vitro* effect of the red alga *Hydropuntia cornea* (J. Agardh) on the respiratory burst activity of sole (*Solea senegalensis*, Kaup 1858) phagocytes. Aquac. Res. 38: 1411–1418.

Diler, I., A.A. Tekinay, D. Guroy, B.K. Guroy and M. Soyuturk. 2007. Effects of *Ulva rigida* on the growth, feed intake and body composition of common Carp, *Cyprinus carpio* L. J. Biol. Sci. 7: 305–308.

Dornbos, D.L. and R.E. Mullen. 1992. Soybean seed protein and oil contents and fatty acid composition adjustments by drought and temperature. J. Am. Oil. Chem. Soc. 69: 228–231.

Ergün, S., M. Soyutürk, B. Güroy, D. Güroy and D. Merrifield. 2009. Influence of *Ulva* meal on growth, feed utilization, and body composition of juvenile Nile tilapia (*Oreochromis niloticus*) at two levels of dietary lipid. Aquacult. Int. 17: 355–361.

Fabrowska, J., B. Łeska, G. Schroeder, B. Messyasz and M. Pikosz. 2015. Biomass and extracts of algae as material for cosmetics. pp. 681–700. *In*: Kim, S.-K. and Chojnacka, K. (ed.). Marine Algae Extracts – Processes, Products and Applications, Vol. II. Wiley,-VCH, Germany.

Faheem, M. and K.P. Lone. 2005. Oxidative stress and histopathologic biomarkers of exposure to bisphenol-A in the freshwater fish, *Ctenopharyngodon idella*. Environ. Pollut. 178: 41–51.

FAO. 2018. The state of world fisheries and aquaculture. Food and Agriculture Organization of the United Nations, Rome, 227 pp.

Felix, S., P.H. Robins and A. Rajeev. 2004. Immune enhancement assessment of dietry incorporated marine alga *Sargassum wightii* (Phaeophyceae/Punctariales) in tiger shrimp *Penaeus monodon* (Crustacia/Penaeidae) through prophenoloxidase (proPO) systems. Indian J. Mar. Sci. 33: 361–364.

Fleurence, J. 1999. Seaweed proteins: biochemical, nutritional aspects and potential uses. Trends Food Sci. Technol. 10: 25–28.

Fleurence, J. 2017. Seaweed proteins. pp. 198–213. *In*: Yada, R.Y. (ed.). Proteins in Food Processing, 2nd Ed. Woodhead Publishing, Elsevier, U.K.

Fleurence, J., G. Gutbier, S. Mabeau and C. Leray. 1994. Fatty acids from 11 marine macroalgae of the French Brittany coast. J. Appl. Phycol. 6: 527–532.

Fleurence, J., M. Ele Morançais, J. Dumay, P. Decottignies, V. Turpin, M. Munier, N. Garcia-Bueno and P. Jaouen. 2012. What are the prospects for using seaweed in human nutrition and for marine animals raised through aquaculture? Trends Food Sci. Technol. 27: 57–61.

Floreto, E.A.T., S. Teshima and M. Ishikawa. 1996. The Effects of seaweed diets on the growth, lipid and acids of juveniles of the white sea Urchin *Tripneustes gratila*. Fish. Sci. 62: 589–593.

Francavilla, M., M. Franchi, M. Monteleone and C. Caroppo. 2013. The red seaweed *Gracilaria gracilis* as a multi products source. Mar. Drugs 11: 3754–3776.

Freitas, A.C., D. Rodrigues, T.A.P. Rocha-Santos, A.M.P. Gomes and A.C. Duarte. 2012. Marine biotechnology advances towards applications in new functional foods. Biotechnol. Adv. 30: 1506–1515.

Fu, Y.-W., W.-Y. Hou, S.-T. Yeh, C.-H. Li and J.-C. Chen. 2007. The immunostimulatory effects of hot-water extract of *Gelidium amansii* via immersion, injection and dietary administrations on white shrimp *Litopenaeus vannamei* and its resistance against *Vibrio alginolyticus*. Fish Shellfish Immunol. 22: 673–685.

Fujiki, K., D.-H. Shin, M. Nakao and T. Yanot. 1997. Effects of kappa-carrageenan on the non-specific defense system of Carp *Cyprinus carpio*. Fish. Sci. 63: 934–938.

Fujiwara-Arasaki, T., N. Mino and M. Kuroda. 1984. The protein value in human nutrition of edible marine algae in Japan. Hydrobiologia 116–117: 513–516.

Gao, J., S. Koshio, M. Ishikawa, S. Yokoyama, R.E.P. Mamauag and Y. Han. 2012. Effects of dietary oxidized fish oil with vitamin E supplementation on growth performance and reduction of lipid peroxidation in tissues and blood of red sea bream Pagrus major. Aquac. 356–357: 73–79.

Gibson, G.R., H.M. Probert, J. Van Loo, R.A. Rastall and M.B. Roberfroid. 2004. Dietary modulation of the human colonic microbiota: updating the concept of prebiotics. Nutr. Res. Rev. 17: 259.

Gora, A.H., N.P. Sahu, S. Sahoo, S. Rehman, S.A. Dar, I. Ahmad and D. Agarwal. 2018. Effect of dietary *Sargassum wightii* and its fucoidan-rich extract on growth, immunity, disease resistance and antimicrobial peptide gene expression in *Labeo rohita*. Int. Aquat. Res. 10: 115–131.

Halver, J.E. and R.W. Hardy. 2002. Fish nutrition, 3rd Ed. Academic Press, Elsevier Science, California, USA, 500 pp.

Hardjani, D.K., G. Suantika and P. Aditiawati. 2017. Nutritional profile of red seaweed *Kappaphycus alvarezii* after fermentation using *Saccharomyces cerevisiae* as a feed supplement for white shrimp *Litopenaeus vannamei* nutritional profile of fermented red seaweed. J. Pure. Appl. Microbiol. 11: 1637–1645.

Hardy, R.H. 2001. Nutritional deficiencies in commercial aquaculture: likelihood, onset, and identification. pp. 131–147. *In*: Lim, C. and Webste, C. (eds.). Nutrition and Fish Health. Food Products Press, N.Y.

Heidarieh, M., M. Soltani, A.H. Tamimi and M.H. Toluei. 2011. Comparative effect of raw fiber (Vitacel) and alginic acid (Ergosan) on growth performance, immunocompetent cell population and plasma lysozyme content of Giant Sturgeon (*Huso huso*). Turkish J. Fish. Aquat. Sci. 11: 445–450.

Heo, S., S. Cha, K. Lee, S.K. Cho and Y. Jeon. 2005. Antioxidant activities of Chlorophyta and Phaeophyta from Jeju Island. Algae 20: 251–260.

Herman, E.M. and M.A. Schmidt. 2016. The potential for engineering enhanced functional-feed soybeans for sustainable aquaculture feed. Front. Plant. Sci. 7: 1–6.

Hernández, A., B. García García, M.J. Jordán and M.D. Hernández. 2014. Natural antioxidants in extruded fish feed: protection at different storage temperatures. Anim. Feed Sci. Technol. 195: 112–119.

Hindu, S.V., N. Chandrasekaran, A. Mukherjee and J. Thomas. 2018. A review on the impact of seaweed polysaccharide on the growth of probiotic bacteria and its application in aquaculture. Aquacult. Int. 23: 543–597.

Holdt, S.L. and S. Kraan. 2011. Bioactive compounds in seaweed: functional food applications and legislation. J. Appl. Phycol. 23: 543–597.

Hoseinifar, S.H., S. Yousefi, G. Capillo, H. Paknejad, M. Khalili, A. Tabarraei, H. Van Doan, N. Spanò and C. Faggio. 2018. Mucosal immune parameters, immune and antioxidant defence related genes expression and growth performance of zebrafish (*Danio rerio*) fed on *Gracilaria gracilis* powder. Fish Shellfish Immunol. 83: 232–237.

Huang, X., H. Zhou and H. Zhang. 2006. The effect of *Sargassum fusiforme* polysaccharide extracts on vibriosis resistance and immune activity of the shrimp, *Fenneropenaeus chinensis*. Fish Shellfish Immunol. 20: 750–757.

Hutson, K.S., L. Mata, N.A. Paul and R. de Nys. 2012. Seaweed extracts as a natural control against the monogenean ectoparasite, *Neobenedenia* sp., infecting farmed barramundi (*Lates calcarifer*). Int. J. Parasitol. 42: 1135–1141.

Huynh, T.-G., S.-T. Yeh, Y.-C. Lin, J.-F. Shyu, L.-L. Chen and J.-C. Chen. 2011. White shrimp *Litopenaeus vannamei* immersed in seawater containing *Sargassum hemiphyllum* var. *chinense* powder and its extract showed increased immunity and resistance against *Vibrio alginolyticus* and white spot syndrome virus. Fish Shellfish Immunol. 31: 286–293.

andIsnansetyo, A., H.M. Irpani, T.A. Wulansari and N. Kasanah. 2015. Oral administration of alginate from a tropical brown seaweed, *Sargassum* sp. to enhance non-specific defense in Walking Catfish (*Clarias* sp.). Aquac. Indones. 15.

Isnansetyo, A., A. Fikriyah, N. Kasanah and Murwantoko. 2016. Non-specific immune potentiating activity of fucoidan from a tropical brown algae (Phaeophyceae), *Sargassum cristaefolium* in tilapia (*Oreochromis niloticus*). Aquacult. Int. 24: 465–477.

Javed, M., M.I. Ahmad, N. Usmani and M. Ahmad. 2017. Multiple biomarker responses (serum biochemistry, oxidative stress, genotoxicity and histopathology) in *Channa punctatus* exposed to heavy metal loaded waste water. Sci. Rep. 7: 1–11.

Kanimozhi, S., M. Krishnaveni, B. Deivasigmani, T. Rajasekar and P. Priyadarshni. 2013. Immunostimulation effects of *Sargassum whitti* on *Mugil cephalus* against *Pseudomonas fluorescence*. Int. J. Curr. Microbiol. App. Sci. 2(7): 93–103.

Kanjana, K., T. Radtanatip, S. Asuvapongpatana, B. Withyachumnarnkul and K. Wongprasert. 2011. Solvent extracts of the red seaweed *Gracilaria fisheri* prevent *Vibrio harveyi* infections in the black tiger shrimp *Penaeus monodon*. Fish Shellfish Immunol. 30: 389–396.

Karuppiah, V. 2014. Evaluation of antibacterial activity and immunostimulant of red seaweed *Chondrococcus hornemanni* (Kuetzing, 1847) against marine ornamental fish pathogens. J. Coast. Life Med. 2: 61–69.

Kasanah, N., W. Amelia, A. Mukminin, Triyanto and A. Isnansetyo. 2018. Antibacterial activity of Indonesian red algae *Gracilaria edulis* against bacterial fish pathogens and characterization of active fractions. Nat. Prod. Res. 7: 1–5.

Kaur, M. 2017. Oxidative stress response in liver, kidney and gills of *Ctenopharyngodon Idellus* (Cuvier & Valenciennes) exposed to *Chlorpyrifos*. MOJ Biology and Medicine 1: 103–112.

Kendel, M., G. Wielgosz-Collin, S. Bertrand, C. Roussakis, N. Bourgougnon and G. Bedoux. 2015. Lipid composition, fatty acids and sterols in the seaweeds *Ulva armoricana*, and *Solieria chordalis* from Brittany (France): an analysis from nutritional, chemotaxonomic, and antiproliferative activity perspectives. Mar. Drugs 13: 5606–5628.

Kibenge, F.S.B., M.G. Godoy, M. Fast, S. Workenhe and M.J.T. Kibenge. 2012. Countermeasures against viral diseases of farmed fish. Antiviral Res. 95: 257–281.

Kitikiew, S., J.-C. Chen, D.F. Putra, Y.-C. Lin, S.-T. Yeh and C.-H. Liou. 2013. Fucoidan effectively provokes the innate immunity of white shrimp *Litopenaeus vannamei* and its resistance against experimental *Vibrio alginolyticus* infection. Fish Shellfish Immunol. 34: 280–290.

Kosanić, M., B. Ranković and T. Stanojković. 2015. Biological activities of two macroalgae from Adriatic coast of Montenegro. Saudi J. Biol. Sci. 22: 390–397.

Kraan, S. 2012. Algal polysaccharides, novel applications and outlook. pp. 489–532. *In*: Chang, C.F. (ed.). Carbohydrates – Comprehensive Studies on Glycobiology and Glycotechnology. InTech Open Science, Croatia, Rijeka.

Kumar, A., K. Kumar, P.K. Pandey, R.P. Raman, K.P. Prasad and S. Roy. 2014. Growth and hemato-immunological response to dietary ι-carrageenan in *Labeo rohita* (Hamilton, 1822) juveniles. Isr. J. Aquac. 66.2014.971.

Kumar, M., P. Kumari, N. Trivedi, M.K. Shukla, V. Gupta, C.R.K. Reddy and B. Jha. 2011. Minerals, PUFAs and antioxidant properties of some tropical seaweeds from Saurashtra coast of India. J. Appl. Phycol. 23: 797–810.

Kumari, P., M. Kumar, V. Gupta, C.R.K. Reddy and B. Jha. 2010. Tropical marine macroalgae as potential sources of nutritionally important PUFAs. Food Chem. 120: 749–757.

Lahaye, M. and A. Robic. 2007. Structure and functional properties of Ulvan, a polysaccharide from green seaweeds. Biomacromolecules 8: 1765–1774.

Lauzon, Q.D. and A. Serrano. 2015. Ulvan extract from *Enteromorpha intestinalis* enhances immune responses in *Litopenaeus vannamei* and *Penaeus monodon* juveniles mudcrab view project enhancing imbao fisheries (*Anodontia* sp.) in Davao gulf view project. ABAH Bioflux 7(3): 153–161.

Lee, R.E. 2008. Phycology. Cambridge University Press, U.K. 560 pp.

Lee, W.W., G. Ahn, J.Y. Oh, S.M. Kim, N. Kang, E.A. Kim, K.N. Kim, J.B. Jeong and Y.J. Jeon. 2016a. A prebiotic effect of *Ecklonia cava* on the growth and mortality of olive flounder infected with pathogenic bacteria. Fish Shellfish Immunol. 51: 313–320.

Lee, W.W., J.Y. Oh, E.A. Kim, N. Kang, K.N. Kim, G. Ahn and Y.J. Jeon. 2016b. A prebiotic role of *Ecklonia cava* improves the mortality of *Edwardsiella tarda*-infected zebrafish models via regulating the growth of lactic acid bacteria and pathogen bacteria. Fish Shellfish Immunol. 54: 620–628.

Li, Y.X. and S.K. Kim. 2011. Utilization of seaweed derived ingredients as potential antioxidants and functional ingredients in the food industry: an overview. Food Sci. Biotechnol. 20: 1461–1466.

Lin, Y.-H., Y.-C. Su and W. Cheng. 2017. Simple heat processing of brown seaweed *Sargassum cristaefolium* supplementation in diet can improve growth, immune responses and survival to *Vibrio alginolyticus* of white shrimp, *Litopenaeus vannamei*. J. Mar. Sci. Tech. 25: 242–248.

Liu, M., P.E. Hansen and X. Lin. 2011. Bromophenols in marine algae and their bioactivities. Mar. Drugs 9: 1273–1292.

Lopes, G., C. Sousa, L.R. Silva, E. Pinto, P.B. Andrade, J. Bernardo, T. Mouga and P. Valentão. 2012. Can phlorotannins purified extracts constitute a novel pharmacological alternative for microbial infections with associated inflammatory conditions? PloS One. 7: e31145.

Lourenço, S.O., E. Barbarino, J.C. De-paula, L. Otávio, S. Pereira and U.M.L. Marquez. 2002. Amino acid composition, protein content and calculation of nitrogen-to-protein conversion factors for 19 tropical seaweeds. Phycological Res. 50: 233–241.

Lu, Y., X.-P. Liang, M. Jin, P. Sun, H.-N. Ma, Y. Yuan and Q.-C. Zhou. 2016. Effects of dietary vitamin E on the growth performance, antioxidant status and innate immune response in juvenile yellow catfish (Pelteobagrus fulvidraco). Aquaculture 464: 609–617.

Lundebye, A.K., H. Hovea, A. Mågea, V.J.B. Bohneb and K. Hamrea. 2010. Levels of synthetic antioxidants (ethoxyquin, butylated hydroxytoluene and butylated hydroxyanisole) in fish feed and commercially farmed fish. Food Addit. Contam. Part A Chem. Anal. Control. Expo. Risk Assess. 27: 1652–1657.

Macartain, P., C.I.R. Gill, M. Brooks, R. Campbell and I.R. Rowland. 2007. Nutritional value of edible seaweeds. Nutr. Rev. 65: 535–543.

Magnoni, L.J., J.A. Martos-Sitcha, A. Queiroz, J.A. Calduch-Giner, J.F.M. Gonçalves, C.M.R. Rocha, H.T. Abreu, J.W. Schrama, R.O.A. Ozorio and J. Pérez-Sánchez. 2017. Dietary supplementation of heat-treated *Gracilaria* and *Ulva* seaweeds enhanced acute hypoxia tolerance in gilthead seabream (*Sparus aurata*) Biol. Open Bio. 024299.

Makkar, H., S. Giger-Reverdin, H.P.S. Makkar, G. Tran, V. Heuzé, M. Lessire, F. Ois Lebas and P. Ankers. 2015. Seaweeds for livestock diets: a review reducing dietary CP in broilers diet view project by products valorization in animal nutrition. Anim. Feed Sci. Technol. 212: 1–17.

Manilal, A., J. Selvin and S. George. 2012. *In vivo* therapeutic potentiality of red seaweed, *Asparagopsis* (Bonnemaisoniales, Rhodophyta) in the treatment of Vibriosis in *Penaeus monodon* Fabricius. Saudi J. Biol. Sci. 19: 165–175.

Manilal, A., S. Sujith, G.S. Kiran and C. Shakir. 2009. *In vivo* antiviral activity of polysaccharide from the Indian green alga, *Acrosiphonia orientalis* (J. Agardh): potential implication in shrimp disease management. World J. Fish and Mar. Sci. 1: 278–282.

Mar Costa, M., B. Novoa and A. Figueras. 2008. Influence of β-glucans on the immune responses of carpet shell clam (*Ruditapes decussatus*) and Mediterranean mussel (*Mytilus galloprovincialis*). Fish Shellfish Immunol. 24: 498–505.

Marinho, G., C. Nunes, I. Sousa-Pinto, R. Pereira, P. Rema and L.M.P. Valente. 2013. The IMTA-cultivated Chlorophyta *Ulva* spp. as a sustainable ingredient in Nile tilapia (*Oreochromis niloticus*) diets. J. Appl. Phycol. 25: 1359–1367.

Marinho-Soriano, E., M.R. Camara, T. de M. Cabral and M.A. do A. Carneiro. 2007. Preliminary evaluation of the seaweed *Gracilaria cervicornis* (Rhodophyta) as a partial substitute for the industrial feeds used in shrimp (*Litopenaeus vannamei*) farming. Aquacult. Res. 38: 182–187.

Marino, F., G. Di Caro, C. Gugliandolo, A. Spanò, C. Faggio, G. Genovese, M. Morabito, A. Russo, D. Barreca, F. Fazio and A. Santulli. 2016. Preliminary study on the *in vitro* and *in vivo* effects of *Asparagopsis taxiformis* bioactive phycoderivates on teleosts. Front. Physiol. 7: 1–11.

Marrion, O., A. Schwertz, J. Fleurence, J.L. Guéant and C. Villaume. 2003. Improvement of the digestibility of the proteins of the red alga *Palmaria palmata* by physical processes and fermentation. Nahrung. 47: 339–344.

Martirosyan, D.M. and J. Singh. 2015. A new definition of functional food by FFC: what makes a new definition unique? Funct. Food Health Dis. 5: 209–223.

Meena, D.K., P. Das, S. Kumar, S. Mandal, A. Prusty, A. Singh, M. Akhtar, S. Kumar, Á.S. K. Singh, A.K. Kumar, A.A.K. Pal, A.S.C. Mukherjee, S.C. Mandal, A.K. Prusty and M.S. Akhtar. 2013. Beta-glucan: an ideal immunostimulant in aquaculture (a review). Fish Physiol. Biochem. 39: 431–457.

Mendoza-Rodriguez, M.G., C. Pohlenz and D.M. Gatlin. 2017. Supplementation of organic acids and algae extracts in the diet of red drum *Sciaenops ocellatus*: immunological impacts. Aquacult. Res. 48: 1778–1786.

Miles, D.J., J. Polchana, J.H. Lilley, S. Kanchanakhan, K.D. Thompson and A. Adams. 2001. Immunostimulation of striped snakehead *Channa striata* against epizootic ulcerative syndrome. Aquaculture 195: 1–15.

Miranda, M., M. Lopez-Alonso and M. Garcia-Vaquero. 2017. Macroalgae for functional feed development: applications in aquaculture, ruminant and swine feed industries. pp. 133–154. *In*: Newton, P. (ed.). Seaweeds: Biodiversity, Environmental Chemistry and Ecological Impacts. NOVA Science Publishers, N.Y.

Mišurcová, L., S. Kráčmar, B. Klejdus and J. Vacek. 2010. Nitrogen content, dietary fiber, and digestibility in algal food products. Czech J. Food Sci. 28: 27–35.

Mona, M.H., E.-S.T. Rizk, W.M. Salama and M.L. Younis. 2015. Efficacy of probiotics, prebiotics, and immunostimulant on growth performance and immunological parameters of *Procambarus clarkii* juveniles. J. Basic Appl. Zool. 69: 17–25.

Montgomery, W.L. and S.D. Gerking, 1980. Marine macroalgae as foods for fishes: an evaluation of potential food quality. Environ. Biol. Fish. 5: 143.

Moroney, N.C., A.H. Wan, A. Soler-Vila, R.D. Fitzgerald, M.P. Johnson and J.P. Kerry. 2015. Inclusion of *Palmaria palmata* (red seaweed) in Atlantic salmon diets: effects on the quality, shelf-life parameters and sensory properties of fresh and cooked salmon fillets. J. Sci. Food Agric. 95: 897–905.

Moroney, N.C., A.H.L. Wan, A. Soler-Vila, M.N. O'Grady, R.D. Fitzgerald, M.P. Johnson and J.P. Kerry. 2017. Influence of green seaweed (*Ulva rigida*) supplementation on the quality and shelf life of Atlantic salmon fillets. J. Aquat. Food Prod. T. 26: 1175–1188.

Morshedi, V., M. Nafisi Bahabadi, E. Sotoudeh, M. Azodi and M. Hafezieh. 2018. Nutritional evaluation of *Gracilaria pulvinata* as partial substitute with fish meal in practical diets of barramundi (*Lates calcarifer*). J. Appl. Phycol. 30: 619–628.

Moutinho, S., F. Linares, J.L. Rodríguez, V. Sousa and L.M.P. Valente. 2018. Inclusion of 10% seaweed meal in diets for juvenile and on-growing life stages of Senegalese sole (*Solea senegalensis*). J. Appl. Phycol. 1–13.

Mulvaney, W.J., P.C. Winberg and L. Adams. 2013. Comparison of macroalgal (*Ulva* and *Grateloupia* spp.) and formulated terrestrial feed on the growth and condition of juvenile abalone. J. Appl. Phycol. 25: 815–824.

Mustafa, M.G. and H. Nakagawa. 1995. A review: dietary benefits of algae as an additive in fish feed. Isr. J. Aquac. 47: 155–162.

Mustafa, M., S. Wakamatsu, T. Takeda, T. Umino and H. Nakagawa. 1995. Effect of algae as a feed additive on growth performance in red sea bream, *Pagrus major*. Trace Nutrients 61: 67–72.

Nakagawa, H., T. Umino and Y. Tasaka. 1997. Usefulness of *Ascophyllum* meal as a feed additive for red sea bream, *Pagrus major*. Aquaculture 151: 275–281.

Narasimhan, M.K., S.K. Pavithra, V. Krishnan and M. Chandrasekaran. 2013. *In vitro* analysis of antioxidant, antimicrobial and antiproliferative activity of *Enteromorpha antenna*, *Enteromorpha linza* and *Gracilaria corticata* extracts. Jundishapur J. Nat. Pharm. Prod. 8: 151–159.

Narayani, C.G.S., M. Arulpriya, P. Ruban, K. Anantharaj and R. Srinivasan. 2011. *In vitro* antimicrobial activities of seaweed extracts against human pathogens. J. Pharm. Res. 4: 2076–2077.

Nasaran, D.S. and V.A.J. Huxley. 2013. Effect of chosen immunostimulant induced immunological changes in common Carp (*Cyprinus carpio*). J. Theor. Exp. Biol. 10: 67–73.

Natify, W., M. Droussi, N. Berday, A. Araba and M. Benabid. 2015. Effect of the seaweed *Ulva lactuca* as a feed additive on growth performance, feed utilization and body composition of Nile tilapia (*Oreochromis niloticus* L.). Int. J. Agr. Agri. Res. 7: 85–92.

Neveux, N., J.J. Bolton, A. Bruhn and M. Ras. 2018. The bioremediation potential of seaweeds: recycling nitrogen, phosphorus, and other waste products. pp. 217–239. *In*: Barre, L. and Bates, S.S. (eds.). Blue Biotechnology: Production and Use of Marine Molecules. Wiley-VCH, Weihheim, Germany.

Niu, J., X. Chen, X. Lu, S.-G. Jiang, H.-Z. Lin, Y.-J. Liu, Z. Huang, J. Wang, Y. Wang and L.-X. Tian. 2015. Effects of different levels of dietary wakame (*Undaria pinnatifida*) on growth, immunity and intestinal structure of juvenile *Penaeus monodon*. Aquaculture 435: 78–85.

Niu, J., S.-W. Xie, H.-H. Fang, J.-J. Xie, T.-Y. Guo, Y.-M. Zhang, Z.-L. Liu, S.-Y. Liao, J.-Y. He, L.-X. Tian and Y.-J. Liu. 2018. Dietary values of macroalgae *Porphyra haitanensis* in *Litopenaeus vannamei* under normal rearing and WSSV challenge conditions: effect on growth, immune response and intestinal microbiota. Fish Shellfish Immunol. 81: 135–149.

Norambuena, F., K. Hermon, V. Skrzypczyk, J.A. Emery, Y. Sharon, A. Beard and G.M. Turchini. 2015. Algae in fish feed: performances and fatty acid metabolism in juvenile atlantic salmon. PLoS One 10: e0124042.

Novais, C., J. Campos, A.R. Freitas, M. Barros, E. Silveira, T.M. Coque, P. Antunes and L. Peixe. 2018. Water supply and feed as sources of antimicrobial-resistant *Enterococcus* spp. in aquacultures of rainbow trout (*Oncorhyncus mykiss*), Portugal. Sci. Total Environ. 625: 1102–1112.

NRC. 2011. Nutrient Requirements of Fish and Shrimp. National Academies Press, Washington, D.C., 392 pp.

Olmos-Soto, J., Paniagua-Michel, J.d.J., Lopez, L. and Ochoa, L. (2015). Functional Feeds in Aquaculture. pp. 1303–1319. *In*: Kim, S.-K. (ed.). Springer Handbook of Marine Biotechnology. Springer, Berlin, Heidelberg.

Omnes, M.H., J. Le Goasduff, H. Le Delliou, N. Le Bayon, P. Quazuguel and J.H. Robin. 2017. Effects of dietary tannin on growth, feed utilization and digestibility, and carcass composition in juvenile European seabass (*Dicentrarchus labrax* L.). Aquaculture Reports 6: 21–27.

Omont, A., E. Quiroz-Guzman, D. Tovar-Ramirez and A. Peña-Rodríguez. 2018. Effect of diets supplemented with different seaweed extracts on growth performance and digestive enzyme activities of juvenile white shrimp *Litopenaeus vannamei*. J. Appl. Phycol. 31: 1433–1442.

Paiva, L., E. Lima, R.F. Patarra, A.I. Neto and J. Baptista. 2014. Edible Azorean macroalgae as source of rich nutrients with impact on human health. Food Chem. 164: 128–135.

Pallaoro, M.F., F. do Nascimento Vieira and L. Hayashi. 2016. *Ulva lactuca* (Chlorophyta Ulvales) as co-feed for Pacific white shrimp. J. Appl. Phycol. 28: 3659-3665.

Pan, C.H., Y.H. Chien and Y.J. Wang. 2011. Antioxidant defence to ammonia stress of characins (*Hyphessobrycon eques* Steindachner) fed diets supplemented with carotenoids. Aquac. Nutr. 17: 258–266.

Pearce, E.N., M. Andersson and M.B. Zimmermann. 2013. Global iodine nutrition: where do we stand in 2013? Thyroid 23: 523–528.

Peixoto, M.J., E. Salas-Leitón, L.F. Pereira, A. Queiroz, F. Magalhães, R. Pereira, H. Abreu, P.A. Reis, J.F.M. Gonçalves and R.O. de A. Ozório. 2016. Role of dietary seaweed supplementation on growth performance, digestive capacity and immune and stress responsiveness in European seabass (*Dicentrarchus labrax*). Aquaculture Reports 3: 189–197.

Peixoto, M.J., E. Salas-Leitón, F. Brito, L.F. Pereira, J.C. Svendsen, T. Baptista, R. Pereira, H. Abreu, P.A. Reis, J.F.M. Gonçalves and R.O. de Almeida Ozório. 2017. Effects of dietary *Gracilaria* sp. and *Alaria* sp. supplementation on growth performance, metabolic rates and health in meagre (*Argyrosomus regius*) subjected to pathogen infection. J. Appl. Phycol. 29: 433–447.

Peixoto, M.J., L. Magnoni, J.F.M. Gonçalves, R.H. Twijnstra, A. Kijjoa, R. Pereira, A.P. Palstra and R.O.A. Ozório. 2018. Effects of dietary supplementation of *Gracilaria* sp. extracts on fillet quality, oxidative stress, and immune responses in European seabass (*Dicentrarchus labrax*). J. Appl. Phycol. 1–10.

Percival, E. 1979. The polysaccharides of green, red and brown seaweeds: their basic structure, biosynthesis and function. Eur. J. Phycol. 14: 103–117.

Pereira, R., L.M.P. Valente, I. Sousa-Pinto and P. Rema. 2012. Apparent nutrient digestibility of seaweeds by rainbow trout (*Oncorhynchus mykiss*) and Nile tilapia (*Oreochromis niloticus*). Algal Res. 1: 77–82.

Philis, G., E.O. Gracey, L.C. Gansel, A.M. Fet and C. Rebours. 2018. Comparing the primary energy and phosphorus consumption of soybean and seaweed-based aquafeed proteins – a material and substance flow analysis. J. Clean Prod. 200: 1142–1153.

Pillay, T.V.R. and M.N. Kutty. 2005. Aquaculture – Principles and Practices, 2nd ed. Blackwell Publishing, Oxford, U.K., 640 pp.

Purbomartono, C., A. Husin, D. Priyambodo and E.N. Anggraeni. 2018. Non specific immune potentiating evaluation of brown seaweed of *Padina* sp. in Catfish and Pangasius. Adv. Sci. Lett. 24: 80–83.

Qi, Z., H. Liu, B. Li, Y. Mao, Z. Jiang, J. Zhang and J. Fang. 2010. Suitability of two seaweeds, *Gracilaria lemaneiformis* and *Sargassum pallidum*, as feed for the abalone *Haliotis discus* hannai Ino. Aquaculture 300: 189–193.

Quezada-Rodríguez, P.R. and E.J. Fajer-Ávila. 2017. The dietary effect of ulvan from *Ulva clathrata* on hematological-immunological parameters and growth of tilapia (*Oreochromis niloticus*). J. Appl. Phycol. 29: 423–431.

Raa, J. 2000. The use of immune-stimulants in fish and shellfish feeds. pp. 47–56. *In*: Cruz-Suárez, L.E., Ricque-Marie, D., Tapia-Salazar, M., Olvera-Novoa, M.A. and Civera-Cerecedo, R. (eds.). Avances En Nutrición Acuícola V. Memorias Del V Simposium Internacional de Nutrición Acuícola. 19-22 Nov. 2000. Merida, Yucatan, Mexico.

Rabiei, R., S.M. Phang, P.E. Lim, A. Salleh, J. Sohrabipour, D. Ajdari and G.A. Zarshenas. 2016. Productivity, biochemical composition and biofiltering performance of agarophytic seaweed, *Gelidium elegans* (Red algae) grown in shrimp hatchery effluents in Malaysia. Iranian J. Fish. Sci. 15: 53–74.

Radhika, D.D. and A. Mohaideen. 2016. Effect of utilization of seaweeds as an immunostimulant in *Aeromonas hydrophila* infected *Cyprinus carpio*. World J. Pharm. Pharm. Sci. 5: 851–857.

Ragaza, J., S. Koshio, R.E. Mamauag, M. Ishikawa, S. Yokoyama and S. Villamor. 2013. Dietary supplemental effects of red seaweed *Eucheuma denticulatum* on growth performance, carcass composition and blood chemistry of juvenile Japanese flounder, *Paralichthys olivaceus*. Aquac. Res. 46: 647–657.

Rajapakse, N. and S.-K. Kim. 2011. Nutritional and digestive health benefits of seaweed. pp. 17–28. *In*: Kim, S.-K. (ed.). Advances in Food and Nutrition Research. Academic Press, London, U.K.

Rajendran, P., P.A. Subramani and D. Michael. 2016. Polysaccharides from marine macroalga, *Padina gymnospora* improve the nonspecific and specific immune responses of *Cyprinus carpio* and protect it from different pathogens. Fish Shellfish Immunol. 58: 220–228.

Rayman, M. 2017. Selenium intake and status in health & disease. Free Radic. Biol. Med. 112: 5.

Reverter, M., N. Bontemps, D. Lecchini, B. Banaigs and P. Sasal. 2014. Use of plant extracts in fish aquaculture as an alternative to chemotherapy: current status and future perspectives. Aquaculture 433: 50–61.

Reyes-Cerpa, S., E. Vallejos-Vidal, M. José Gonzalez-Bown, J. Morales-Reyes, D. Pérez-Stuardo, D. Vargas, M. Imarai, V. Cifuentes, E. Spencer, A. María Sandino and F.E. Reyes-López. 2018. Effect of yeast (*Xanthophyllomyces dendrorhous*) and plant (Saint John's wort, lemon balm, and rosemary) extract based functional diets on antioxidant and immune status of Atlantic salmon (*Salmo salar*) subjected to crowding stress. Fish Shellfish Immunol. 74: 250–259.

Rodrigues, D., G. Walton, S. Sousa, T.A.P. Rocha-Santos, A.C. Duarte, A.C. Freitas and A.M.P. Gomes. 2016. *In vitro* fermentation and prebiotic potential of selected extracts from seaweeds and mushrooms. LWT – Food Science and Technology 73: 131–139.

Romero, J., C. Feijoó and P. Navarrete. 2012. Antibiotics in aquaculture – use, abuse and alternatives. pp. 159–198. *In*: Carvalho, E. (ed.). Health and Environment in Aquaculture. InTech Open Science, Cróacia.

Rudtanatip, T., S.A. Lynch, K. Wongprasert and S.C. Culloty. 2018. Assessment of the effects of sulfated polysaccharides extracted from the red seaweed Irish moss *Chondrus crispus* on the immune-stimulant activity in mussels *Mytilus* spp. Fish Shellfish Immunol. 75: 284–290.

Rudtanatip, T., S. Asuvapongpatana, B. Withyachumnarnkul and K. Wongprasert. 2014. Sulfated galactans isolated from the red seaweed *Gracilaria fisheri* target the envelope proteins of white spot syndrome virus and protect against viral infection in shrimp haemocytes. J. Gen. Virol. 95: 1126–1134.

Sakthivel, M., B. Deivasigamani, T. Rajasekar, S. Kumaran and K. Alagappan. 2015. Immunostimulatory effects of polysaccharide compound from seaweed *Kappaphycus alvarezii* on Asian seabass (*Lates calcarifer*) and its resistance against *Vibrio parahaemolyticus*. J. Mar. Biol. Oceanogr. 4: 1–9.

Sanz-Pintos, N., J. Pérez-Jiménez, A.H. Buschmann, J.R. Vergara-Salinas, J.R. Pérz-Correa and F. Saura-Calixto. 2017. Macromolecular antioxidants and dietary fiber in edible seaweeds. J. Food Sci. 82: 289–295.

Sarojini, Y. and K. Uma Devi. 2014. The marine macro algae as potential source of nutritionally important polyunsaturated fatty acids. Pharm. Lett. 6: 348–351.

Saura-Calixto, F. 2011. Dietary fiber as a carrier of dietary antioxidants: an essential physiological function. J. Agric. Food Chem. 12: 43–49.

Schleder, D.D., L.G.B. Peruch, M.A. Poli, T.H. Ferreira, C.P. Silva, E.R. Andreatta, L. Hayashi and F. do Nascimento Vieira. 2018. Effect of brown seaweeds on Pacific white shrimp growth performance, gut morphology, digestive enzymes activity and resistance to white spot virus. Aquaculture 495: 359–365.

Schmid, M., L.G.K. Kraft, L.M. Van Der Loos, G.T. Kraft, P. Virtue, P.D. Nichols and C.L. Hurd. 2018. Southern Australian seaweeds: a promising resource for omega-3 fatty acids. Food Chem. 265: 70–77.

Serrano Jr., A.E., R.S. Declarador, M.G.C. Sedanza and B.L.M. Tumbokon. 2017. *Rhizoclonium riparium* protein concentrate can replace soybean meal in the diet of the Pacific white shrimp *Litopenaeus vannamei* post larvae. Isr. J. Aquac. 9: 1391.

Selvin, J., A. Manilal, S. Sujith, G. Seghal Kiran and A. Premnath Lipton. 2011. Efficacy of marine green alga *Ulva fasciata* extract on the management of shrimp bacterial diseases. Lat. Am. J. Aquat. Res. 39: 197–204.

Shanab, S.M.M., E.A. Shalaby and E.A. El-Fayoumy. 2011. *Enteromorpha compressa* exhibits potent antioxidant activity. J. Biomed. Biotech. Volume 2011: Article ID 726405, 11 pp.

Shields, R.J. and I. Lupatsch. 2012. Algae for aquaculture and animal feeds. Technikfolgenabschätzung – Theorie und Praxis 21: 23–37.

Shpigel, M., L. Guttman, L. Shauli, V. Odintsov, D. Ben-Ezra and S. Harpaz. 2017. *Ulva lactuca* from an integrated multi-trophic aquaculture (IMTA) biofilter system as a protein supplement in gilthead seabream (*Sparus aurata*) diet. Aquaculture 481: 112–118.

Silva, D.M., L.M.P. Valente, I. Sousa-Pinto, R. Pereira, M.A. Pires, F. Seixas and P. Rema. 2015. Evaluation of IMTA-produced seaweeds (*Gracilaria*, *Porphyra*, and *Ulva*) as dietary ingredients in Nile tilapia, *Oreochromis niloticus* L., juveniles. Effects on growth performance and gut histology. J. Appl. Phycol. 27: 1671–1680.

Siva Kumar, K. and S.V. Rajagopal. 2011. Radical scavenging activity of green algal species. J. Pharm. Res. 4: 723–725.

Skjermo, J., T. Defoortt, M. Dehasque, T. Espevik, Y. Olsen, G. Skjak-Briek, P. Sorgeloost and O. Vadstein. 1995. Immunostimulation of juvenile turbot (*Scophthalmus maximus* L.) using an alginate with high mannuronic acid content administered via the live food organism Artemia. Fish Shellfish Immunol. 5: 531–534.

Soler-Vila, A., S. Coughlan, M.D. Guiry and S. Kraan. 2009. The red alga *Porphyra dioica* as a fish-feed ingredient for rainbow trout (*Oncorhynchus mykiss*): effects on growth, feed efficiency, and carcass composition. J. Appl. Phycol. 21: 617–624.

Suantika, G., M.L. Situmorang, P. Aditiawati, A. Khakim, S. Suryanaray, S.S. Nori, S. Kumar and F. Putri. 2017. Effect of red Seaweed *Kappaphycus alvarezii* on growth, salinity stress tolerance and vibriosis resistance in Shrimp *Litopenaeus vannamei* Hatchery. J. Fish. Aquat. Sci. 12: 127–133.

Targett, N.M. and T.M. Arnold. 1998. Minireview-predicting the effects of brown algal phlorotannins on marine herbivores in tropical and temperate oceans. J. Phycol. 34: 195–205.

Telles, C.B.S., C. Mendes-Aguiar, G.P. Fidelis, A.P. Frasson, W.O. Pereira, K.C. Scortecci, R.B.G. Camara, L.T.D.B. Nobre, L.S. Costa, T. Tasca and H.A.O. Rocha. 2018. Immunomodulatory effects and antimicrobial activity of heterofucans from *Sargassum filipendula*. J. Appl. Phycol. 30: 569–578.

Thanigaivel, S., S. Vijayakumar, A. Mukherjee, N. Chandrasekaran and J. Thomas. 2014. Antioxidant and antibacterial activity of *Chaetomorpha antennina* against shrimp pathogen *Vibrio parahaemolyticus*. Aquaculture 433: 467–475.

Thanigaivel, S., N. Chandrasekaran, A. Mukherjee and J. Thomas. 2015. Investigation of seaweed extracts as a source of treatment against bacterial fish pathogen. Aquaculture 448: 82–86.

Thanigaivel, S., N. Chandrasekaran, A. Mukherjee and J. Thomas. 2016. Seaweeds as an alternative therapeutic source for aquatic disease management. Aquaculture 464: 529–536.

Thanigaivel, S., S.K. Bhullar, N. Chandrasekaran, A. Mukherjee, J. Thomas and M. Ramalingam. 2018. Antibacterial Activity of *Sargasum longifolium* – Polycaprolactone Nanobiocomposite for Fish Pathogen. J. Bionanosci. 12: 417–421.

Traifalgar, R.F. 2017. Development of immunostimulant for mud crab, *Scylla serrata*. pp. 52–62. *In*: Quinitio, E.T., Parado-Estepa, F.D. and Coloso, R.M. (eds.). Philippines: In the forefront of the mud crab industry development. Proceedings of the 1st National Mud Crab Congress, 16–18 November 2015, Iloilo City, Philippines.

Turchini, G.M., B.E. Torstensen and W.K. Ng. 2009. Fish oil replacement in finfish nutrition. Rev. Aquacult. 1: 10–57.

United Nations, Department of Economic and Social Affairs, Population Division. 2017. World Population Prospects: The 2017 Revision, Data Booklet.

Valentão, P., P. Trindade, D. Gomes, P. Guedes de Pinho, T. Mouga and P.B. Andrade. 2010. *Codium tomentosum* and *Plocamium cartilagineum*: chemistry and antioxidant potential. Food Chem. 119: 1359–1368.

Valente, L.M.P., A. Gouveia, P. Rema, J. Matos, E.F. Gomes and I.S. Pinto. 2006. Evaluation of three seaweeds *Gracilaria bursa-pastoris*, *Ulva rigida* and *Gracilaria cornea* as dietary ingredients in European sea bass (Dic*entrarchus labrax*) juveniles. Aquaculture 252: 85–91.

Valente, L.M.P., P. Rema, V. Ferraro, M. Pintado, I. Sousa-Pinto, L.M. Cunha, M.B. Oliveira and M. Araújo. 2015. Iodine enrichment of rainbow trout flesh by dietary supplementation with the red seaweed *Gracilaria vermiculophylla*. Aquaculture 446: 132–139.

Vallejos-Vidal, E., F. Reyes-López, M. Teles and S. MacKenzie. 2016. The response of fish to immunostimulant diets. Fish Shellfish Immunol. 56: 34–69.

Van Doan, H., W. Tapingkae, T. Moonmanee and A. Seepai. 2016. Effects of low molecular weight sodium alginate on growth performance, immunity, and disease resistance of tilapia, Oreochromis niloticus. Fish Shellfish Immunol. 55: 186–194.

Vatsos, I.N. and C. Rebours. 2015. Seaweed extracts as antimicrobial agents in aquaculture. J. Appl. Phycol. 27: 2017–2035.

Vieira, E.F., C. Soares, S. Machado, M. Correia, M.J. Ramalhosa, M.T. Oliva-teles, A. Paula Carvalho, V.F. Domingues, F. Antunes, T.A.C. Oliveira, S. Morais and C. Delerue-Matos. 2018. Seaweeds from the Portuguese coast as a source of proteinaceous material: total and free amino acid composition profile. Food Chem. 269: 264–275.

Viera, M.P., G.C. de Viçose, J.L. Gómez-Pinchetti, A. Bilbao, H. Fernandez-Palacios and M.S. Izquierdo. 2011. Comparative performances of juvenile abalone (*Haliotis tuberculata coccinea* Reeve) fed enriched vs non-enriched macroalgae: effect on growth and body composition. Aquaculture 319: 423–429.

Viera, M.P., G. Courtois de Viçose, L. Robaina and M.S. Izquierdo. 2015. First development of various vegetable-based diets and their suitability for abalone *Haliotis tuberculata coccinea* Reeve. Aquaculture 448: 350–358.

Villamil, L., S. Infante Villamil, G. Rozo and J. Rojas. 2018. Effect of dietary administration of kappa carrageenan extracted from *Hypnea musciformis* on innate immune response, growth, and survival of Nile tilapia (*Oreochromis niloticus*). Aquacult. Int. 1–10.

Vollstad, D., J. Bøgwald, O. Gåserød and R.A. Dalmo. 2006. Influence of high-M alginate on the growth and survival of Atlantic cod (*Gadus morhua* L.) and spotted wolffish (*Anarhichas minor* Olafsen) fry. Fish Shellfish Immunol. 20: 548–561.

Wan, A.H.L., A. Soler-Vila, D. O'Keeffe, P. Casburn, R. Fitzgerald and M.P. Johnson. 2016. The inclusion of *Palmaria palmata* macroalgae in Atlantic salmon (*Salmo salar*) diets: effects on growth, haematology, immunity and liver function. J. Appl. Phycol. 28: 3091–3100.

Wassef, E.A., A.-F.M. El-Sayed and E.M. Sakr. 2013. *Pterocladia* (Rhodophyta) and *Ulva* (Chlorophyta) as feed supplements for European seabass, *Dicentrarchus labrax* L., fry. J. Appl. Phycol. 25: 1369–1376.

Watson, R.A., G.B. Nowara, K. Hartmann, B.S. Green, S.R. Tracey and C.G. Carter. 2015. Marine foods sourced from farther as their use of global ocean primary production increases. Nat. Commun. 6: 1–6.

Welker, T.L. and J.L. Congleton. 2009. Effect of dietary α-tocopherol + ascorbic acid, selenium, and iron on oxidative stress in sub-yearling Chinook salmon (*Oncorhynchus tshawytscha* Walbaum). J. Anim. Physiol. Anim. Nutr. 93: 15–25.

Wong, K.H., P.C.K. Cheung and K.H. Wong. 2000. Nutritional evaluation of some subtropical red and green seaweeds: Part I – proximate composition, amino acid profiles and some physico-chemical properties. Food Chem. 71: 475–482.

World Health Organization. 2011. Tackling antibiotic resistance from a food safety perspective in Europe. WHO Regional Office for Europe, Denmark.

Xia, S., P. Zhao, K. Chen, Y. Li, S. Liu, L. Zhang and H. Yang. 2012. Feeding preferences of the sea cucumber *Apostichopus japonicus* (Selenka) on various seaweed diets. Aquaculture 344–349: 205–209.

Xu, S., L. Zhang, Q. Wu, X. Liu, S. Wang, C. You and Y. Li. 2011. Evaluation of dried seaweed *Gracilaria lemaneiformis* as an ingredient in diets for teleost fish *Siganus canaliculatus*. Aquacult. Int. 19: 1007–1018.

Xuan, X., X. Wen, S. Li, D. Zhu and Y. Li. 2013. Potential use of macro-algae *Gracilaria lemaneiformis* in diets for the black sea bream, *Acanthopagrus schlegelii*, juvenile. Aquaculture 412–413: 167–172.

Yang, Q., R. Yang, M. Li, Q. Zhou, X. Liang and Z.C. Elmada. 2014. Effects of dietary fucoidan on the blood constituents, anti-oxidation and innate immunity of juvenile yellow catfish (*Pelteobagrus fulvidraco*). Fish Shellfish Immunol. 41: 264–270.

Yeganeh, S. and M. Adel. 2018. Effects of dietary algae (*Sargassum ilicifolium*) as immunomodulator and growth promoter of juvenile great sturgeon (*Huso huso* Linnaeus, 1758). J. Appl. Phycol. 1–10.

Yeh, S.-T., C.-S. Lee and J.-C. Chen. 2006. Administration of hot-water extract of brown seaweed *Sargassum duplicatum* via immersion and injection enhances the immune resistance of white shrimp *Litopenaeus vannamei*. Fish Shellfish Immunol. 20: 332–345.

Yone, Y., M. Furuichi and K. Urano. 1986. Effects of dietary wakame *Undaria pinnatifida* and *Ascophyllum nodosum* supplements on growth, feed efficiency, and proximate compositions of liver and muscle of red sea bream. Nippon Suisan Gakkaishi 52: 1465–1488.

Yu, S., A. Blennow, M. Bojko, F. Madsen, C.E. Olsen and S.B. Engelsen. 2002. Physico-chemical characterization of floridean starch of red algae. Strach 54: 66–74.

Zahra, A., S. Sukenda and D. Wahjuningrum. 2017. Extract of seaweed *Gracilaria verrucosa* as immunostimulant to controlling white spot disease in Pacific white shrimp *Litopenaeus vannamei*. JAI. 16: 174.

Zaporozhets, T.S., N.N. Besednova, T.A. Kuznetsova, T.N. Zvyagintseva, I.D. Makarenkova, S.P. Kryzhanovsky and V.G. Melnikov. 2014. The prebiotic potential of polysaccharides and extracts of seaweeds. Russ. J. Mar. Biol. 40: 1–9.

Zeraatpisheh, F., F. Firouzbakhsh and K. Khalili. 2018. Effects of the macroalga *Sargassum angustifolium* hot water extract on hematological parameters and immune responses in rainbow trout (*Oncohrynchus mykiss*) infected with Yersinia rukeri. J. Appl. Phycol. 30: 2029–2037.

Zhang, Z., T. Swain, J. Bøgwald, R.A. Dalmo and J. Kumari. 2009. Bath immunostimulation of rainbow trout (*Oncorhynchus mykiss*) fry induces enhancement of inflammatory cytokine transcripts, while repeated bath induce no changes. Fish Shellfish Immunol. 26: 677–684.

Zhu, D., X. Wen, X. Xuan, S. Li and Y. Li. 2016. The green alga *Ulva lactuca* as a potential ingredient in diets for juvenile white spotted snapper *Lutjanus stellatus* Akazaki. J. Appl. Phycol. 28: 703–711.

Zubia, M., D. Robledo and Y. Freile-Pelegrin. 2007. Antioxidant activities in tropical marine macroalgae from the Yucatan Peninsula, Mexico. J. Appl. Phycol. 19: 449–458.

10

Effects of Feeding with Seaweeds on Ruminal Fermentation and Methane Production

Ana Rita J. Cabrita[1*], **Inês M. Valente**[1,2], **Hugo M. Oliveira**[3], **António J.M. Fonseca**[1] **and Margarida R.G. Maia**[1]

[1] REQUIMTE, LAQV, ICBAS, Instituto de Ciências Biomédicas de Abel Salazar, Universidade do Porto, Rua de Jorge Viterbo Ferreira n.º 228, 4050-313 Porto, Portugal
[2] REQUIMTE, LAQV, Departamento de Química e Bioquímica, Faculdade de Ciências, Universidade do Porto, Rua do Campo Alegre 687, 4169-007 Porto, Portugal
[3] INL, International Iberian Nanotechnology Laboratory, Avenida Mestre José Veiga s/n, 4715-330 Braga, Portugal

1 Introduction

Livestock production is rapidly developing, motivated by the flourishing human population and consequent demand for animal products. Global meat and dairy production is expected to increase to 465 million tonnes and 1043 million tonnes, respectively, in 2050, over twice the annual meat and dairy produced in 2000. Increasing amounts of feed supplies will be required to meet this estimated demand for animal products. Hence, for sustainable development of the livestock sector, it is essential to enlarge the feedstock base through the search for novel feeds, increased use of underexploited feeds, or development of additives that improve resource use efficiency. In this context, a renewed interest in seaweeds for animal feed applications has emerged over the last decades.

Seaweeds are primary producers thriving on coastal environments under natural light and temperature conditions. Seaweed production in aquaculture represents the advantage of not competing with land that can be used for the cultivation of food and fodder crops, and of not requiring freshwater. In addition, the wild and cultivated stocks of seaweed may make an important contribution to the sequestration of atmospheric carbon dioxide (Duarte et al. 2017).

In coastal regions, seaweeds have been used in livestock feeding since ancient times, particularly in periods of feed shortage (Makkar et al. 2016). Depending on their market cost and availability, seaweeds have been assessed as prebiotic promoter (Ramnani et al. 2012) or as feed ingredient (Machado et al. 2015), when included at low or high rates, respectively. Seaweeds offer a potential resource for use as animal feed (Makkar et al. 2016) as they contain high levels of micronutrients and small molecules, including carotenoids and phlorotannins. The red and green seaweed varieties can also be protein-rich, depending on the area and season of harvest (Gojon-Baez et al. 1998, Makkar et al. 2016). Seaweeds contain large amounts of polysaccharides (up to 76% of dry matter

*Corresponding author: arcabrita@icbas.up.pt

(DM); Holdt and Kraan 2011), markedly structural cell wall polysaccharides, and to a lesser extent mucopolysaccharides and non-structural polysaccharides (Murata and Nakazoe 2001, Kumar et al. 2008). In light of their polysaccharide diversity and complexity, the livestock species best suited to be fed on seaweeds may be herbivorous animals, particularly ruminants.

2 Ruminant Digestive System

Ruminants are herbivorous animals with a unique digestive adaptation: the development of three forestomachs (reticulum, rumen, and omasum) and the characteristic feature of rumination (i.e., regurgitation and new mastication) of ingested feeds. There are several ruminant species worldwide, with different anatomical and feeding behaviour characteristics that have evolved in adaptation to their habitats. Nevertheless, in all species, the feeds undergo a microbial fermentation in the major forestomach, the rumen, before entering the true stomach, the abomasum. The rumen is a huge fermentation vat that hosts a complex symbiotic microbial consortium composed of bacteria, protozoa, archaea, fungi, and bacteriophages. This microbial consortium degrades most of the plant complex structural and non-structural polysaccharides, proteins, and lipids to monomers, and further ferments them mainly to volatile fatty acids, ammonia, carbon dioxide, and methane (Fig. 1).

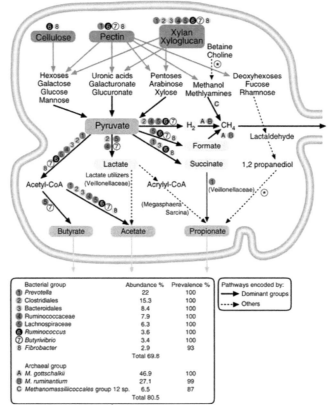

Figure 1. Simplified illustration showing the degradation and metabolism of plant structural carbohydrates by the dominant bacterial and archaeal groups identified in the Global Rumen Census project (Henderson et al. 2015) using information from metabolic studies and analysis of the reference genomes. Abundance represents the mean relative percentage for that genus-level group in samples that contain that group, while prevalence represents the prevalence of that genus-level group in all samples (n = 684). *The conversion of choline to trimethylamine, and propanediol to propionate, generates toxic intermediates that are contained within bacterial microcompartments. Adapted, with permission, from Seshadri et al. (2018) under CC BY 4.0 licence.

Color version at the end of the book

Volatile fatty acids produced in the rumen can fulfil up to 70% of the host energetic needs, playing the role of glucose in monogastric animals (Nagaraja et al. 1997). In addition, part of the microbial population leaves the rumen along with the unfermented or partially fermented feed and enters the abomasum, where the enzymatic digestion begins, being an extensive source of protein for the host.

The rumen microbiome includes a diversity of microorganisms with a wide range of enzymatic capabilities, including cellulolytic, hemicellulolytic, amylolytic, proteolytic, lipolytic, and methanogenic, which often overlap in the degradation of feeds and/or in the metabolization of other microbial fermentation products (Hungate 1966). This intricate microbial population differs among ruminant species and within species, but the existence of a core microbiome has been reported (Lettat and Benchaar 2013, Petri et al. 2014). Indeed, Henderson et al. (2015) accessed the geographical distribution and diversity of 742 samples of rumen contents from different animals from 35 countries and concluded that over 90% of the samples presented 30 of the most abundant bacterial groups (Fig. 2).

The dominant archaeal groups were also similar around the world. The two most common clades of methanogens (*Methanobrevibacter gottschalkii* and *Methanobrevibacter ruminantium*) were present in most of the samples and accounted for 74% of total archaeal population. This study concluded that diet, and not animal species or animal distribution, was the main factor to influence the rumen microbial population composition (Henderson et al. 2015). Conversely, a recent study evaluated the postprandial microbial community composition of 18 cows and demonstrated that the diurnal oscillatory patterns of the main bacterial and archaeal population relative abundance changed 3- to 5-fold, independently of individual animal variability and diet (Shaani et al. 2018).

It was suggested that the core bacteriome of the rumen was composed of seven bacterial groups: *Prevotella*, *Butyrivibrio*, and *Ruminococcus*, and unclassified Lachnospiraceae, Ruminococcaceae,

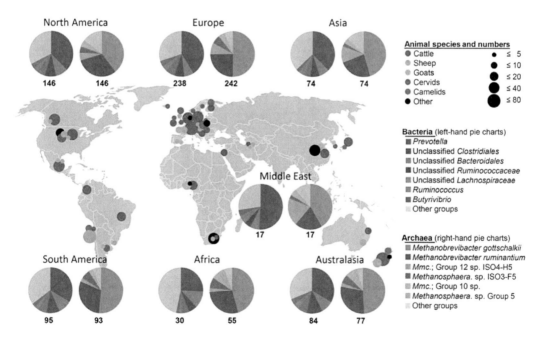

Figure 2. Bacterial and archaeal community compositions of rumen samples from different ruminant species and regions. Numbers below pie charts represent the number of samples for which data were obtained. The most abundant bacteria and archaea are named in clockwise order starting at the top of the pie chart. Mmc. Methanomassiliicoccales. Reproduced, with permission, from Henderson et al. (2015) under CC BY 4.0 licence.

Bacteroidales, and Clostridiales (Henderson et al. 2015). Due to the lack of characterized cultures, their function is not entirely understood, but major enzymatic activity is expected to be cellulolytic and hemicellulolytic (Wallace 2008, Firkins and Yu 2015). Therefore, the rumen microbiome might provide the host animal the ability to use seaweeds by degrading their complex and diverse polysaccharides. However, the use of seaweeds in ruminant feeding is constrained because of the paucity of information concerning the specific nutritive value and variability of each species, and the existence of substances that could be challenging to the digestive system of land animals.

3 Effect of Seaweeds on Ruminant Feeding

The use of seaweeds to feed livestock has been long known, with the first reports dating from Ancient Greece and from Icelandic sagas. In times of feed scarcity, ruminants, particularly sheep, grazed on fresh seaweed on the coast or were fed dried seaweed in barns, occasionally or for long periods of time (Evans and Critchley 2014, Makkar et al. 2016). However, in the 20th century, the inclusion level of seaweeds was recommended to be, at most, 10%, as higher incorporation levels would have detrimental effects on the animals (Chapman and Chapman 1980, Evans and Critchley 2014). Recently, a new interest emerged in seaweeds as alternative feeds and valuable sources of organic minerals (Cabrita et al. 2016), and of complex polysaccharides, pigments and polyunsaturated fatty acids with health-promoting properties (Makkar et al. 2016, Cabrita et al. 2017). Seaweed inclusion at 2% was reported to exert nutraceutical properties on monogastrics, such as prebiotic activity, increased immune competence, and improved productivity (Evans and Critchley 2014). Nutraceutical effects of seaweeds on ruminant animals are not expected to differ from those on monogastric animals. In addition, the complex microbial ecosystem of ruminants may allow the host to use seaweed compounds as valuable nutritive compounds, even at high inclusion levels.

3.1 Digestibility and degradability of seaweeds in ruminants

Published data on *in vivo* and *in vitro* digestibility and *in sacco* degradability of seaweeds in ruminants is limited (Makkar et al. 2016). The DM degradability of *Alaria esculenta, Laminaria digitata, Pelvetia canaliculata* (Phaeophyceae), *Mastocarpus stellatus, Palmaria palmata,* and *Pyropia/Porphyra* sp. (Rhodophyta), of *Acrosiphonia* sp. and *Ulva* sp. (Chlorophyta) (Tayyab et al. 2016), and of *Macrocystis pyrifera* and *Sargassum* spp. (Phaeophyceae) (Gojon-Baez et al. 1998) were evaluated. After 96 h of rumen incubation, the DM degradability of these seaweed species varied between 279 g per kg DM for *M. stellatus* and 946 g per kg DM for *L. digitata* (Gojon-Baez et al. 1998, Tayyab et al. 2016). This broad range of values is also apparent in *in vitro* digestibility studies. Indeed, several *in vitro* studies reported organic matter (OM) digestibility from 150 g per kg DM for *Fucus serratus* (Phaeophyceae) to values as high as 970 g per kg DM for *Saccharina latissima* (Phaeophyceae) (Greenwood et al. 1983a, b, Ventura and Castanon 1998, Hansen et al. 2003, Machado et al. 2014, El-Waziry et al. 2015, Kinley and Fredeen 2015). In the study of Machado et al. (2016a), an inclusion higher than 10% of *Asparagopsis taxiformis* (Rhodophyta) or *Oedogonium* (Chlorophyta) decreased *in vitro* OM digestibility of Rhode grass hay. Conversely, the *in vivo* trials of Castro et al. (2009) and Marin et al. (2009) did not find negative effects on digestibility with the inclusion of *Macrocystis pyrifera* or *Sargassum* sp. up to 30% DM.

Data on energy digestibility of seaweeds is very scarce. Arieli et al. (1993) suggested that the energetic value of *Ulva lactuca* (Chlorophyta), which had energy digestibility of 600 g per kg DM, is similar to that of a medium-quality hay. Similarly, Cabrita et al. (2017) found values of *in vivo* energy digestibility of 523 and 558 g per kg DM respectively for *Gracilaria vermiculophyla* (Rhodophyta) and *Ulva rigida* (Chlorophyta).

From the results available, a high variability of the digestibility values between and within the various seaweed species is evident and can be explained, at least partially, by the different composition of the cell wall of the different species. The cellulose and hemicellulose contents of most seaweed species of interest for animal feeding are 2-10% and 9% dry weight, respectively, whereas lignin was only found in *Ulva* sp. at 3% of DM (Holdt and Kraan 2011). However, a direct comparison of results between studies is complicated because of the use of different methodologies to determine digestibility, differences among the length of *in vitro* or *in situ* incubations, different level of seaweed inclusion, and the nature of the basal substrate.

Seaweed structural and non-structural polysaccharides are species-specific. This specificity and absence of most of the complex polysaccharides in terrestrial plants may pose a hurdle to their fermentation by the rumen microbiome, and further digestion and utilization by the host animal.

The cell wall matrix of green algae contains highly complex sulphated hetero-polysaccharides in which the major sugars are glucuronic acid, rhamnose, arabinose, and galactose, in a diversity of combinations (Percival 1979, Bikker et al. 2016). Ulvans comprise highly charged sulphated polyelectrolytes, with the main monomer sugars being rhamnose, uronic acid, and xylose. Ulvans also contain a common disaccharide (aldobiuronic acid), and the sugar iduronic acid (Holdt and Kraan 2011). Although green algae ulvans are potentially hydrolysable to oligosaccharides with biological activity (Andrieux et al. 1998), ulvan lyases have only been isolated in Proteobacteria and Flavobacteria species from marine environments (Barbeyron et al. 2011, Collén et al. 2011, Kopel et al. 2016, Salinas and French 2017, Konasani et al. 2018).

Alginic acid is the main polysaccharide found in brown algae (Davis et al. 2003). This linear cell wall polysaccharide is composed of 1,4-linked β-D-mannuronic acid and α-L-guluronic acid residues (Haug et al. 1967). Fucoidans and fucans are polysaccharides mainly composed of sulphated L-fucose that are broadly found in the cell walls of brown seaweed, but absent in green and red algae and in higher plants (Berteau and Mulloy 2003). Laminarin is a relatively low molecular weight storage polysaccharide (β-1,3 glucan) generally found in brown algae, where it can reach 35% of the dried weight (Holdt and Kraan 2011). Mannitol is a sugar alcohol present in many species of brown algae, especially in *Laminaria* and *Ecklonia* (Holdt and Kraan 2011). To varying extents, carbohydrates from brown seaweed can be hydrolysed by the ruminal population, leading to the production of methane and acetic acid (Williams et al. 2013). The ability to hydrolyse mannitol has been found to differ between seaweed-fed and grass-fed animals (Ahmed et al. 2013), suggesting an adaptation of the rumen microbial population to mannitol. Similar adaptation may be required for effective brown seaweed polysaccharide fermentation by the microbial ecosystem of the rumen.

Polysaccharides in red algae include agars, carrageenans, xylans, floridean starch, water-soluble sulphated galactan, as well as porphyrin (a complex galactan) and mucopolysaccharides (Falshaw et al. 1998, Murata and Nakazoe 2001, Kumar et al. 2008, O'Sullivan et al. 2010, Holdt and Kraan 2011). Even though galactans can be hydrolysed by microbes from marine environments, the enzymes that catalyse it are less frequent, or even absent, in bacteria that hydrolyse polysaccharides from higher plants (Hehemann et al. 2010).

Differences in seaweed digestibility among studies could be explained by their different nutrient composition, particularly of complex polysaccharides, but also by the ability of the microbial ecosystem of the ruminant to adapt to this particular feed (Makkar et al. 2016). Additionally, we may consider that secondary metabolites produced by some species of seaweeds that exert health benefits (anti-bacterial, anti-viral, antioxidant, and anti-inflammatory; Bach et al. 2008, O'Sullivan et al. 2010) might also have a detrimental effect on rumen fermentation, particularly on fibre degradation (Wang et al. 2008).

3.2   Rumen microbiome adaptation to seaweeds

The use of seaweeds as feed has been closely related to times of feed scarcity in some coastal geographies around the world. The exception is North Ronaldsay, the northernmost of the Orkney Islands (Scotland), where sheep have been fed almost exclusively on seaweeds. A dyke was built on this island, and the sheep population is kept by the shore all year round, except for fertile ewes that are grazed on grass during the lambing season (Hansen et al. 2003). Although some pasture is available, sheep prefer to feed on fresh seaweeds washed ashore or found in intertidal beds (Hansen et al. 2003) and often ignore the pasture (Orpin et al. 1985).

The North Ronaldsay sheep were observed to eat a variety of seaweed species (*P. palmata, A. esculenta, A. nodosum, Fucus* sp. and *Laminaria* spp.), but, due to apparent preference and availability, *Laminaria* spp. (*L. digitata, Laminaria hyperborea* and *S. latissima*) (Phaeophyceae) account for up to 90% of their diet (Hansen et al. 2003). In face of the unusual diet, Orpin et al. (1985) investigated the microbial population of North Ronaldsay sheep exclusively fed on seaweeds and that of the same breed fed on pasture. These authors found great differences in the protozoal and bacterial populations between the two groups of animals, but not in total bacterial numbers. The rumen population of sheep fed on brown seaweed was found to be dominated by ciliate protozoa (e.g., *Dasytricha ruminantium, Entodinium* species) and bacteria, particularly *Streptococcus bovis, Selenomonas ruminantium, Butyrivibrio fibrisolvens* and lactate-utilizing species (Greenwood et al. 1983a, Orpin et al. 1985). Spirochetes and an unidentified filamentous bacterium, observed using electronic microscopy, were suggested to be of major significance in the fermentation of seaweed (Greenwood et al. 1983a). Seaweed-fed sheep presented high numbers of the bacteria fermenting non-structural polysaccharides, but not of bacteria fermenting structural polysaccharides (Orpin et al. 1985). Unlike in grass-fed sheep, cellulolytic bacteria were undetectable by growth in cellulose-containing medium from rumen fluid of seaweed-fed sheep, suggesting that these animals have adapted to a diet with scarce amount of cellulose (Orpin et al. 1985).

The adaptation of rumen microbiome to seaweeds was also suggested by the fermentation ability of bacteria isolated from grass-fed and seaweed-fed rumen fluid. Indeed, of the culturable bacteria from seaweed-fed sheep, 99% grew on mannitol, 71% grew on laminarin, and 13% grew on alginate and on fucoidan, while those from pasture-fed animals were considerably lower (0%, 32%, 2%, and 0%, respectively; Orpin et al. 1985). Degradation of alginates seems to be dependent on polyphenols present and the ratio of calcium and sodium alginate salt forms (Mohen, Horn et al. 1997). Williams et al. (2013) further evaluated the rumen bacterial population of seaweed-fed North Ronaldsay sheep. These authors isolated rumen bacteria and crude enzyme extracts with ability to hydrolyse brown seaweed polysaccharides, e.g., laminarin, alginate, carrageenan, and cellulose. Fucoidan hydrolysis was also reported but to a lesser extent (Williams et al. 2013). These findings reinforce the ability of the rumen microbiome to adapt to uncommon seaweed complex polysaccharides. Although data on the adaptation of rumen microbiota to green and red seaweeds is absent, we cannot discard this possibility. Therefore, the gradual increase of the levels of seaweeds in the diet may drive the rumen microbe's adaptation, enhancing energy availability from these complex carbohydrates (Makkar et al. 2016).

4 Effect of Seaweeds on Ruminal Fermentation and Enteric Methane Production

4.1 Livestock and greenhouse gas emissions

In recent decades there has been a great public concern about climate change and sustainable livestock production. The livestock sector is estimated to be responsible for 14.5% of total greenhouse gas emissions (Gerber et al. 2013), either directly or indirectly. Direct emissions result from rumen fermentation, breathing, and urinary and faecal excretions (Jungbluth et al. 2001). On the other hand, indirect emissions result from feed production, manure management, farm operations, animal transport, and processing of livestock products (Mosier et al. 1998; Fig. 3). Globally, livestock contribute to 44% of the world's anthropogenic methane, 53% of anthropogenic nitrous oxide and 5% of anthropogenic carbon dioxide emissions. Among livestock, the animals that most contribute

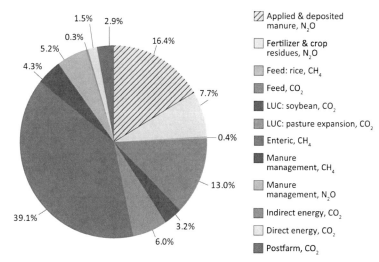

Figure 3. Global emissions from livestock supply chains. Reproduced, with permission, from Gerber et al. (2013).

Color version at the end of the book

to greenhouse gas emissions (65% of the total) are ruminants, particularly beef and dairy cattle (Gerber et al. 2013). Methane constitutes more than 40% of the total greenhouse gas emissions from ruminants (Rojas-Downing et al. 2017, Niu et al. 2018). In addition to an estimated half-life of 12.4 years (Cottle et al. 2011), the high warming potential of methane (25 times that of carbon dioxide) indicates its considerable impact on global greenhouse gas emissions and climate change (Moss et al. 2000, Beauchemin et al. 2008). The enteric fermentation of ruminants is estimated to be responsible for 6.4% of the overall methane produced by anthropogenic activity each year (Ripple et al. 2013).

4.2 Enteric production of methane by ruminant animals

Rumen fermentation of feed is carried out by a complex microbial ecosystem composed by bacteria, protozoa, archaea, fungi, and bacteriophages. In the rumen, ingested feed is digested and fermented in a two-stage process (McAllister and Newbold 2008, Krause et al. 2013): (i) enzymatic degradation with release of sugars, amino acids, glycerol, and fatty acids; and (ii) microbial fermentation of those compounds mainly to volatile fatty acids, ammonia, hydrogen, carbon dioxide and methane (Morgavi et al. 2010). The main volatile fatty acids produced are acetic acid, propionic acid and butyric acid, although formic acid, lactic acid, succinic acid and branched-chain fatty acids are also formed. These volatile fatty acids are transported across the rumen papillae and omasal walls into the circulatory system and transported to the liver, where they are used by the host animal as energy sources (Chesson and Forsberg 1997). Ammonia can be absorbed across the rumen wall, flowed into the omasum or taken up by the rumen microorganisms, along with amino acids and oligopeptides from dietary proteins and non-protein N, and incorporated into microbial protein (Wallace et al. 1997). After enzymatic digestion, this microbial protein will be a source of amino acids for the host animal. Hydrogen mostly results from the microbial fermentation of structural and non-structural polysaccharides of feeds. Clearance of metabolic hydrogen is essential to prevent the accumulation of reducing equivalents, which would otherwise inhibit the microbial fermentation of substrates (Sharp et al. 1998). Disposal of hydrogen can occur either through volatile fatty acid production, namely of propionic acid, through biohydrogenation of dietary unsaturated fatty acid, or, predominantly, through the reduction of carbon dioxide to methane in the reductive methanogenesis pathway by methanogenic archaea (Janssen 2010, Morgavi et al. 2010). Carbon dioxide produced during microbial fermentation of feeds constitutes the main gas in the rumen (up to 65%; Ellis et al.

1991), and it can be either eructed or transported to the lungs and exhaled. Most methane is directly eructed from the rumen headspace, but 10% to 15% is exhaled or emitted via flatus, as a result of hindgut fermentation of digesta (Huhtanen et al. 2015).

The archaeal population found in the rumen is mainly composed of methanogens, which are responsible for the enteric methane production. Rumen methanogens represent a small portion of the microbiome but play an important role in rumen function by disposal of excess reducing power (Janssen and Kirs 2008). Several genera of methanogenic archaea have been identified in the rumen, including *Methanobrevibacter*, the most abundant methanogen, *Methanomicrobium*, *Methanosphaera, Methanosarcina, Methanobacterium*, and members of the rumen cluster C (Wright et al. 2004, Wright et al. 2006, Janssen and Kirs 2008, Williams et al. 2009, Hook et al. 2010, Singh et al. 2010, St-Pierre and Wright 2013). Methanogens of the rumen cluster C and of the Thermoplasmatales-affiliated lineage C are now part of the recently proposed order Methanoplasmatales, which mainly contains uncultured archaea (Cersosimo and Wright 2015). The relative abundance of these methanogens in the microbiome varies among individual animals, ruminant species and diets, but a core archaeal population seems to exist (Fig. 4; Henderson et al. 2015), independently of the effect of diet or individual host animal (Kong et al. 2013).

Substrates for enteric rumen methanogenesis include hydrogen, carbon dioxide, methylamines, methanol, formic acid, and acetic acid (Cersosimo and Wright 2015). The main pathway is the hydrogenotrophic (by carbon dioxide reduction; Fig. 5), followed by the methylotrophic pathway (by methyl group reduction), and sparsely by the acetoclastic pathway (by acetic acid reduction; Morgavi et al. 2010, Huws et al. 2018). These alternative pathways (methylotrophic and acetoclastic) may alter the overall stoichiometry of methane production and shift the hydrogen sink in the rumen (Ungerfeld 2015).

Hydrogenotrophic methanogens are the most abundant and include the genera *Methanobrevibacter*, whereas the less abundant methylotrophic methanogens include the genera *Methanosarcina* and *Methanosphaera*, and the family Methanomassiliicoccaceae of the Methanoplasmatales order (Huws et al. 2018). Friedman et al. (2017) reported a high activity of methylotrophic methanogens in young ruminants, putatively of Methanosarcinales, whereas in

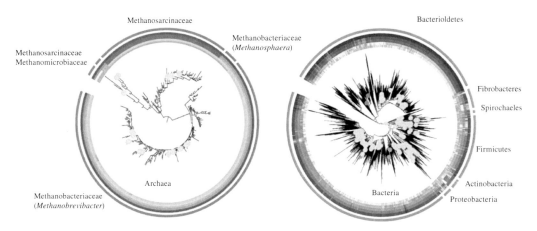

Figure 4. Archaeal (left) and bacterial (right) community composition data from Global Rumen Census project (Henderson et al. 2015) overlaid with the 16S rRNA gene sequences (yellow dots) from the Hungate 1000 project (Seshadri et al. 2018). The coloured rings around the trees represent the taxonomic classifications of each operational taxonomic unit from the Ribosomal Database Project (from the innermost to the outermost): genus, family, order, class and phylum. The strength of the colour is indicative of the percentage similarity of the operational taxonomic units to a sequence in the Ribosomal Database Project of that taxonomic level. Reproduced, with permission, from Seshadri et al. (2018) under CC BY 4.0 licence.

Color version at the end of the book

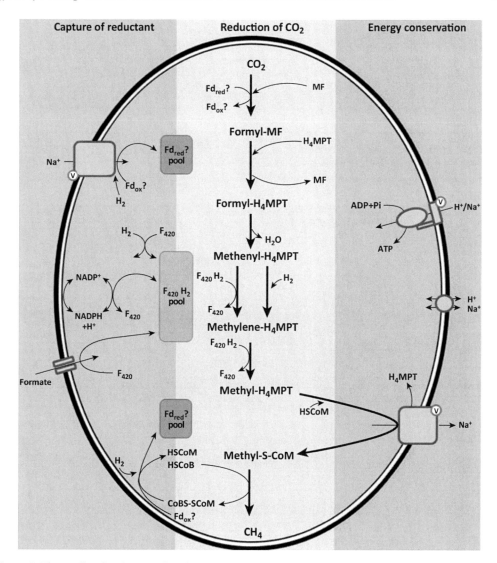

Figure 5. The predicted pathways of methane (CH_4) production in *Methanobrevibacter ruminantium* without cytochromes. The diagram is divided into three parts, showing the capture of reductant, the reduction of carbon dioxide (CO_2), and conservation of energy at the methyl transfer step. The main reactions are shown by thick arrows and cofactor participation is indicated by thin arrows. Protons are not shown, and the overall reaction is not balanced. Membrane-located proteins are shown by light brown boxes and potential vaccine is labelled with a circled V. Abbreviations: CoMS-SCoB, coenzyme B-coenzyme M heterodisulfide; F_{420}, coenzyme F_{420} oxidized; $F_{420}H_2$, coenzyme F_{420} reduced; Fd_{ox}?, unknown oxidized ferredoxin; Fd_{red}?, unknown reduced ferredoxin; H_4MPT, tetrahydromethanopterin; HSCoB, reduced coenzyme B; HSCoM, reduced coenzyme M; MF, methanofuran. Reproduced, with permission, from Leahy et al. (2010) and Hill et al. (2016) under CC BY 4.0 licence.

mature ruminants the hydrogenotrophs predominate. Methanosarcinales were also identified as acetoclastic producers (Morgavi et al. 2010).

Regardless of the metabolic pathway involved, methanogenesis final step converges in the reduction of coenzyme M methyl group to methane catalysed by methyl-coenzyme M reductase (Ellermann et al. 1988, Ermler et al. 1997, Hedderich and Whitman 2013).

Symbiotic relationships in the rumen have been reported between methanogens and protozoa, and less frequently between methanogens and fungi. This symbiosis involves interspecies hydrogen transfer, with archaea being associated either intracellularly or extracellularly, or both (Sharp et al. 1998) and may account for over one third of total enteric methane production (Finlay et al. 1994). Common ciliate protozoa associated with methanogens include the genera *Entodinium*, *Polyplastron*, *Epidinium*, and *Ophryoscolex* (Sharp et al. 1998), whereas among flagellated fungi methanogens were found to be associated with *Neocallimastix frontalis* (Joblin et al. 2002). The orders Methanobacteriales and Methanomicrobiales are those most frequently associated with protozoa (Sharp et al. 1998).

4.3 Mitigation of methane production

Although enteric methane production presents the advantage of reducing the excess reducing power caused by hydrogen, which could otherwise inhibit the fermentation of feed, it presents the disadvantage of reducing the energy and carbon available for formation of volatile fatty acids, which are essential for host nutrition (Morgavi et al. 2010). Methanogenesis is estimated to lead to a loss of 2-12% of the gross energy of the host animal (Johnson and Johnson 1995, Ellis et al. 2007). Improvement of feed utilization, diet digestibility, and ultimately livestock productivity might be achieved if the metabolic hydrogen is used in volatile fatty acid production or in alternative biochemical pathways thermodynamically more favourable (Grainger and Beauchemin 2011, Mitsumori et al. 2012). In addition, growing concern about the environmental impact and energy economics of rumen methane emissions has compelled researchers to search for new strategies that could mitigate this scenario with or without the negligible deterioration of animal performance (Hook et al. 2010) and lead to a more sustainable animal production.

Several strategies have been proposed to reduce enteric methane emissions. These include the modulation of forage type and forage-to-concentrate ratio, lipid supplementation, supplementation of dietary additives (chemicals and plant secondary metabolites), defaunation, immunization, phage therapy, and genetic selection (Martin et al. 2010, Cottle et al. 2011, Kumar et al. 2014, Patra 2016).

Nutritional management offers an efficient short-term strategy to reduce enteric methane emissions either by directly affecting methanogens or enhancing alternative hydrogen uses. For example, the inclusion of forages with high digestibility rates, the inclusion of ensiled forages, and the increased proportion of energy feeds (containing cereal grains and oil meals) in the animal diet reduce methane emissions from the animal (Beauchemin et al. 2008, Dourmad et al. 2008, Martin et al. 2010, Cottle et al. 2011, Patra 2012). The evaluation of forage digestibility and preservation with respect to methanogenesis is scarce and requires further research, whereas the effect of concentrate level is well established (Martin et al. 2010). The replacement of complex structural polysaccharides by easily digested non-structural polysaccharides leads to a modification of the rumen microbiome towards lower abundance of microorganisms with fibrolytic activity, lower production of hydrogen, and lower bacterial and archaeal diversity (Huws et al. 2018). However, the high grain and low roughage strategy may lead to digestive disorders such as sub-acute rumen acidosis (Gonzalez et al. 2012). Furthermore, it offers the additional disadvantage of using high-energy grains and oil meals, such as soybeans, wheat and maize, that are also human food sources.

Lipid supplementation was shown to be an effective strategy to reduce rumen methanogenesis, although its effect is mostly dependent on its fatty acid profile. A meta-analysis by Martin et al. (2010) observed a 3.8% methane reduction per percentage unit of lipid supplementation, which increased to 7.3% when the lipid was rich in medium-chain fatty acids (mainly from coconut oil) and to 4.8% when it was rich in polyunsaturated fatty acids (mainly from soybean oil, sunflower oil, and linseed oil). Both medium-chain and polyunsaturated fatty acids were reported to reduce the abundance and activity of methanogens (Machmuller et al. 2003, Soliva et al. 2003, Lillis et al. 2011). Polyunsaturated fatty acids may also provide an alternative hydrogen sink through biohydrogenation, even though it only accounts for 1-2% of the hydrogen consumed (Nagaraja et al. 1997).

Dietary additives such as ionophores (e.g., monensin) have been reported to reduce enteric methane production (Beauchemin et al. 2008), although their effect is transient (McGinn et al. 2004, Cottle et al. 2011). Organic acids, such as malate and fumarate, were reported to effectively mitigate methane production *in vitro* (Newbold et al. 2005) but responses to *in vivo* supplementation are contradictory (Martin et al. 2010). Halogenated compounds such as bromoethanosulphonate and 3-nitrooxypropanol are potent inhibitors of methyl-coenzyme M reductase, the enzyme involved in the last step of methane formation by rumen archaea (Gräwert et al. 2014, Martínez-Fernández et al. 2014, Duin et al. 2016). Unlike bromoethanosulphonate, which was considered a toxic compound (Gräwert et al. 2014), 3-nitrooxypropanol successfully reduced methane production without negatively affecting the host animal (Hristov et al. 2015, Jayanegara et al. 2018). Anti-methanogenic effects of plant secondary metabolites such as tannins, flavonoids, essential oils and saponins are related to their interactions with enzymes and antimicrobial effects, namely protozoa inhibition (Wallace et al. 2002). These effects vary according to plant secondary metabolite molecular structure, which may lead to a simultaneous decrease in feed digestibility (Patra and Saxena 2010, Goel and Makkar 2012, Patra 2016) or become inactive (Newbold et al. 1997).

Methanogens may also be inhibited by the use of antibiotics, bacteriophage therapy, defaunation, and vaccination (Cottle et al. 2011, Patra 2016). However, the majority of the attempts to inhibit methanogens in the rumen failed, or had only limited success, because of toxicity to the host animal, adaptation of the archaeome, low efficacy or poor selectivity (McAllister and Newbold 2008).

Genetic selection might be the most sustainable strategy to reduce enteric methane emission from ruminants (Pickering et al. 2015). Indeed, ruminants emitting low levels of methane tend to be more productive and more productive animals tend to emit less methane (Patra 2016).

4.4 Seaweeds as modulators of rumen methane production and fermentation

Identification of effective rumen enteric mitigation strategies with minimal or absent effects on the rumen microbial fermentation, and with persistent performance, are needed. In this context, seaweeds have been playing a growing and emerging role in the potential modulation of rumen methanogenesis. The effects of seaweed dietary inclusion on fermentation and particularly on methane emissions have been investigated, but results have been contradictory.

Dubois et al. (2013) studied the effects of different levels of inclusion of two green algae (*Caulera taxifolia, Oedogonium*) and one brown alga (*Cystoseira trinodis*) on methane production. The authors observed that only *C. trinodis* decreased methane production with no difference between the two doses tested (3.85% and 7.41%, OM basis). In another study, *Furcellaria* spp., *Chondrus crispus* (Rhodophyta) and "stormtoss seaweed" (a mixture with variable proportions of *C. crispus*, *Saccharina longicruris*—formerly *Laminaria longicruris*—and *Fucus vesiculosus*) significantly reduced methane production without negatively affecting the rumen fermentation (Kinley and Fredeen 2015).

Belanche et al. (2016b) observed no changes in methane production by the inclusion of *L. digitata*, only finding a decrease in methane production when *Ascophyllum nodosum* was incubated at the highest dose (2.0 g per L) in 24 h batch incubations. The inclusion of *A. nodosum* at 2.0 g per L also showed anti-protozoal properties, decreasing protozoal activity by 23%. Conversely, *L. digitata* had no effect on protozoal activity and increased volatile fatty acid production. When incubated in RUSITEC fermenters, either *L. digitata* or *A. nodosum* at 5% (DM basis; equivalent to 1.56 g per L) left methane and total volatile fatty acid production unaffected (Belanche et al. 2016a). The absence of effects of *A. nodosum* in the latter study can be explained by the lower concentration of seaweeds used (1.56 g per L), or by the greater dilution of the rumen fluid. On the other hand, results with *Laminaria* suggest that this seaweed had no negative antimicrobial effect and favoured rumen fermentation (Belanche et al. 2016a).

The accumulation of phlorotannins by brown algae is an adaptive strategy of defence against stress conditions and herbivory (Li et al. 2011). Highest concentrations of phlorotannins are found in seaweed species that live at the mid-tide level (e.g., *A. nodosum*) and lowest concentrations in seaweeds living on the low shore (e.g., *L. digitata*; Connan et al. 2004). Like terrestrial tannins, pholorotannins are able to form protein-phenol complexes, but their ability to form carbohydrate-phenol complexes and to reduce rumen enteric methane production is less evident. Wang et al. (2009) suggested that most *A. nodosum* protein would complex to phenols through oxidative and nucleophilic reactions, thus escaping microbial fermentation (Wang et al. 2009). However, results from Belanche et al. (2016a) suggest that the phlorotannin content in *L. digitata* is insufficient to have negative effects on feed degradability. Additionally, the formation of pholorotannin-enzyme complexes could be the main factor that limits cellulase activity in *A. nodosum*, and consequently fibre degradability. Conversely, enzyme inhibition is unlikely to affect *L. digitata* fermentation as a higher xylanase and carboxymethylcellulase activity per unit of enzyme was described (Belanche et al. 2016a). Therefore, the antimicrobial action of brown seaweeds might underlie the formation of pholorotannin-protein complexes in the cell wall, reducing the excretion of extracellular enzymes, and of nutrient hydrolysis and absorption. The structure and biodiversity of the archaeome and the distribution of the main protozoal groups were not affected by brown seaweeds in the study of Belanche et al. (2016a). These results suggest a nonspecific antimicrobial action of brown seaweeds against methanogens and protozoa or that the inclusion rate used was low and failed to affect the structure of these microbial communities.

In 24 h incubations by Molina-Alcaide et al. (2017), methane production was differently affected by the seven seaweeds tested, though these conclusions were taken without control experiments. *Palmaria palmata* had the highest volatile fatty acids and methane production and the lowest acetate to propionate ratio, and *Porphyra/Pyropia* sp. had the second highest methane production of all the seaweeds studied (Molina-Alcaide et al. 2017). Lee et al. (2018) evaluated the effects of increasing amounts of *Gelidium amansii*, a red alga commonly found in the shallow coasts of many East Asian countries, in rumen fermentation, observing increased total volatile fatty acids and methane production. Increased methane production may have been due, at least partially, to an increase in the protozoan population and of *Ruminococcus flavefaciens*, a fibrolytic bacteria that normally produces succinic acid as a major fermentation product together with acetic and formic acids, hydrogen and carbon dioxide (Lee et al. 2018).

Machado et al. (2014) demonstrated that seaweeds can effectively reduce *in vitro* methane production as all 20 species of tropical seaweed studied had similar or lower methane production compared to a positive control of decorticated cottonseed meal. The highest effects were obtained with *Dictyota* (Phaeophyceae) and *Asparagopsis* (Rhodophyta), which reduced methane production by 92.2% and 98.9% after 72 h, respectively (Machado et al. 2014). However, *Dictyota* (Phaeophyceae) and *Asparagopsis* (Rhodophyta) also resulted in the lowest concentration of total volatile fatty acids and the highest concentration of propionic acid, thus suggesting that the rumen fermentation was affected (Machado et al. 2014). These authors concluded that the reduced methane production was associated with secondary metabolites produced by seaweeds, as no relationship was found between the chemical composition of seaweed species and gas parameters. Indeed, *Asparagopsis* and *Dictyota* are rich in secondary metabolites with strong antimicrobial properties (Paul et al. 2006). *Dictyota* produces isoprenoids (Blunt et al. 2013) and *Asparagopsis* produces halogenated low molecular weight compounds, including bromine- and iodine-containing haloforms, as well other halogenated methanes and ethanes in lesser quantities (Paul et al. 2006) that may inhibit the rumen microbiome (Genovese et al. 2009). Machado et al. (2016b) identified bromoform as the most abundant natural product in the biomass of *Asparagopsis* followed by dibromochloromethane, bromochloroacetic acid and dibromoacetic acid. However, only bromoform was concluded to be present in sufficient quantities in the biomass at 2% OM to elicit a significant effect on methane production. The methane blocking mechanism seems to be related with the inhibition of the cobamine-dependent methyl transferase step in the reduction of carbon dioxide (Wood et al. 1968, Johnson et al. 1972).

This mechanism has been originally described for bromochloromethane (Lanigan 1972), which has a chemical similarity with the bromoform. Other structural analogues of coenzyme-M, such as 2-bromoethanesulfonate and 2-chloroethanesulfonate, competitively inhibit the methyl transfer reactions essential for the biosynthesis of methane (Liu et al. 2011). Nevertheless, it is quite likely that new and/or complementary inhibition mechanisms can be identified and described, regarding the diversity of chemical compounds present in seaweeds (Greff et al. 2014).

Machado et al. (2018) evaluated the impact of *A. taxiformis* supplementation on the relative abundance of methanogens and microbial community structure during *in vitro* batch fermentation. A reduction of methane production by over 99% was found when *A. taxiformis* was added at 2% OM or the halogenated methane analogue bromoform at 5 µM, compared to the control (basal substrate only). The decrease in methane production was related with a reduced abundance of the main orders of methanogens in adult ruminants (Methanobacteriales, Methanomassiliicoccales and Methanomicrobiales; Machado et al. 2018).

Since the major efforts in the study of the effect of seaweeds in methanogenesis aim at the mitigation of methane emissions without influencing the overall fermentation efficiency, one of the crucial questions is related to the inclusion levels of seaweeds. Kinley et al. (2016a) studied increasing inclusion rates of *A. taxiformis* on enteric methane production throughout 72 h fermentations (Fig. 6). The inclusion of more than 2% of *A. taxiformis* eliminated methane production, without negative impact on substrate digestibility for seaweed inclusion up to 5%. With 2% inclusion, acetic acid decreased and propionic acid and, to a lesser extent, butyric acid increased. Machado et al. (2016a) aimed to identify the optimal doses of *A. taxiformis* and *Oedogonium* sp., individually and in combination, that would decrease the production of methane while minimizing adverse effects on fermentation. *Asparagopsis taxiformis* was highly effective in decreasing the production of methane with a reduction of 99% at doses as low as 2% OM basis, but this level also decreased the production of volatile fatty acids, and increased the proportion of propionic acid, butyric acid, valeric acid, and isovaleric acid, suggesting that alternative fermentation processes took place. *Oedogonium* sp.

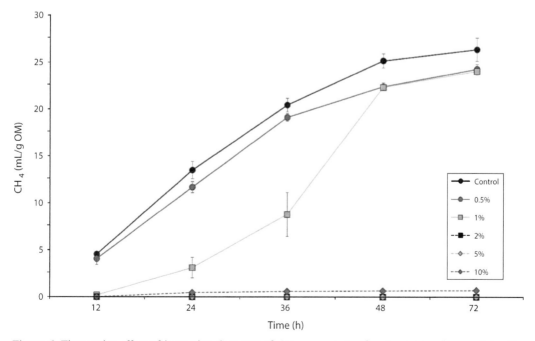

Figure 6. Time series effect of increasing dose rate of *Asparagopsis taxiformis* on mean (± s.e.m.) *in vitro* methane production (mL/g organic matter). Control was a high-quality Rhodes grass hay. Reproduced, with permission, from Kinley et al. (2016a).

was less effective as methane production was only reduced with doses ≥ 50% OM. A combination of 2% OM of *A. taxiformis* and 25% and 50% OM of *Oedogonium* sp. suppressed the production of methane independently of the inclusion rate of *Oedogonium* sp. Similarly, Kinley et al. (2016b) reported that the effects on fermentation of the combination of seven seaweeds with *A. taxiformis* were dominated by presence of *A. taxiformis* at 2% and no further benefits were demonstrated (Fig. 7). An inclusion rate of 2% OM caused a near-complete abatement of methane production and a 12-25% reduction in total volatile fatty acid production with minimal impact on substrate degradability.

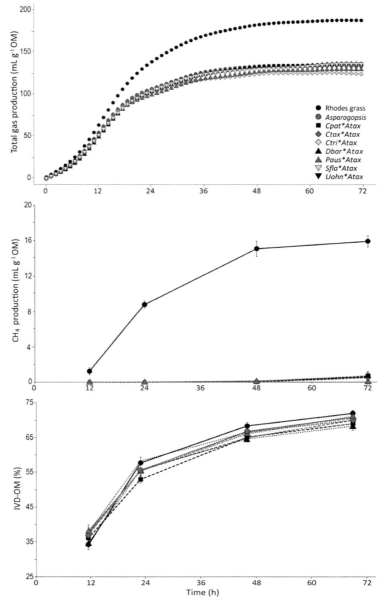

Figure 7. The effect of inclusion of seven different seaweeds on gas production and substrate digestibility over 72 h of *in vitro* fermentation with rumen fluid. From top down: Total gas production (TGP); methane (CH_4) production; and apparent *in vitro* digestibility (IVD-OM). The Rhodes grass control substrate was equal in all fermentations, the *Asparagopsis* control was included at a concentration of 2% of substrate organic matter, and the other seven seaweeds were included at 5%. No ±SE is presented for TGP because SE was smaller than the symbols. Reproduced, with permission, from Kinley et al. (2016b) under CC BY 4.0 licence.

Li et al. (2018) investigated the effect of five dietary inclusion levels of *A. taxiformis* (0%, 0.5%, 1%, 2% and 3% OM basis) on methane production from sheep offered a high-fibre diet for 72 d. The inclusion of *A. taxiformis* resulted in a dose-dependent reduction in methane production, with up to 80% reduction with 3% inclusion relatively to the control group. Consistently with earlier *in vitro* studies, *A. taxiformis* inclusion promoted a lower concentration of total volatile fatty acids and acetic acid, but a higher propionic acid concentration, suggesting that dietary inclusion of *A. taxiformis* directs hydrogen to propionic acid formation, probably because reductive propionogenesis is more favourable than acetogenesis in the presence of excess hydrogen (Mitsumori et al. 2012).

Earlier studies clearly showed that *A. taxiformis* is an effective antimethanogen in ruminants. However, in all the studies referred to, freeze-dried seaweed was used without considering other post-harvest processing methods. In this context, Vucko et al. (2017) evaluated *in vitro* effects of different processing methods of *A. taxiformis* in a factorial design [rinsing (unrinsed/dip-rinsed/submerged), freezing (frozen/not frozen) and drying (freeze-dried/kiln-dried/dehydrated)] on methane production and on the concentration of bromoform in the biomass. Treatments that were frozen and freeze-dried, regardless of rinsing, were the most effective, completely inhibiting the production of methane. Of these, the unrinsed treatment had higher concentration of bromoform (4.39 mg per g DM basis) than the rinsed treatments (2.0-3.2 mg per g DM basis). The next most effective group that completely inhibited the production of methane included treatments that were unrinsed, and kiln-dried or dehydrated without freezing. These treatments had lower but still considerable concentrations of bromoform (1.0-2.0 mg per g DM basis; Vucko et al. 2017). All other treatments inhibited methane production but not completely and presented bromoform concentrations lower than 1.0 mg per g DM basis. Therefore, it is suggested that the threshold for complete inhibition of methane production *in vitro* is 1 mg per g DM basis of bromoform in *A. taxiformis,* when included at 2% of OM (Vucko et al. 2017).

The broad range of inclusion levels and its effects on the fermentation process may be also related with the interaction of seaweed and basal substrate. Although the studies have been performed in controlled conditions, there is still a lack of systematic information about the effect of seaweeds in methane production when combined with different basal substrates. Maia et al. (2016) evaluated for the first time the effects of five seaweeds (*Ulva* sp.—Chlorophyta, *Laminaria ochroleuca* and *S. latissima*—Phaeophyceae, and *Gigartina* sp. and *Gracilaria vermiculophylla*—Rhodophyta) on methane production and ruminal fermentation parameters when incubated *in vitro* at 25% (DM basis) with two substrates (meadow hay and maize silage) for 24 h. Methane production was reduced by green and red algae, the highest effects being observed with *G. vermiculophylla* and *Gigartina* sp. (Rhodophyta) (38.2% and 35.8% reduction, respectively). Compared to the control, brown seaweeds had no effect on methanogenesis (Maia et al. 2016). These authors demonstrated that the seaweed effects on methane production depended on the basal substrate used (Fig. 8). *Ulva* sp., *Gigartina* sp. and *G. vermiculophylla* decreased methane production when added to meadow hay, but with maize silage only *G. vermiculophylla* reduced methanogenesis. Supplementation of *L. ochroleuca* and *S. latissima* to meadow hay produced methane levels similar to the control, but methanogenesis was promoted when *L. ochroleuca* was added to maize silage (Maia et al. 2016).

The varying effects of basal substrate are also suggested in earlier results, as the supplementation of *Oedogonium* (0.2 g OM basis) to different basal substrates (1 g OM basis) was found to decrease methane at different rates, by nearly 40% (Dubois et al. 2013), 30% (Machado et al. 2014) or 15% (Machado et al. 2016a), when Rhodes grass, Finders grass or Rhodes grass hay, respectively, was used as basal substrate. The interaction between seaweed and basal substrate is clearly underexploited and it will certainly be a driver for future studies, considering that basal substrates define the path of volatile fatty acid production, and consequently the nature and efficiency of the fermentation process.

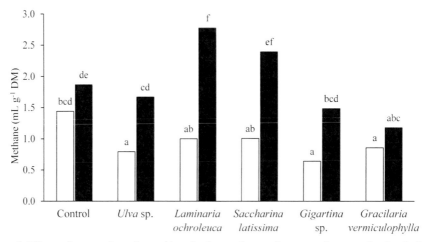

Figure 8. Effects of seaweed species and basal substrate interaction on methane production (mL g^{-1} dry matter) after 24 h of *in vitro* incubation. Meadow hay (□), maize silage (■). Mean values with different superscript letters were significantly different (P < 0.05). Reproduced, with permission, from Maia et al. (2016) under CC BY 4.0 licence.

5 Conclusion

Seaweeds are primary producers that could constitute alternative feed ingredients for a sustainable livestock production. In light of the complexity of seaweed polysaccharides, ruminant animals are the species most suitable for being fed seaweeds because of the symbiotic microbial population established in the rumen. Although the rumen microbiome can ferment seaweeds and provide energy to the host animal, a high variability of digestibility values is evident among and within seaweed species. These differences might be explained not only by the different nutrient composition of the algae, in particular complex polysaccharides and secondary metabolites, but also by the adaptation of the rumen microbial ecosystem of the animal to this particular feed.

Moreover, reducing enteric methanogenesis is challenging, and a strategy to be adopted needs to be sustainable, practical, and economically viable, ensuring functional capacity of rumen microbiota for fermentation of the basal diets and to improve animal productivity. The effect of seaweeds and its metabolites is still an open-ended research topic. Since most of the recent findings of the effect of seaweeds on methane reduction are supported by short-term *in vitro* studies, it is important to confirm the current observations not only in long-term *in vitro* studies but also with *in vivo* studies. This is a pivotal aspect to answer some of the major questions about the effect of methane reduction in population of rumen archaea: (i) the direct effect of seaweeds in the populations and/or activity of rumen methanogenic archaea, and (ii) the potential adaptation of the rumen microbial ecosystem to the long-term use of diets containing seaweeds. Regarding the first question, no relationship is reported between the abundance of rumen methanogenic archaea and methane production (Molina-Alcaide et al. 2017), which makes unclear the role of seaweeds in the growth of archaea. The same rationale is valid for the potential accumulation of hydrogen that is not eliminated through methanogenesis, the destiny of which is still unclear. The second question is more complex, and it is related with the extremely high capacity of adaptation of the rumen microbial ecosystem to new environmental conditions that could be created with the mid- to long-term use of seaweeds as a feed additive. The answer to this question can only be provided by information generated from long-term *in vivo* studies that are currently unavailable.

The use of seaweeds as a feed additive with a positive impact in the mitigation of methane emissions should be viewed as a new and complementary strategy in the current toolbox for methane mitigation based on the use of feed additives (Hook et al. 2010, Rojas-Downing et al. 2017). This includes different options such as dietary composition, defaunation based on chemical additives, and the use of plant compounds or lipids. Nevertheless, all of these strategies have potential drawbacks that may negatively affect the performance of the feeding strategy. In this context, seaweeds could play a completely new role due to the unique association of methane mitigation with an added nutritional value. Furthermore, the chemical diversity of bioactive compounds found in seaweeds (O'Sullivan et al. 2010) may contribute to improving the global health status of the animals, expanding the potential use of seaweeds as a feed additive in ruminant nutrition.

References Cited

Ahmed, S., A. Minuti and P. Bani. 2013. *In vitro* rumen fermentation characteristics of some naturally occurring and synthetic sugars. Ital. J. Anim. Sci. 12(3): e57.

Andrieux, C., A. Hibert, A.-M. Houari, M. Bensaada, F. Popot and O. Szylit. 1998. *Ulva lactuca* is poorly fermented but alters bacterial metabolism in rats inoculated with human faecal flora from methane and non-methane producers. J. Sci. Food Agric. 77(1): 25–30.

Arieli, A., D. Sklan and G. Kissil. 1993. A note on the nutritive-value of *Ulva lactuca* for ruminants. Anim. Prod. Sci. 57: 329–331.

Bach, S.J., Y. Wang and T.A. McAllister. 2008. Effect of feeding sun-dried seaweed (*Ascophyllum nodosum*) on fecal shedding of *Escherichia coli* O157:H7 by feedlot cattle and on growth performance of lambs. Anim. Feed Sci. Technol. 142(1–2): 17–32.

Barbeyron, T., Y. Lerat, J.-F. Sassi, S. Le Panse, W. Helbert and P.N. Collén. 2011. *Persicivirga ulvanivorans* sp. nov., a marine member of the family *Flavobacteriaceae* that degrades ulvan from green algae. Int. J. Syst. Evol. Microbiol. 61(8): 1899–1905.

Beauchemin, K.A., M. Kreuzer, F. O'Mara and T.A. McAllister. 2008. Nutritional management for enteric methane abatement: a review. Aust. J. Exp. Agric. 48(2): 21–27.

Belanche, A., E. Jones, I. Parveen and C.J. Newbold. 2016a. A metagenomics approach to evaluate the impact of dietary supplementation with *Ascophyllum nodosum* or *Laminaria digitata* on rumen function in Rusitec fermenters. Front. Microbiol. 7: 299.

Belanche, A., E. Ramos-Morales and C.J. Newbold. 2016b. *In vitro* screening of natural feed additives from crustaceans, diatoms, seaweeds and plant extracts to manipulate rumen fermentation. J. Sci. Food Agric. 96(9): 3069–3078.

Berteau, O. and B. Mulloy. 2003. Sulfated fucans, fresh perspectives: structures, functions, and biological properties of sulfated fucans and an overview of enzymes active toward this class of polysaccharide. Glycobiology 13(6): 29R–40R.

Bikker, P., M.M. van Krimpen, P. van Wikselaar, B. Houweling-Tan, N. Scaccia, J.W. van Hal, W.J.J. Huijgen, J.W. Cone and A.M. Lopez-Contreras. 2016. Biorefinery of the green seaweed *Ulva lactuca* to produce animal feed, chemicals and biofuels. J. Appl. Phycol. 28(6): 3511–3525.

Blunt, J.W., B.R. Copp, R.A. Keyzers, M.H. Munro and M.R. Prinsep. 2013. Marine natural products. Nat. Prod. Rep. 30(2): 237–323.

Cabrita, A.R.J., M.R.G. Maia, H.M. Oliveira, I. Sousa-Pinto, A.A. Almeida, E. Pinto and A.J.M. Fonseca. 2016. Tracing seaweeds as mineral sources for farm-animals. J. Appl. Phycol. 28(5): 3135–3150.

Cabrita, A.R.J., A. Correia, A.R. Rodrigues, P.P. Cortez, M. Vilanova and A.J.M. Fonseca. 2017. Assessing *in vivo* digestibility and effects on immune system of sheep fed alfalfa hay supplemented with a fixed amount of *Ulva rigida* and *Gracilaria vermiculophylla*. J. Appl. Phycol. 29(2): 1057–1067.

Castro, N.M., M.C. Valdez, A.M. Alvarez, R.N.A. Ramirez, I.S. Rodriguez, H.H. Contreras and L.S. Garcia. 2009. The kelp *Macrocystis pyrifera* as nutritional supplement for goats. Rev. Cient.-Fac. Cien. V. 19(1): 63–70.

Cersosimo, L.M. and A.-D.G. Wright. 2015. Rumen methanogens. pp. 143–150. *In*: Puniya, A.K., Singh, R. and Kamra, D.N. (eds.). Rumen Microbiology: From Evolution to Revolution. Springer India, New Delhi.

Chapman, V.J. and D.J. Chapman. 1980. Seaweeds and their uses. Springer Netherlands, Dordrecht, 334 pp.

Chesson, A. and C.W. Forsberg. 1997. Polysaccharide degradation by rumen microorganisms. pp. 329–381. *In*: Hobson, P.N. and Stewart, C.S. (eds.). The Rumen Microbial Ecosystem. Chapman and Hall, London, UK.

Collén, P.N., J.-F. Sassi, H. Rogniaux, H. Marfaing and W. Helbert. 2011. Ulvan lyases isolated from the Flavobacteria *Persicivirga ulvanivorans* are the first members of a new polysaccharide lyase family. J. Biol. Chem. 286(49): 42063–42071.

Connan, S., F. Goulard, V. Stiger, E. Deslandes and E. Ar Gall. 2004. Interspecific and temporal variation in phlorotannin levels in an assemblage of brown algae. Bot. Mar. 47(5): 410.

Cottle, D.J., J.V. Nolan and S.G. Wiedemann. 2011. Ruminant enteric methane mitigation: a review. Anim. Prod. Sci. 51(6): 491–514.

Davis, T.A., B. Volesky and A. Mucci. 2003. A review of the biochemistry of heavy metal biosorption by brown algae. Water Res. 37(18): 4311–4330.

Dourmad, J.Y., C. Rigolot and H. van der Werf. 2008. Emission of greenhouse gas, developing management and animal farming systems to assist mitigation. Proceedings of the Livestock and Global Climate Change 36–39.

Duarte, C.M., J. Wu, X. Xiao, A. Bruhn and D. Krause-Jensen. 2017. Can seaweed farming play a role in climate change mitigation and adaptation? Front. Mar. Sci. 4(100): 1–8.

Dubois, B., N.W. Tomkins, R.D. Kinley, M. Bai, S. Seymour, N.A. Paul and R. de Nys. 2013. Effect of tropical algae as additives on rumen *in vitro* gas production and fermentation characteristics. Am. J. Plant. Sci. 4: 34–43.

Duin, E.C., T. Wagner, S. Shima, D. Prakash, B. Cronin, D.R. Yáñez-Ruiz, S. Duval, R. Rümbeli, R.T. Stemmler, R.K. Thauer and M. Kindermann. 2016. Mode of action uncovered for the specific reduction of methane emissions from ruminants by the small molecule 3-nitrooxypropanol. Proc. Natl. Acad. Sci. 113(22): 6172–6177.

El-Waziry, A., A. Al-Haidary, A. Okab, E. Samara and K. Abdoun. 2015. Effect of dietary seaweed (*Ulva lactuca*) supplementation on growth performance of sheep and on *in vitro* gas production kinetics. Turk. J. Vet. Anim. Sci. 39(1): 81–86.

Ellermann, J., R. Hedderich, R. Bocher and R.K. Thauer. 1988. The final step in methane formation. Eur. J. Biochem. 172(3): 669–677.

Ellis, J.E., P.S. McIntyre, M. Saleh, A.G. Williams and D. Lloyd. 1991. Influence of CO_2 and low concentrations of O_2 on fermentative metabolism of the rumen ciliate *Dasytricha ruminantium*. Microbiology 137(6): 1409–1417.

Ellis, J.L., E. Kebreab, N.E. Odongo, B.W. McBride, E.K. Okine and J. France. 2007. Prediction of methane production from dairy and beef cattle. J. Dairy Sci. 90(7): 3456–3466.

Ermler, U., W. Grabarse, S. Shima, M. Goubeaud and R.K. Thauer. 1997. Crystal structure of methyl-coenzyme M reductase: the key enzyme of biological methane formation. Science 278(5342): 1457.

Evans, F.D. and A.T. Critchley. 2014. Seaweeds for animal production use. J. Appl. Phycol. 26(2): 891–899.

Falshaw, R., R.H. Furneaux and D.E. Stevenson. 1998. Agars from nine species of red seaweed in the genus *Curdiea* (Gracilariaceae, Rhodophyta). Carbohydr. Res. 308(1): 107–115.

Finlay, B.J., G. Esteban, K.J. Clarke, A.G. Williams, T.M. Embley and R.P. Hirt. 1994. Some rumen ciliates have endosymbiotic methanogens. FEMS Microbiol. Lett. 117(2): 157–161.

Firkins, J.L. and Z. Yu. 2015. Ruminant Nutrition Symposium: how to use data on the rumen microbiome to improve our understanding of ruminant nutrition. J. Anim. Sci. 93(4): 1450–1470.

Friedman, N., E. Jami and I. Mizrahi. 2017. Compositional and functional dynamics of the bovine rumen methanogenic community across different developmental stages. Environ. Microbiol. 19(8): 3365–3373.

Genovese, G., L. Tedone, M.T. Hamann and M. Morabito. 2009. The Mediterranean red alga *Asparagopsis*: a source of compounds against *Leishmania*. Mar. Drugs 7(3): 361–366.

Gerber, P.J., H. Steinfeld, B. Henderson, A. Mottet, C. Opio, J. Dijkman, A. Falcucci and G. Tempio. 2013. Tackling climate change through livestock – A global assessment of emissions and mitigation opportunities. Food and Agriculture Organization of the United Nations (FAO), Rome, Italy, 139 pp.

Goel, G. and H.P.S. Makkar. 2012. Methane mitigation from ruminants using tannins and saponins. Trop. Anim. Health Prod. 44(4): 729–739.

Gojon-Baez, H.H., D.A. Siqueiros-Beltrones and H. Hernandez-Contreras. 1998. *In situ* ruminal digestibility and degradability of *Macrocystis pyrifera* and *Sargassum* spp. in bovine livestock. Cienc. Mar. 24(4): 463–481.

Gonzalez, L.A., X. Manteca, S. Calsamiglia, K.S. Schwartzkopf-Genswein and A. Ferret. 2012. Ruminal acidosis in feedlot cattle: interplay between feed ingredients, rumen function and feeding behavior (a review). Anim. Feed Sci. Technol. 172(1–2): 66–79.

Grainger, C. and K.A. Beauchemin. 2011. Can enteric methane emissions from ruminants be lowered without lowering their production? Anim. Feed Sci. Technol. 166–167(0): 308–320.

Gräwert, T., H.-P. Hohmann, M. Kindermann, S. Duval, A. Bacher and M. Fischer. 2014. Inhibition of methyl-CoM reductase from *Methanobrevibacter ruminantium* by 2-bromoethane sulfonate. J. Agric. Food Chem. 62(52): 12487–12490.

Greenwood, Y., F.J. Hall, C.G. Orpin and I.W. Paterson. 1983a. Microbiology of seaweed digestion in Orkney sheep. J. Physiol.-London 343(Oct): 121.

Greenwood, Y., C.G. Orpin and I.W. Paterson. 1983b. Digestibility of seaweeds in Orkney sheep. J. Physiol.-London 343(Oct): 120.

Greff, S., M. Zubia, G. Genta-Jouve, L. Massi, T. Perez and O.P. Thomas. 2014. Mahorones, highly brominated cyclopentenones from the red alga *Asparagopsis taxiformis*. J. Nat. Prod. 77(5): 1150–1155.

Hansen, H.R., B.L. Hector and J. Feldmann. 2003. A qualitative and quantitative evaluation of the seaweed diet of North Ronaldsay sheep. Anim. Feed Sci. Technol. 105(1): 21–28.

Haug, A., B. Larsen and O. Smidsrod. 1967. Studies on sequence of uronic acid residues in alginic acid. Acta Chem. Scand. 21: 691–704.

Hedderich, R. and W.B. Whitman. 2013. Physiology and biochemistry of the methane-producing archaea. pp. 635–662. *In*: Rosenberg, E., DeLong, E.F., Lory, S.. Stackebrandt, E. and Thompson, F. (eds.). The Prokaryotes: Prokaryotic Physiology and Biochemistry. Springer Berlin Heidelberg, Berlin, Heidelberg.

Hehemann, J.-H., G. Correc, T. Barbeyron, W. Helbert, M. Czjzek and G. Michel. 2010. Transfer of carbohydrate-active enzymes from marine bacteria to Japanese gut microbiota. Nature 464: 908.

Henderson, G., F. Cox, S. Ganesh, A. Jonker, W. Young, Global Rumen Census collaborators and P.H. Janssen. 2015. Rumen microbial community composition varies with diet and host, but a core microbiome is found across a wide geographical range. Sci. Rep. 5: 14567.

Hill, J., C. McSweeney, A.-D.G. Wright, G. Bishop-Hurley and K. Kalantar-zadeh. 2016. Measuring methane production from ruminants. Trends Biotechnol. 34(1): 26–35.

Holdt, S.L. and S. Kraan. 2011. Bioactive compounds in seaweed: functional food applications and legislation. J. Appl. Phycol. 23(3): 543–597.

Hook, S.E., A.-D.G. Wright and B.W. McBride. 2010. Methanogens: methane producers of the rumen and mitigation strategies. Archaea 2010: 945785.

Hristov, A.N., J. Oh, F. Giallongo, T.W. Frederick, M.T. Harper, H.L. Weeks, A.F. Branco, P.J. Moate, M.H. Deighton, S.R.O. Williams, M. Kindermann and S. Duval. 2015. An inhibitor persistently decreased enteric methane emission from dairy cows with no negative effect on milk production. Proc. Natl. Acad. Sci. 112(34): 10663.

Huhtanen, P., E.H. Cabezas-Garcia, S. Utsumi and S. Zimmerman. 2015. Comparison of methods to determine methane emissions from dairy cows in farm conditions. J. Dairy Sci. 98(5): 3394–3409.

Hungate, R.E. 1966. The Rumen and its Microbes. Academic Press, New York, 544 pp.

Huws, S.A., C.J. Creevey, L.B. Oyama, I. Mizrahi, S.E. Denman, M. Popova, R. Muñoz-Tamayo, E. Forano, S.M. Waters, M. Hess, I. Tapio, H. Smidt, S.J. Krizsan, D.R. Yáñez-Ruiz, A. Belanche, L. Guan, R.J. Gruninger, T.A. McAllister, C.J. Newbold, R. Roehe, R.J. Dewhurst, T.J. Snelling, M. Watson, G. Suen, E.H. Hart, A.H. Kingston-Smith, N.D. Scollan, R.M. do Prado, E.J. Pilau, H.C. Mantovani, G.T. Attwood, J.E. Edwards, N.R. McEwan, S. Morrisson, O.L. Mayorga, C. Elliott and D.P. Morgavi. 2018. Addressing global ruminant agricultural challenges through understanding the rumen microbiome: past, present, and future. Front. Microbiol. 9: 2161.

Janssen, P.H. 2010. Influence of hydrogen on rumen methane formation and fermentation balances through microbial growth kinetics and fermentation thermodynamics. Anim. Feed Sci. Technol. 160(1–2): 1–22.

Janssen, P.H. and M. Kirs. 2008. Structure of the archaeal community of the rumen. Appl. Environ. Microbiol. 74(12): 3619–3625.

Jayanegara, A., K.A. Sarwono, M. Kondo, H. Matsui, M. Ridla, E.B. Laconi and Nahrowi. 2018. Use of 3-nitrooxypropanol as feed additive for mitigating enteric methane emissions from ruminants: a meta-analysis. Ital. J. Anim. Sci. 17(3): 650–656.

Joblin, K.N., H. Matsui, G.E. Naylor and K. Ushida. 2002. Degradation of fresh ryegrass by methanogenic co-cultures of ruminal fungi grown in the presence or absence of *Fibrobacter succinogenes*. Curr. Microbiol. 45(1): 46–53.

Johnson, E.D., A.S. Wood, J.B. Stone and E.T. Moran Jr. 1972. Some effects of methane inhibition in ruminants (steers). Can. J. Anim. Sci. 52(4): 703–712.

Johnson, K.A. and D.E. Johnson. 1995. Methane emissions from cattle. J. Anim. Sci. 73(8): 2483–2492.

Jungbluth, T., E. Hartung and G. Brose. 2001. Greenhouse gas emissions from animal houses and manure stores. Nutr. Cycl. in Agroecosys. 60(1): 133–145.

Kinley, R.D. and A.H. Fredeen. 2015. *In vitro* evaluation of feeding North Atlantic stormtoss seaweeds on ruminal digestion. J. Appl. Phycol. 27(6): 2387–2393.

Kinley, R.D., R. de Nys, M.J. Vucko, L. Machado and N.W. Tomkins. 2016a. The red macroalgae *Asparagopsis taxiformis* is a potent natural antimethanogenic that reduces methane production during *in vitro* fermentation with rumen fluid. Anim. Prod. Sci. 56(3): 282–289.

Kinley, R.D., M.J. Vucko, L. Machado and N.W. Tomkins. 2016b. *In vitro* evaluation of the antimethanogenic potency and effects on fermentation of individual and combinations of marine macroalgae. Am. J. Plant. Sci. 7: 14.

Konasani, V.R., C. Jin, N.G. Karlsson and E. Albers. 2018. A novel ulvan lyase family with broad-spectrum activity from the ulvan utilisation loci of *Formosa agariphila* KMM 3901. Sci. Rep. 8(1): 14713.

Kong, Y., Y. Xia, R. Seviour, R. Forster and T.A. McAllister. 2013. Biodiversity and composition of methanogenic populations in the rumen of cows fed alfalfa hay or triticale straw. FEMS Microbiol. Ecol. 84(2): 302–315.

Kopel, M., W. Helbert, Y. Belnik, V. Buravenkov, A. Herman and E. Banin. 2016. New family of ulvan lyases identified in three isolates from the Alteromonadales order. J. Biol. Chem. 291(11): 5871–5878.

Krause, D.O., T.G. Nagaraja, A.D.G. Wright and T.R. Callaway. 2013. BOARD-INVITED REVIEW: Rumen microbiology: leading the way in microbial ecology. J. Anim. Sci. 91(1): 331–341.

Kumar, C.S., P. Ganesan, P.V. Suresh and N. Bhaskar. 2008. Seaweeds as a source of nutritionally beneficial compounds – a review. J. Food Sci. Tech. Mys. 45(1): 1–13.

Kumar, S., P.K. Choudhury, M.D. Carro, G.W. Griffith, S.S. Dagar, M. Puniya, S. Calabro, S.R. Ravella, T. Dhewa, R.C. Upadhyay, S.K. Sirohi, S.S. Kundu, M. Wanapat and A.K. Puniya. 2014. New aspects and strategies for methane mitigation from ruminants. Appl. Microbiol. Biotechnol. 98(1): 31–44.

Lanigan, G. 1972. Metabolism of pyrrolizidine alkaloids in the ovine rumen. IV. Effects of chloral hydrate and halogenated methanes on rumen methanogenesis and alkaloid metabolism in fistulated sheep. Aust. J. Agric. Res. 23(6): 1085–1091.

Leahy, S.C., W.J. Kelly, E. Altermann, R.S. Ronimus, C.J. Yeoman, D.M. Pacheco, D. Li, Z. Kong, S. McTavish, C. Sang, S.C. Lambie, P.H. Janssen, D. Dey and G.T. Attwood. 2010. The genome sequence of the rumen methanogen *Methanobrevibacter ruminantium* reveals new possibilities for controlling ruminant methane emissions. PLoS One 5(1): e8926.

Lee, S.J., N.H. Shin, J.S. Jeong, E.T. Kim, S.K. Lee, I.D. Lee and S.S. Lee. 2018. Effects of *Gelidium amansii* extracts on *in vitro* ruminal fermentation characteristics, methanogenesis, and microbial populations. Asian-Australas. J. Anim. Sci. 31(1): 71–79.

Lettat, A. and C. Benchaar. 2013. Diet-induced alterations in total and metabolically active microbes within the rumen of dairy cows. Plos One 8(4): e60978.

Li, X., H.C. Norman, R.D. Kinley, M. Laurence, M. Wilmot, H. Bender, R. de Nys and N. Tomkins. 2018. *Asparagopsis taxiformis* decreases enteric methane production from sheep. Anim. Prod. Sci. 58(4): 681–688.

Li, Y.-X., I. Wijesekara, Y. Li and S.-K. Kim. 2011. Phlorotannins as bioactive agents from brown algae. Process Biochem. 46(12): 2219–2224.

Lillis, L., B. Boots, D.A. Kenny, K. Petrie, T.M. Boland, N. Clipson and E.M. Doyle. 2011. The effect of dietary concentrate and soya oil inclusion on microbial diversity in the rumen of cattle. J. Appl. Microbiol. 111(6): 1426–1435.

Liu, H., J. Wang, A. Wang and J. Chen. 2011. Chemical inhibitors of methanogenesis and putative applications. Appl. Microbiol. Biotechnol. 89(5): 1333–1340.

Machado, L., R.D. Kinley, M. Magnusson, R. Nys and N.W. Tomkins. 2015. The potential of macroalgae for beef production systems in Northern Australia. J. Appl. Phycol. 27(5): 2001–2005.

Machado, L., M. Magnusson, N.A. Paul, R. de Nys and N. Tomkins. 2014. Effects of marine and freshwater macroalgae on *in vitro* total gas and methane production. PLoS One 9(1): e85289.

Machado, L., M. Magnusson, N.A. Paul, R. Kinley, R. de Nys and N. Tomkins. 2016a. Dose-response effects of *Asparagopsis taxiformis* and *Oedogonium* sp. on *in vitro* fermentation and methane production. J. Appl. Phycol. 28(2): 1443–1452.

Machado, L., M. Magnusson, N.A. Paul, R. Kinley, R. de Nys and N. Tomkins. 2016b. Identification of bioactives from the red seaweed *Asparagopsis taxiformis* that promote antimethanogenic activity *in vitro*. J. Appl. Phycol. 28(5): 3117–3126.

Machado, L., N. Tomkins, M. Magnusson, D.J. Midgley, R. de Nys and C.P. Rosewarne. 2018. *In vitro* response of rumen microbiota to the antimethanogenic red macroalga *Asparagopsis taxiformis*. Microb. Ecol. 75(3): 811–818.

Machmuller, A., C.R. Soliva and M. Kreuzer. 2003. Effect of coconut oil and defaunation treatment on methanogenesis in sheep. Reprod. Nutr. Dev. 43(1): 41–55.

Maia, M.R.G., A.J.M. Fonseca, H.M. Oliveira, C. Mendonça and A.R.J. Cabrita. 2016. The potential role of seaweeds in the natural manipulation of rumen fermentation and methane production. Sci. Rep. 6: 32321.

Makkar, H.P.S., G. Tran, V. Heuzé, S. Giger-Reverdin, M. Lessire, F. Lebas and P. Ankers. 2016. Seaweeds for livestock diets: a review. Anim. Feed Sci. Technol. 212: 1–17.

Marin, A., M. Casas-Valdez, S. Carrillo, H. Hernandez, A. Monroy, L. Sangines and F. Perez-Gil. 2009. The marine algae *Sargassum* spp. (Sargassaceae) as feed for sheep in tropical and subtropical regions. Rev. Biol. Trop. 57(4): 1271–1281.

Martin, C., D.P. Morgavi and M. Doreau. 2010. Methane mitigation in ruminants: from microbe to the farm scale. Animal 4(3): 351–365.

Martínez-Fernández, G., L. Abecia, A. Arco, G. Cantalapiedra-Hijar, A.I. Martín-García, E. Molina-Alcaide, M. Kindermann, S. Duval and D.R. Yáñez-Ruiz. 2014. Effects of ethyl-3-nitrooxy propionate and 3-nitrooxypropanol on ruminal fermentation, microbial abundance, and methane emissions in sheep. J. Dairy Sci. 97(6): 3790–3799.

McAllister, T.A. and C.J. Newbold. 2008. Redirecting rumen fermentation to reduce methanogenesis. Aust. J. Exp. Agric. 48(1–2): 7–13.

McGinn, S.M., K.A. Beauchemin, T. Coates and D. Colombatto. 2004. Methane emissions from beef cattle: effects of monensin, sunflower oil, enzymes, yeast, and fumaric acid. J. Anim. Sci. 82(11): 3346–3356.

Mitsumori, M., T. Shinkai, A. Takenaka, O. Enishi, K. Higuchi, Y. Kobayashi, I. Nonaka, N. Asanuma, S.E. Denman and C.S. McSweeney. 2012. Responses in digestion, rumen fermentation and microbial populations to inhibition of methane formation by a halogenated methane analogue. Br. J. Nutr. 108(3): 482–491.

Mohen, E., S. Horn and K. Ostgaard. 1997. Alginate degradation during anaerobic digestion of Laminaria hyperborea stipes. J. Appl. Phycol. 9(2): 157–166.

Molina-Alcaide, E., M.D. Carro, M.Y. Roleda, M.R. Weisbjerg, V. Lind and M. Novoa-Garrido. 2017. *In vitro* ruminal fermentation and methane production of different seaweed species. Anim. Feed Sci. Technol. 228: 1–12.

Morgavi, D.P., E. Forano, C. Martin and C.J. Newbold. 2010. Microbial ecosystem and methanogenesis in ruminants. Animal 4(7): 1024–1036.

Mosier, A., C. Kroeze, C. Nevison, O. Oenema, S. Seitzinger and O. van Cleemput. 1998. Closing the global N_2O budget: nitrous oxide emissions through the agricultural nitrogen cycle. Nutr. Cycl. Agroecosys. 52(2): 225–248.

Moss, A.R., J.-P. Jouany and J. Newbold. 2000. Methane production by ruminants: its contribution to global warming. Ann. Zootech. 49(3): 231–253.

Murata, M. and J.-i. Nakazoe. 2001. Production and use of marine algae in Japan. JARQ-Jpn. Agr. Res. Q. 35(4): 281–290.

Nagaraja, T.G., C.J. Newbold, C.J. van Nevel and D.I. Demeyer. 1997. Manipulation of ruminal fermentation. pp. 523–632. *In*: Hobson, P.N. and Stewart C.S. (eds.). The Rumen Microbial Ecosystem. Springer Netherlands, Dordrecht.

Newbold, C.J., S.M. El Hassan, J. Wang, M.E. Ortega and R.J. Wallace. 1997. Influence of foliage from African multipurpose trees on activity of rumen protozoa and bacteria. Br. J. Nutr. 78(2): 237–249.

Newbold, C.J., S. Lopez, N. Nelson, J.O. Ouda, R.J. Wallace and A.R. Moss. 2005. Propionate precursors and other metabolic intermediates as possible alternative electron acceptors to methanogenesis in ruminal fermentation *in vitro*. Br. J. Nutr. 94(1): 27–35.

Niu, M., E. Kebreab, A.N. Hristov, J. Oh, C. Arndt, A. Bannink, A.R. Bayat, A.F. Brito, T. Boland, D. Casper, L.A. Crompton, J. Dijkstra, M.A. Eugene, P.C. Garnsworthy, M.N. Haque, A.L.F. Hellwing, P. Huhtanen, M. Kreuzer, B. Kuhla, P. Lund, J. Madsen, C. Martin, S.C. McClelland, M. McGee, P.J. Moate, S. Muetzel, C. Munoz, P. O'Kiely, N. Peiren, C.K. Reynolds, A. Schwarm, K.J. Shingfield, T.M. Storlien, M.R. Weisbjerg, D.R. Yanez-Ruiz and Z. Yu. 2018. Prediction of enteric methane production, yield, and intensity in dairy cattle using an intercontinental database. Glob. Chang. Biol. 24(8): 3368–3389.

O'Sullivan, L., B. Murphy, P. McLoughlin, P. Duggan, P.G. Lawlor, H. Hughes and G.E. Gardiner. 2010. Prebiotics from marine macroalgae for human and animal health applications. Mar. Drugs 8(7): 2038–2064.

Orpin, C.G., Y. Greenwood, F.J. Hall and I.W. Paterson. 1985. The rumen microbiology of seaweed digestion in Orkney sheep. J. Appl. Phycol. 58(6): 585–596.

Patra, A.K. 2012. Enteric methane mitigation technologies for ruminant livestock: a synthesis of current research and future directions. Environ. Monit. Assess. 184(4): 1929–1952.

Patra, A.K. 2016. Recent advances in measurement and dietary mitigation of enteric methane emissions in ruminants. Front. Vet. Sci. 3: 39.

Patra, A.K. and J. Saxena. 2010. A new perspective on the use of plant secondary metabolites to inhibit methanogenesis in the rumen. Phytochemistry 71(11): 1198–1222.

Paul, N.A., R. de Nys and P.D. Steinberg. 2006. Chemical defence against bacteria in the red alga *Asparagopsis armata*: linking structure with function. Mar. Ecol. Prog. Ser. 306: 87–101.

Percival, E. 1979. The polysaccharides of green, red and brown seaweeds: their basic structure, biosynthesis and function. Br. Phycol. J. 14(2): 103–117.

Petri, R.M., T. Schwaiger, G.B. Penner, K.A. Beauchemin, R.J. Forster, J.J. McKinnon and T.A. McAllister. 2014. Characterization of the core rumen microbiome in cattle during transition from forage to concentrate as well as during and after an acidotic challenge. PLoS One 8(12): e83424.

Pickering, N.K., V.H. Oddy, J. Basarab, K. Cammack, B. Hayes, R.S. Hegarty, J. Lassen, J.C. McEwan, S. Miller, C.S. Pinares-Patiño and Y. de Haas. 2015. Animal board invited review: genetic possibilities to reduce enteric methane emissions from ruminants. Animal 9(9): 1431–1440.

Ramnani, P., R. Chitarrari, K. Tuohy, J. Grant, S. Hotchkiss, K. Philp, R. Campbell, C. Gill and I. Rowland. 2012. *In vitro* fermentation and prebiotic potential of novel low molecular weight polysaccharides derived from agar and alginate seaweeds. Anaerobe 18(1): 1–6.

Ripple, W.J., P. Smith, H. Haberl, S.A. Montzka, C. McAlpine and D.H. Boucher. 2013. Ruminants, climate change and climate policy. Nat. Clim. Chang. 4: 2.

Rojas-Downing, M.M., A.P. Nejadhashemi, T. Harrigan and S.A. Woznicki. 2017. Climate change and livestock: impacts, adaptation, and mitigation. Clim. Risk Manag. 16: 145–163.

Salinas, A. and C.E. French. 2017. The enzymatic ulvan depolymerisation system from the alga-associated marine flavobacterium *Formosa agariphila*. Algal Res. 27: 335–344.

Seshadri, R., S.C. Leahy, G.T. Attwood, K.H. Teh, S.C. Lambie, A.L. Cookson, E.A. Eloe-Fadrosh, G.A. Pavlopoulos, M. Hadjithomas, N.J. Varghese, D. Paez-Espino, Hungate project collaborators, R. Perry, G. Henderson, C.J. Creevey, N. Terrapon, P. Lapebie, E. Drula, V. Lombard, E. Rubin, N.C. Kyrpides, B. Henrissat, T. Woyke, N.N. Ivanova and W.J. Kelly. 2018. Cultivation and sequencing of rumen microbiome members from the Hungate 1000 Collection. Nat. Biotechnol. 36: 359.

Shaani, Y., T. Zehavi, S. Eyal, J. Miron and I. Mizrahi. 2018. Microbiome niche modification drives diurnal rumen community assembly, overpowering individual variability and diet effects. ISME. J. 12(10): 2446–2457.

Sharp, R., C.J. Ziemer, M.D. Stern and D.A. Stahl. 1998. Taxon-specific associations between protozoal and methanogen populations in the rumen and a model rumen system. FEMS Microb. Ecol. 26(1): 71–78.

Singh, K.M., P.R. Pandya, S. Parnerkar, A.K. Tripathi, U. Ramani, P.G. Koringa, D.N. Rank, C.G. Joshi and R.K. Kothari. 2010. Methanogenic diversity studies within the rumen of Surti buffaloes based on methyl coenzyme M reductase A (mcrA) genes point to Methanobacteriales. Pol. J. Microbiol. 59(3): 175–178.

Soliva, C.R., I.K. Hindrichsen, L. Meile, M. Kreuzer and A. Machmuller. 2003. Effects of mixtures of lauric and myristic acid on rumen methanogens and methanogenesis *in vitro*. Lett. Appl. Microbiol. 37(1): 35–39.

St-Pierre, B. and A.D.G. Wright. 2013. Diversity of gut methanogens in herbivorous animals. Animal 7: 49–56.

Tayyab, U., M. Novoa-Garrido, M.Y. Roleda, V. Lind and M.R. Weisbjerg. 2016. Ruminal and intestinal protein degradability of various seaweed species measured *in situ* in dairy cows. Anim. Feed Sci. Technol. 213: 44–54.

Ungerfeld, E.M. 2015. Shifts in metabolic hydrogen sinks in the methanogenesis-inhibited ruminal fermentation: a meta-analysis. Front.Microbiol. 6: 538.

Ventura, M.R. and J.I.R. Castanon. 1998. The nutritive value of seaweed (*Ulva lactuca*) for goats. Small Rumin. Res. 29(3): 325–327.

Vucko, M.J., M. Magnusson, R.D. Kinley, C. Villart and R. de Nys. 2017. The effects of processing on the *in vitro* antimethanogenic capacity and concentration of secondary metabolites of *Asparagopsis taxiformis*. J. Appl. Phycol. 29(3): 1577–1586.

Wallace, R.J. 2008. Gut microbiology – broad genetic diversity, yet specific metabolic niches. Animal 2(5): 661–668.

Wallace, R.J., N.R. McEwan, F.M. McIntosh, B. Teferedegne and C.J. Newbold. 2002. Natural products as manipulators of rumen fermentation. Asian-Australas. J. Anim. Sci. 15(10): 1458–1468.

Wallace, R.J., R. Onodera and M.A. Cotta. 1997. Metabolism of nitrogen-containing compounds. pp. 283–328. *In*: Hobson, P.N. and Stewart, C.S. (eds.). The Rumen Microbial Ecosystem. Chapman and Hall, London, UK.

Wang, Y., Z. Xu, S.J. Bach and T.A. McAllister. 2008. Effects of phlorotannins from *Ascophyllum nodosum* (brown seaweed) on *in vitro* ruminal digestion of mixed forage or barley grain. Anim. Feed Sci. Technol. 145(1–4): 375–395.

Wang, Y., Z. Xu, S.J. Bach and T.A. McAllister. 2009. Sensitivity of *Escherichia coli* to seaweed (*Ascophyllum nodosum*) phlorotannins and terrestrial tannins. Asian-Australas. J. Anim. Sci. 22(2): 238–245.

Williams, A.G., S. Withers and A.D. Sutherland. 2013. The potential of bacteria isolated from ruminal contents of seaweed-eating North Ronaldsay sheep to hydrolyse seaweed components and produce methane by anaerobic digestion *in vitro*. Microb. Biotechnol. 6(1): 45–52.

Williams, Y.J., S. Popovski, S.M. Rea, L.C. Skillman, A.F. Toovey, K.S. Northwood and A.-D.G. Wright. 2009. A vaccine against rumen methanogens can alter the composition of archaeal populations. Appl. Environ. Microbiol. 75(7): 1860–1866.

Wood, J.M., F.S. Kennedy and R.S. Wolfe. 1968. Reaction of multihalogenated hydrocarbons with free and bound reduced vitamin B_{12}. Biochemistry 7(5): 1707–1713.

Wright, A.D.G., A.F. Toovey and C.L. Pimm. 2006. Molecular identification of methanogenic archaea from sheep in Queensland, Australia reveal more uncultured novel Archaea. Anaerobe 12(3): 134–139.

Wright, A.D.G., A.J. Williams, B. Winder, C.T. Christophersen, S.L. Rodgers and K.D. Smith. 2004. Molecular diversity of rumen methanogens from sheep in western Australia. Appl. Environ. Microbiol. 70(3): 1263–1270.

Index

About the Editors

Leonel Pereira

MARE – Marine and Environmental Sciences Centre, Department of Life Sciences, Faculty of Sciences and Technology, University of Coimbra, 3000-456 Coimbra, Portugal

Leonel Pereira has a degree in Biology (scientific branch) and a PhD in Biology (Cell Biology specialty) from the Faculty of Science and Technology of the University of Coimbra (Portugal), where he is currently Professor. In addition to teaching at this university, he is also an Investigator in the Marine and Environmental Sciences Centre. His interests are mainly focused on marine biodiversity (algae), marine biotechnology (bioactive compounds of macroalgae) and marine ecology (environmental assessment). Since 2008, he has been the author and editor of the electronic publication MACOI—Portuguese Seaweeds Website (www.uc.pt/seaweeds). He has authored more than 20 books and book chapters, has published more than 40 scientific articles in international journals, and is the editor of 5 books published by international publishers, and he has presented more than a hundred lectures and oral communications in various national and international scientific events. He received the Francisco de Holanda Prize (Honorable Mention) in 1998 and, more recently, the King D. Carlos Award (18th edition), and the CHOICE Award 2016 for Outstanding Academic Title: *Edible Seaweeds of the World*, CRC Press.

Kiril Bahcevandziev

CERNAS—the Research Centre for Natural Resources, Environment and Society, and IIA—Institute of Applied Research, Polytechnic Institute of Coimbra, Coimbra Agriculture College, Bencanta, 3045-601 Coimbra, Portugal.

Kiril Bahcevandziev has a degree in Agriculture (S.S. Cyril and Methodius University, Skopje, FYR of Macedonia, 1984), a master's degree in Agricultural Engineering (University of Lisbon, 1992), and a PhD in Agricultural Engineering (University of Lisbon, 2003). He is a member of the Portuguese Society for Plant Physiology, European Association for Research on Plant Breeding (EUCARPIA), International Society for Horticultural Science, American Society for Horticultural Science, and Portuguese Society for Horticulture.

He is also member of the Editorial Board of University Bulletins and Reviewer of the *Annals of Agricultural and Environmental Sciences*, *Plant Cell, Tissue & Organ Culture*, *Plant Biotechnology Journal*, *World Journal of Agricultural Research*, and *African Journal of Plant Science*.

Nilesh H. Joshi

Fisheries Research Station, Junagadh Agricultural University, Okha Port, Okha, Dist. Dev Bhoomi Dwarka, Gujarat 361350, India

Nilesh H. Joshi is Editorial Board Member of the *Journal of Marine Science: Research & Development* and *Journal of Marine Biology & Oceanography.* He is a life member of the Seaweed Research and Utilization Association of India and of the Marine Biological Association of India, and a member of the Indian Science Congress, Kolkata.

Color Plate Section

Chapter 1

Figure 1. Main species used as agricultural fertilizer: Chlorophyta – a) *Rhizoclonium linum*, b) *Codium tomentosum*, c) *Ulva* sp.; Phaeophyceae – d) *Ascophyllum nodosum*, e) *Bifurcaria bifurcata*, f) *Fucus serratus*, g) *Fucus vesiculosus*, h) *Himanthalia elongata*, i) *Laminaria digitata*, j) *Laminaria hyperborea*, k) *Pelvetia canaliculata*, l) *Saccharina latissima*, m) *Saccorhiza polyschides*; Rhodophyta – n) *Calliblepharis jubata*, o) *Chondrus crispus*, p) *Cryptopleura ramosa*.

Figure 2. Main species used as agricultural fertilizer: Rhodophyta – a) *Ceramium* sp., b) *Gelidium corneum*, c) *Gigartina pistillata*. d) *Halarachnion ligulatum*, e) *Lithothamnion corallioides*, f) *Lithothamnion glaciale*, g) *Osmundea pinnatifida*, h) *Palmaria palmata*, i) *Phymatolithon calcareum*, j) *Polyneura bonnemaisonii*, k) *Polysiphonia elongata*, l) *Vertebrata thuyoides*.

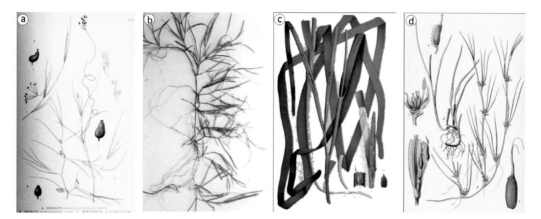

Figure 3. Main species used as agricultural fertilizer: Marine angiosperms – a) *Ruppia* sp., b) *Stuckenia pectinata*, c) *Zostera marina*, d) *Zostera noltei*.

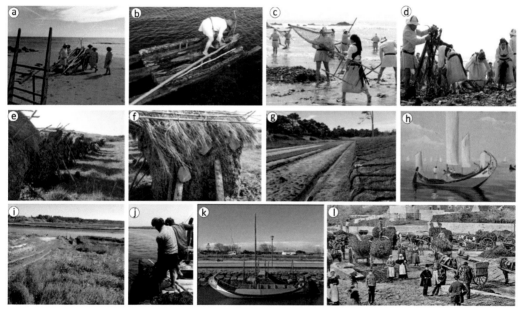

Figure 4. Historical use of algae and marine plants in the European North Atlantic: a) the carrelo (used to transport the raft), and the raft (used to reach the rocks furthest from the beach where the algae cling); b) the jangada (raft); c-d) the instruments for cutting and collecting (foicinhão, croque, engaceira); e-f) the medas or haystacks of sargaço; g) Maceiras or fields of maceira; i) landing dock of moliço (ria of Aveiro); h) mural painting representing the harvest of moliço; j) landing of moliço; k) typical moliceiro boat; l) "vraic" or "wrack", the Jersey terms, were used by many writers to cover all types of seaweed but especially those used for agricultural purposes.

Chapter 3

Figure 2. Some macroalgae products (agriculture, foods, pharmaceuticals).

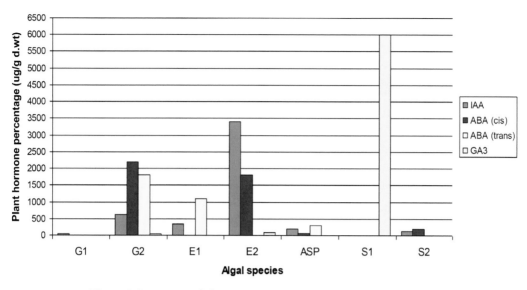

Figure 5. Percentage of plant growth regulators in the macroalgae studied.

Chapter 4

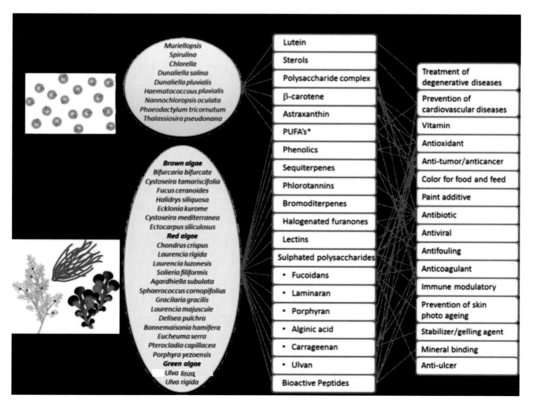

Figure 2. Components of secondary metabolites of marine algae and their possible application.

Chapter 6

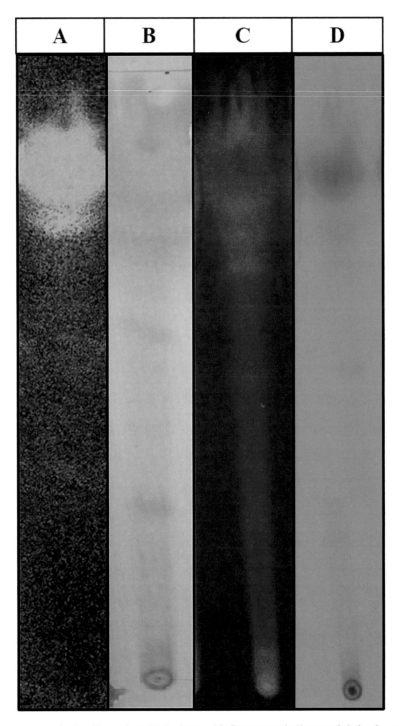

Figure 2. Chromatography in silica gel on TLC plates with fluorescent indicator of *Ochtodes secundiramea* extract eluted at MeOH and dichloromethane (90:10). Destructive analyses. A: Bioautography (*Cladosporium sphaerospermum*). B: Derivatization (p-hydroxybenzaldehyde), nondestructive method. C: observation at 366 nm. D: Observation at 254 nm (Machado 2014).

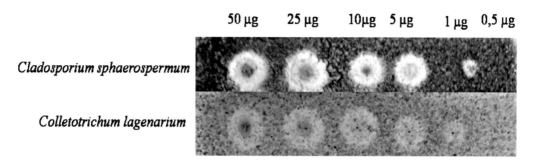

Figure 3. Bioautography of *Ochtodes secundiramea* extract for determination of LD antifungal activity against *Cladosporium sphaerospermum* and *Colletotrichum lagenarium* (Machado 2014, Machado et al. 2014a).

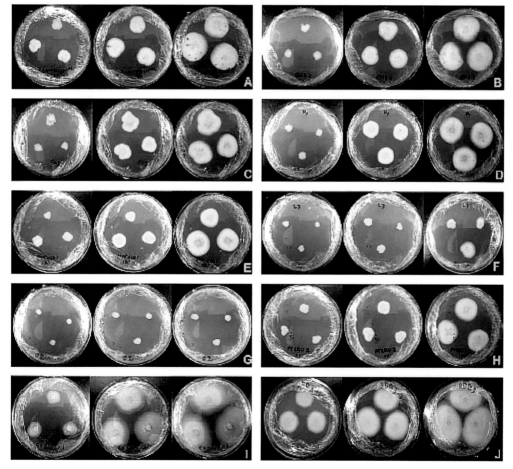

Figure 4. Mycelial growth of *Colletotrichum gloeosporioides* (papaya anthracnose) in response to different seaweed extracts after 2, 4 and 6 days. Conidial suspension was applied on paper disks (Machado et al. 2010). A to H: Crude extracts obtained from different seaweeds. I and J: solvent control and negative control respectively.

Chapter 10

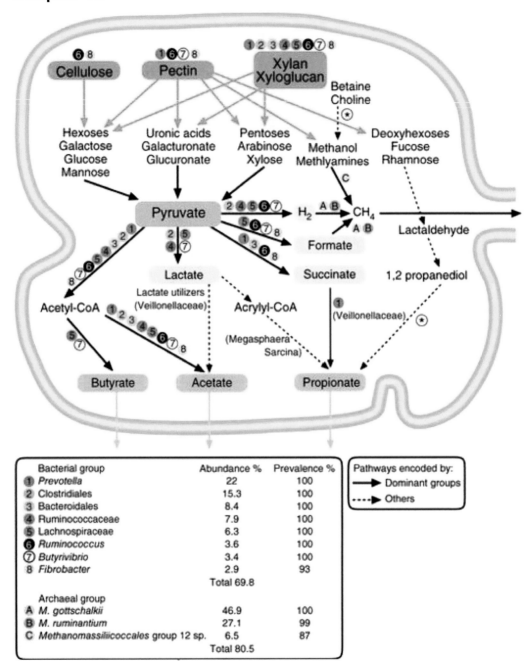

Figure 1. Simplified illustration showing the degradation and metabolism of plant structural carbohydrates by the dominant bacterial and archaeal groups identified in the Global Rumen Census project (Henderson et al. 2015) using information from metabolic studies and analysis of the reference genomes. Abundance represents the mean relative percentage for that genus-level group in samples that contain that group, while prevalence represents the prevalence of that genus-level group in all samples (*n* = 684). *The conversion of choline to trimethylamine, and propanediol to propionate, generates toxic intermediates that are contained within bacterial microcompartments. Adapted, with permission, from Seshadri et al. (2018) under CC BY 4.0 licence.

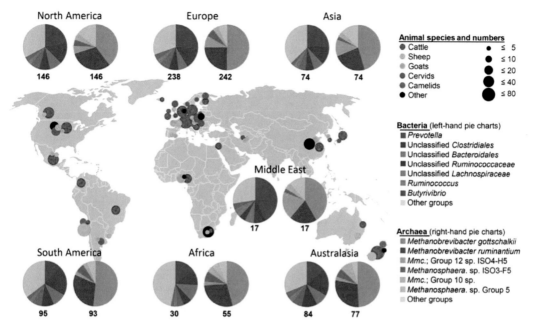

Figure 2. Bacterial and archaeal community compositions of rumen samples from different ruminant species and regions. Numbers below pie charts represent the number of samples for which data were obtained. The most abundant bacteria and archaea are named in clockwise order starting at the top of the pie chart. Mmc. Methanomassiliicoccales. Reproduced, with permission, from Henderson et al. (2015) under CC BY 4.0 licence.

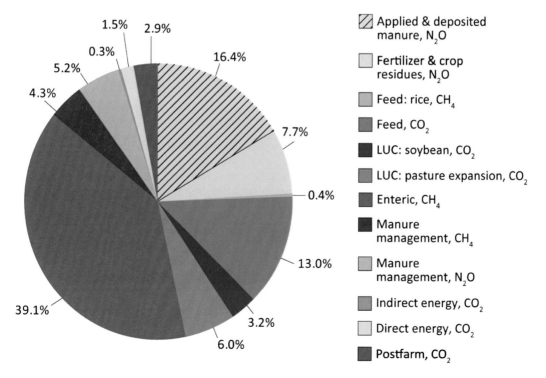

Figure 3. Global emissions from livestock supply chains. Reproduced from Gerber et al. (2013).

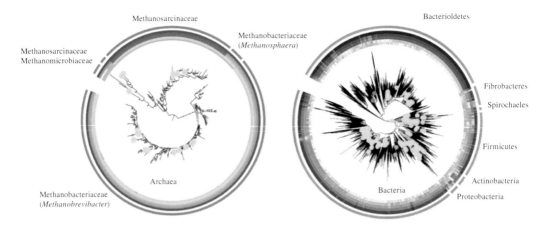

Figure 4. Archaeal (left) and bacterial (right) community composition data from Global Rumen Census project (Henderson et al. 2015) overlaid with the 16S rRNA gene sequences (yellow dots) from the Hungate 1000 project (Seshadri et al. 2018). The coloured rings around the trees represent the taxonomic classifications of each operational taxonomic unit from the Ribosomal Database Project (from the innermost to the outermost): genus, family, order, class and phylum. The strength of the colour is indicative of the percentage similarity of the operational taxonomic units to a sequence in the Ribosomal Database Project of that taxonomic level.

Reproduced, with permission, from Seshadri et al. (2018) under CC BY 4.0 licence.

Printed and bound by CPI Group (UK) Ltd, Croydon, CR0 4YY

22/10/2024

01777333-0006